Design and Verification of Microprocessor Systems
for High-Assurance Applications

David S. Hardin
Editor

Design and Verification of Microprocessor Systems for High-Assurance Applications

 Springer

Editor
David S. Hardin
Rockwell Collins, Inc.
400 Collins Road NE.
Cedar Rapids IA 52498
USA
dshardin@rockwellcollins.com

ISBN 978-1-4419-1538-2   e-ISBN 978-1-4419-1539-9
DOI 10.1007/978-1-4419-1539-9
Springer New York Dordrecht Heidelberg London

Library of Congress Control Number: 2010921140

Printed on acid-free paper

Springer is part of Springer Science+Business Media (www.springer.com)

Preface

Microprocessors are the workhorses of modern high-assurance systems. New cars typically utilize dozens of microprocessors, managing vital functions such as braking, ignition, traction control, and air bags. Most of the critical functions on modern aircraft, including flight control, engine control, flight management, and pilot displays, are microprocessor based. Medical devices, transportation systems, power grids, communications infrastructure, defense systems, and numerous other high-confidence applications are all dependent on microprocessor control.

Over the last 30 years, the microprocessors deployed in high-assurance systems have increased in sophistication from simple 8-bit chips with less than 10,000 transistors to 32- and 64-bit chips with millions of transistors, pipelining, floating point, specialized functional units (e.g., for cryptography), memory management, cache management, and so on. The system software for these systems has likewise increased from simple interrupt handlers to full-fledged operating systems capable of supporting multiple concurrent applications with space and time partitioning. Finally, the application logic on these microprocessor systems has increased from a few hundred lines of assembly code to millions of lines of high-level language code. The means of expressing the application design has changed as well, progressing from schematics and assembly code to high-level languages for both hardware and software. Indeed, many high-assurance applications are now developed using model-based development languages, with the final source code (which may be a mixture of programming language and hardware description language source text) autogenerated from the models.

This begs the question: how are we to trust this complex stack of microprocessor hardware, microcode, operating system kernel, runtime libraries, and application logic? Simulation and testing have shown to be inadequate, as several high-profile design flaws have manifested themselves in the field. How do we even specify the important safety and/or security properties for complex systems, much less assure that implementations maintain them?

In this book, we examine several leading-edge design and verification technologies that have been successfully applied to microprocessor systems at various levels – from arithmetic circuits to microcode to instruction sets to operating systems to applications. We focus on recent hardware, software, and system designs that have actually been built and deployed, and feature systems that have been certified at high Evaluation Assurance Levels, namely the Rockwell Collins AAMP7G microprocessor (EAL7) and the Green Hills INTEGRITY-178B separation kernel (EAL6+). The contributing authors to this book have endeavored to bring forth compelling new material on significant, modern design and verification efforts; many of the results described herein were obtained only within the past year.

This book is intended for practicing computer engineers, computer scientists, professionals in related fields, as well as faculty and students, who have an interest in the intersection of high-assurance design, microprocessor systems, and formal verification, and wish to learn about current developments in the field. It is not intended as a tutorial for any of the aforementioned subjects, for which excellent texts already exist.

The approach we have taken is to treat each subject that we examine in depth. Rather than presenting a mere summary of the work, we provide details: how exactly the design is specified and implemented, how the design is formalized, what the exact correctness properties are, and how the design is shown to meet its specification. Thus, for example, the text describes precisely how a radix-4 SRT divider for a commercial microprocessor is implemented and proven correct. Another chapter details how a complete AES-128 design is refined from an abstract specification traceable back to the FIPS-197 document all the way down to a high-performance hardware-based implementation that is provably equivalent to the FIPS-197 specification. The contributors to this book have made an extraordinary effort to produce descriptions of their work that are as complete and detailed as possible.

Just as important, this book takes the time to derive useful correctness statements from basic principles. The text formally develops the "GWV" family of information flow theorems used in the certifications of the AAMP7G as well as the INTEGRITY-178B kernel, proceeding from a simple model of computing systems (expressed in the language of the PVS theorem prover) called the "calculus of indices", and formally developing the GWVr1 and GWVr2 information flow theorems. The text presents a proof of how the GWV formulation maps to classical noninterference, as well as a proof demonstrating that a system can be shown to uphold the GWVr1 information flow specification via model checking. Another example of development from basic principles can be found in the chapter detailing the refinement frameworks used in the verification of the seL4 microkernel.

Along the way, we delve into a number of "tools of the trade" – theorem provers (e.g., ACL2, HOL4, Isabelle/HOL, PVS), model checkers (BAT, NuSMV, Prover), and equivalence checkers – and show how formal verification toolchains are increasingly able to parse the actual engineering artifacts under analysis, with the result that the formal models are much more detailed and accurate. Another tool trend noted in several chapters is the combination of theorem proving, model checking, symbolic simulation, etc., to produce a final verification result. A notable example of this combination of techniques documented in the text is the process used by Centaur Technology to verify their x86 compatible processors. The book also highlights ways in which ideas from, for example, theorem proving and compiler design, are being combined to produce novel and useful capabilities.

We open with a description of the ACL2 theorem prover, utilized in a number of the succeeding chapters, and then proceeds from the design and verification of basic high-performance hardware found in modern microprocessors (e.g., a divider circuit), to larger functional units (FPUs and cryptographic units), to pipelines, and then on to microcoded functions. The book then addresses the microprocessor at the instruction set level, focusing on machine code proofs, particularly the "decompilation into logic" approach pioneered by users of the HOL4 system. After developing some basic information flow theorems, the book turns its attention to operating system verification, with chapters on recent successes in information flow verification (INTEGRITY-178B) as well as functional correctness (seL4). We then progress to the application level, with a chapter dealing with the specification and checking of software contracts for conditional information flow, targeting the SPARK high-assurance Ada language subset. The book concludes with a description of tools and techniques for model checking information flow properties, applicable to both hardware and software systems that have been implemented using model-based development tools such as Simulink or SCADE.

Looking back over the year or so since I was first contacted by my superb editor at Springer, Charles Glaser, about the possibility of doing a book, I am both surprised and pleased to see that the finished product is quite close to my original, overly ambitious vision (as a colleague, perhaps rightly, called it at the time). To the extent that this book succeeds in describing compelling new results in the design and verification of high-assurance microprocessor systems, I have my contributing authors to thank; conversely, I am solely to blame for any shortcomings in reaching that goal. I would also like to thank my employer, Rockwell Collins, Inc., particularly John Borghese, Ray Kamin, Matt Wilding, and Ray Richards in the Advanced Technology Center, for providing me with time and facilities to work on the book. Several farsighted US government employees also played a significant role in the development of the technologies described herein, and I wish to especially acknowledge the efforts of Bill Legato, Brad Martin, and Mark Vanfleet of DoD, Jahn Luke and Dave Homan of AFRL, as well as Rick Butler and Paul Miner of NASA. Finally, I wish to thank my family and friends for their patience and support.

Cedar Rapids, IA   David S. Hardin

Contents

Contributors

Torben Amtoft Kansas State University, Manhattan, KS, USA, tamtoft@cis.ksu.edu

Sally Browning Galois, Inc., Portland, OR, USA, sally@galois.com

Jared Davis Centaur Technology, Austin, TX, USA, jared@centtech.com

Anthony C. J. Fox University of Cambridge, Cambridge, UK, Anthony.Fox@cl.cam.ac.uk

Michael J. C. Gordon University of Cambridge, Cambridge, UK, mjcg@cl.cam.ac.uk

David A. Greve Rockwell Collins, Inc., Cedar Rapids, IA, USA, dagreve@rockwellcollins.com

David S. Hardin Rockwell Collins, Inc., Cedar Rapids, IA, USA, dshardin@rockwellcollins.com

John Hatcliff Kansas State University, Manhattan, KS, USA, hatcliff@cis.ksu.edu

Jonathan Hoag Kansas State University, Manhattan, KS, USA, jch5588@cis.ksu.edu

Warren A. Hunt, Jr. Centaur Technology, Austin, TX, USA
and
University of Texas at Austin, Austin, TX, USA, hunt@cs.utexas.edu; hunt@centtech.com

Matt Kaufmann University of Texas at Austin, Austin, TX, USA, kaufmann@cs.utexas.edu

Gerwin Klein NICTA, Sydney, NSW, Australia
and
University of New South Wales, Sydney, NSW, Australia, gerwin.klein@nicta.com.au

Guodong Li University of Utah, Salt Lake City, UT, USA, ligd@cs.utah.edu

Panagiotis Manolios Northeastern University, Boston, MA, USA,
pete@ccs.neu.edu

J Strother Moore University of Texas at Austin, Austin, TX, USA,
moore@cs.utexas.edu

Magnus O. Myreen University of Cambridge, Cambridge, UK,
Magnus.Myreen@cl.cam.ac.uk

Scott Owens University of Cambridge, Cambridge, UK,
Scott.Owens@cl.cam.ac.uk

Raymond J. Richards Rockwell Collins, Inc., Cedar Rapids, IA, USA,
rjricha1@rockwellcollins.com

Robby Kansas State University, Manhattan, KS, USA, robby@cis.ksu.edu

Edwin Rodríguez Kansas State University, Manhattan, KS, USA,
edwin@cis.ksu.edu

David M. Russinoff Advanced Micro Devices, Inc., Austin, TX, USA,
david.russinoff@amd.com

Thomas Sewell NICTA, Sydney, NSW, Australia, thomas.sewell@nicta.com.au

Konrad Slind Rockwell Collins, Inc., Bloomington, MN, USA,
klslind@rockwellcollins.com

Anna Slobodova Centaur Technology, Austin, TX, USA, anna@centtech.com

Sudarshan K. Srinivasan North Dakota State University, Fargo, ND, USA,
sudarshan.srinivasan@ndsu.edu

Sol Swords Centaur Technology, Austin, TX, USA, sswords@centtech.com

Lucas G. Wagner Rockwell Collins, Inc., Cedar Rapids, IA, USA,
lgwagner@rockwellcollins.com

Philip Weaver Signali Corp., Portland, OR, USA, pweaver@signalicorp.com

Michael W. Whalen Rockwell Collins, Inc., Bloomington, MN, USA,
mike.whalen@gmail.com

Matthew M. Wilding Rockwell Collins, Inc., Cedar Rapids, IA, USA,
mmwildin@rockwellcollins.com

Simon Winwood NICTA, Sydney, NSW, Australia
and
University of New South Wales, Sydney, NSW, Australia,
simon.winwood@nicta.com.au

ACL2 and Its Applications to Digital System Verification

Matt Kaufmann and J Strother Moore

1 Introduction

Digital systems designs are growing in complexity, with huge state spaces and even larger sets of possible execution paths through those spaces. Traditional simulation-based testing for complex systems covers only relatively few of those execution paths. One solution that is getting increased attention is *formal verification*: the application of mechanized mathematical techniques to verify design properties for *all* execution paths.

In this chapter, we introduce ACL2 – a programming language, logic, and proof development environment – and explore its use in the formal verification of digital systems. Section 2 explores the general problem of proving properties of digital machines. Next, Sect. 3 introduces ACL2. Finally, in Sect. 4, we illustrate how to apply ACL2 to reason about digital system models and programs running on them.

2 Some Basic Decisions

Suppose we wanted to describe an abstract digital machine and to prove properties of it. In what language should we describe the machine? In what mathematical system should we construct our proofs? Different researchers – and this is a topic of ongoing research! – have different answers. In this section we give ours.

If asked "what is a digital system?" in a context requiring a mathematical answer, many designers would think in terms of finite or infinite state machines, state charts, programs, recursive functions, or any of the other equivalent formalization of computational processes. We tend to use state machines or recursive functions modeling some aspects of the system, e.g., functionality, timing, etc. The first question confronting a person wishing to prove theorems about a digital system is in

---

M. Kaufmann (✉)
University of Texas at Austin, Austin, TX, USA
e-mail: kaufmann@cs.utexas.edu

D.S. Hardin (ed.), *Design and Verification of Microprocessor Systems
for High-Assurance Applications*, DOI 10.1007/978-1-4419-1539-9_1,
© Springer Science+Business Media, LLC 2010

what language or languages does one operate? Most readers will immediately think of several programming languages well suited to describing abstract state machines, e.g., VHDL, C, Java, etc., as well as various modeling languages.

But we assume that most readers are less familiar with mathematical proof systems and so we explore that more closely here. A formal *logic* includes a *syntax* describing all well-formed *formulas* together with some *rules of inference* that allow one to deduce "new" formulas from "old" ones. There are many formal logics: propositional calculus, first-order predicate calculus, higher order logic, a plethora of modal logics, etc. A logic can be turned into a *theory* by identifying some formulas as *axioms*. It is conventional to assign *meaning* to formulas so that some are considered *valid* with respect to the axioms and the rules of inference preserve validity: if the "old" formulas used by a rule are valid, so is the new formula produced by the rule. The axioms characterize the properties of the primitive objects and the rules of inference let us deduce additional properties. A *proof* is a derivation of a formula from some axioms using the rules of inference. The new formula is called a *theorem* (or, if it is just a stepping stone used to derive a more interesting formula, a *lemma*) and it is valid. Thus, a way to determine that a formula is valid is to construct a proof of the formula. A piece of software that checks that an alleged proof is indeed a proof is called, naturally enough, a *proof checker*. A piece of software that attempts to discover a proof given an alleged theorem is a *theorem prover*. If we were talking about the game of chess, proof checking is akin to checking that every move in a published game is legal and theorem proving is akin to playing the game against an opponent (the alleged theorem).

Because numbers are so basic in digital machines – they are typically used as data, as addresses, as instructions, etc. – our theory will have to have axioms characterizing the numbers (especially the natural numbers including 0 and 1) and possibly other "atomic" objects relevant to our machine or its description, such as symbols and strings. In addition, our axioms should allow us to build composite structures such as lists or vectors, tables, trees, graphs, etc., because these are typically used in machine descriptions.

All of these objects are *inductively* constructed in the sense that they can be built up by repeatedly applying some basic functions. For example, the naturals are built from 0 by adding 1. Vectors can be built from the empty vector by the operation of adding an element. Tables can be built by adding a row or column (vector) to the empty table, etc. To reason about inductive objects, one must have an inductive rule of inference. The most familiar is: to prove that ψ is valid for all natural numbers n, (a) prove ψ when n is 0, and (b) prove that if ψ holds for the natural number n, then ψ holds for $n+1$. Similar rules of inference can be formulated for vectors, tables, etc.

Because no preexisting theory will contain all the concepts needed to describe our machine, our logical theory must also support the notion of *definition*, allowing us to define new functions and relations. For example, we might need to speak of the "physical address," if any, *associated with* a given "virtual address" in a certain table. The main idea behind a definitional principle is to add one or more new axioms that characterize the properties of some previously undistinguished ("new") symbol, e.g., `virtual-address`. This is called *extending* the theory. It is not possible to add just any axiom to a sound theory while preserving soundness, e.g., one cannot "define" the new function symbol *bad* by the new axiom $bad(x) = 1 + bad(x)$ and expect the new theory to be sound! But there are conditions under which extensions are known to preserve soundness. One is that the new axiom describes a function whose values on all inputs can be algorithmically computed. So adding the axiom $sq(x) = x \times x$ to define the new symbol *sq* ("square") is safe: it is just an abbreviation for something we could have written in the unextended theory. Even recursive definitions can be shown to be safe, provided it is possible to show that you could compute them out for any given values of the inputs. This *termination requirement* excludes our *bad* function; and while termination cannot be decided for all possible rules, it is certainly possible to determine that certain rules always terminate (e.g., those that decrease a natural number by 1 on every iteration and stop when the number reaches 0).

Discouragingly, it is not possible to write a theorem prover that can prove all the theorems (and only theorems) of any interesting theory. This result was established by Gödel in 1931 and was described in many textbooks on mathematical logic, e.g., [55], but was proved for the first time mechanically by Shankar in 1985 [54] using ACL2's predecessor Nqthm [2]. As soon as the theory is rich enough to encompass the natural numbers, equality, addition, and multiplication, it is *undecidable* in two important senses: there are valid formulas for which proofs do not exist, and no algorithm can always terminate and correctly announce whether a formula has or does not have a proof. So while we can always build proof checkers, we cannot build theorem provers that are in any sense *complete*. Sometimes our theorem provers will run forever or stop and announce "I can't prove the alleged theorem but it might be a theorem nevertheless!"

Returning to the question of reasoning about microprocessor designs, if we choose to describe our machine in a conventional hardware design or programming language, e.g., VHDL or C, then we will need a way to convert machine descriptions into logical terms. For example, we might provide a translator from our description language into the logic. This allows the descriptions to be written in familiar terms but can complicate our reasoning by the imposition of an extra layer of abstraction or indirection.

An alternative is to describe the machine in logical terms in the first place. This means we reason about the description itself, but invariably our formal descriptions must be evaluated by engineers and designers who have in mind some actual artifact, and when our descriptions are in unfamiliar terms that evaluation process can be more difficult.

3 ACL2

The name "ACL2" is used to refer to three distinct systems: a functional (side-effect free) programming language, a formal mathematical theory, and an interactive mechanized theorem prover and proof development environment. ACL2 stands for

"*A* Computational *L*ogic for *A*pplicative *C*ommon *L*isp" and hence might have been written "ACL^2."

In this section, we introduce ACL2 and provide a few references, but much more information is available about it. In particular, the reader is invited to visit the ACL2 home page [30], where one can find tutorials, demos, publications, mailing lists, and an extensive hypertext user's manual. The home page also links to home pages of past ACL2 workshops, where one may find many dozens of papers and slide presentations about ACL2 and its applications, many of which are on the topic of digital system verification. Some of that work is also published in journals and conference proceedings, but there are advantages to starting with the workshops Web sites (1) ACL2 Workshop papers are freely available in full on the Web (other that those of the first workshop; see [31]), (2) the Web site often provides supplemental material in the form of ACL2 source material (e.g., "ACL2 books") or exercises, and (3) the reader will learn about the standards and activities of the ACL2 community by browsing the workshop Web sites.

3.1 A Programming Language

As a programming language, ACL2 is an extension of a subset of applicative (side-effect free) Common Lisp [56]. Like Lisp, ACL2 is syntactically untyped: any type of object may be passed to any function. However, there are several runtime types: various types of numbers, characters, strings, symbols, and ordered pairs used to construct lists, trees, and other abstract structures. Predicates in the language allow the programmer to determine, at runtime, what type of objects have been passed to a function and to code accordingly. The "various types of numbers" (and the predicates recognizing them) are natural numbers (`natp`), integers (`integerp`), rationals (`rationalp`), and complex rationals (`acl2-numberp`). The last are complex numbers in which the two parts, real and imaginary, are both rational. While Lisp provides floating point numbers, ACL2 does not. Each type in the sequence above includes the numbers of the preceding type. Examples of each successive type are `23`, `-245`, `22/7`, and `#c(5  3)`, the last being the Lisp way of writing the complex number $5 + 3i$.

Unlike Lisp, ACL2 is first order: functions are not objects. While Higher Order Logics are common, we decided for reasons of efficiency and logical simplicity to disallow functional objects.

The syntax of ACL2 is just the prefix syntax of Lisp and Scheme. For example, `(* n (fact (- n 1)))` could be written more conventionally as $n \times \text{fact}(n - 1)$. The value of `(- n 1)` is the difference between the value of the variable `n` and `1`. The value of `(fact (- n 1))` is the value returned by calling the function `fact` on the value of `(- n 1)`. And the value of `(* n (fact (- n 1)))` is the product of the value of `n` with the value returned by `fact` on the value of `(- n 1)`. We henceforth refrain from repeating "the value of" and just say that `(* n (fact (- n 1)))` is "n times `fact` applied to n minus 1."

Among the primitive data types supported by ACL2, aside from the numbers, are characters (e.g., #\A, #\a, and #\Newline); strings (e.g., "Hello, world!"); the Boolean symbols t and nil denoting *true* and *false*, respectively; other symbols (e.g., LOAD and X); and pairs of objects discussed at greater length below. Various primitive functions allow us to manipulate, compare, and construct such objects. For example, (+ n 1) is the sum of n and 1, and (< x y) is t if x is less than y, and nil otherwise. The predicate equal takes two objects and returns t if they are the same and nil otherwise.

The most basic "method" for constructing composite structures is cons, which takes two objects and returns the ordered pair containing them. The car of such a pair is the first object and the cdr is the second. The predicate consp returns t or nil according to whether its argument is an ordered pair constructed by cons.

Because ACL2 is untyped, any object x can be treated as a *list* denoting a finite sequence of objects. If x is a cons pair whose car is a and whose cdr is d, then x denotes the sequence whose first element is a and whose remaining elements are those denoted by the list d. If x is not a cons pair, it denotes the empty sequence. It is conventional to use nil as the representative of the empty list, though any non-cons will do. When a list is nil-terminated it is said to be a "true-list."

For example, the sequence containing the elements 1, 2, and 3 is denoted by the object constructed by (cons 1 (cons 2 (cons 3 nil))). This object is written (1 2 3) in ACL2.

Given any object v in ACL2, it is possible to write a literal constant in the language that evaluates to that object, namely ' v.

Suppose we wished to define the function sum-list which takes one argument, x, and treats it as a list of numbers, returning the sum of its elements. Then we could define sum-list as follows:

```
(defun sum-list (x)
  (if (consp x)
      (+ (car x)
         (sum-list (cdr x)))
      0))
```

In particular, if x is an ordered pair, then sum-list returns the sum of the car (head) of x and the result of recursively summing the elements of the cdr (rest) of x; otherwise, sum-list returns 0. Thus, (sum-list '(1 2 3)) is 6.

The syntax of ACL2 may be extended by a powerful abbreviational facility called "macros." A *macro* is similar to a function except it operates on the syntax of the language. For example, it is possible to define list as a macro so that (list x(+ y 1) (* y 2)) is just an abbreviation for (cons x (cons (+ y 1)(cons (* y 2) nil))). The idea is that list is defined (as a macro) so that when given the list (x (+ y 1) (* y 2)) it returns the list (cons x (cons (+ y 1) (cons (* y 2) nil))). The syntax of ACL2 is defined so that if a macro is called in an alleged expression, the macro is evaluated on the argument list (*not* on the value of arguments) and the object returned is treated as the expression meant by the macro call. The process is repeated recursively until no macros are called in the expression. It is beyond the scope of this chapter to explain macros in greater detail. Macros allow the full power of recursive computation to be exploited in the syntax; amazing abbreviations can be introduced.

3.2 A Logical Theory

When we refer to ACL2 as a logical theory, we essentially mean first-order predicate calculus with equality, a set of axioms describing certain data types in Common Lisp, a Definitional Principle, an Induction Principle, and several very useful derived rules of inference. We only sketch the logic here. See [26, 27, 32, 34].

The Common Lisp specification [56] describes a variety of data types and the primitive operations on them. Following [56] we identified a set of axioms, some of which are shown below.

Axioms
```
x ≠ nil → (if x y z) = y
x = nil → (if x y z) = z
(car (cons x y)) = x
(cdr (cons x y)) = y
(consp x) = t → (cons (car x) (cdr x)) = x
```

The arrow, "→," is logical implication. The first axiom above may be read "when x is different from nil, then (if x y z) is y," or, alternatively, "(if x y z) is y, provided x is different from nil." The others should make sense now.

So that everything can be written in Lisp, the standard logical operators, "and," "or," "not," "implies," and "iff," are characterized by equality axioms defining the logical symbols in terms of the primitive if–then–else. For example, (and p q) is equivalent to (if p q nil) and is thus true (non-nil) if both p and q are true and false (nil) otherwise.

In order to formulate both a Definitional Principle and an Induction Principle, ACL2 includes the "ordinals." The ordinals are an extension of the natural numbers and can be compared, added, multiplied, and exponentiated. Informally, think of ω as the limit of the sequence 0, 1, 2, ..., i.e., ω is the "least infinite" ordinal. One can imagine an ordinal 1 larger than ω, written as $\omega + 1$, and start to build up an infinite collection of such ordinals by algebraic addition, multiplication, and exponentiation. The limit of all the ordinals one can construct that way is called ε_0 and may be thought of as $\omega^{\omega^{\omega^{\cdots}}}$. In ACL2, we represent all the ordinals below ε_0 in terms of natural numbers and conses similar to the construction in [11] but improved upon in [37]. We also define recursively the "less than" total order o< on such ordinals. We assume the *well foundedness* of o< on ordinals, i.e., there is no infinite strictly decreasing (according to o<) sequence of ordinals.

The ACL2 Definitional Principle permits one to define new function symbols in terms of old ones, including recursive use of the new symbols, provided there is an ordinal measure that decreases according to o< in each recursive call. This principle is logically conservative (nothing new can be proved in the extended logic unless it involves the newly introduction symbol) and thus insures the soundness of the extended theory. Informally, the Definitional Principle insures that every defined function terminates. To use the principle, the user must exhibit an appropriate measure of the size of the arguments, although by default the system uses an often-convenient notion of the "size" of an object.

For example, here is a definition of a list concatenation function that takes two lists, x and y, and returns a list containing all the elements of x followed by all the elements of y, in sequence.

```
(defun app (x y)
  (if (consp x)
      (cons (car x) (app (cdr x) y))
      y))
```

This definition terminates because the size of x decreases on each recursion. Once accepted, it is *executable*: given two concrete lists x and y the answer can be computed directly from the definitions and axioms. (app '(1 2 3) '(4 5 6)) evaluates to (1 2 3 4 5 6).

In duality with the Definitional Principle, the ACL2 Induction Principle permits one, in the induction step of a proof of ψ, to assume any number of instances each of o<-smaller "size," as determined by a user-supplied ordinal-valued measure on the variables in ψ. To use the induction principle one must exhibit the measure, prove that it is ordinal valued, and prove that it decreases under the case analysis and variable substitutions used in the induction steps.

For example, using induction and axioms about the basic data types, ACL2 can automatically prove the theorem

```
(equal (sum-list (app x y))
       (+ (sum-list x) (sum-list y)))
```

by induction on x. The base case is the formula

```
(implies (not (consp x))
         (equal (sum-list (app x y))
                (+ (sum-list x) (sum-list y))))
```

and the induction step is the formula

```
(implies (and (consp x)
              (equal (sum-list (app (cdr x) y))
                     (+ (sum-list (cdr x)) (sum-list y))))
         (equal (sum-list (app x y))
                (+ (sum-list x) (sum-list y))))
```

Both follow easily from the definitions of sum-list and app and the axioms.

The Defchoose Principle permits one to introduce a function symbol that returns an object satisfying a given property, if such an object exists. It is conservative. Using it we can provide the power of full first-order quantification, e.g., it is possible to introduce a function that returns t or nil according to whether there exists an object satisfying a given formula. This power is provided by a mechanism akin to Skolemization and is called defun-sk. Of course, unlike the functions in [56], the symbols introduced by the choose and Skolemization principles cannot be "run" (i.e., applications of such function symbols to literal constants cannot necessarily be reduced to literal constants using the axioms).

The Encapsulation Principle permits one to introduce undefined symbols that are axiomatized to have given properties. To use the principle, one must exhibit functions ("witnesses") that have the given properties. Constrained functions cannot be "run."

Associated with the Encapsulation Principle is a derived rule of inference permitting one to *Functionally Instantiate* any theorem to derive a new theorem, by replacing the function symbols in the old theorem by other function symbols provided one can prove that the "other function symbols" satisfy the same constraints as the replaced symbols. This gives ACL2 a "second-order" feel.

For example, one might introduce the notions of an abstract state and an abstract state transition function, step, constrained to preserve abstract states. Then one might define a function, run, that repeatedly steps a state n times. One might prove that run preserves abstract states and enjoys certain commutativity properties. These theorems follow inductively from the constraints on step. Then one might define a concrete notion of state, a specific state transition function on such concrete states, and a concrete "run" function. In particular, this concrete "run" function might model a microprocessor and be "runnable" in the sense of being executable on literal constants. Using functional instantiation one could then immediately derive that the concrete "run" function preserves the concrete notion of states and enjoys those commutativity properties, provided one can prove that the concrete "step" function preserves the concrete notion of state.

3.3 A Mechanical Theorem Prover and Proof Environment

The ACL2 theorem prover accepts as input an alleged theorem and attempts to find a proof. Also provided as input is a *logical world* which, roughly speaking, lists the axioms, user-supplied definitions, and proved theorems of the session. Additional arguments allow the user to provide hints and other pragmatic advice. The theorem prover either terminates with a success or failure message or else runs until interrupted by the user. When it reports failure, the meaning is simply that no proof was found and not that the formula is not a theorem. However, the prover's output can help the user to find a counterexample if the formula is not valid or to formulate useful lemmas to assist in subsequent proof attempts [28]. ACL2 also includes numerous proof debugging tools [24, 29].

It tries to find proofs by applying a suite of standard proof techniques based largely on the function symbols used in the goal formula. The most common technique is simplification, which replaces the goal formula by a set of supposedly simpler subgoals, e.g., by splitting apart conjoined formulas and by attempting to normalize the terms in each subgoal. This normalization is done by *rewriting* with axioms, definitions, and previously proved theorems.

For example, suppose earlier in the session the user issued the command:

```
(defthm app-right-id
  (implies (true-listp x)
           (equal (app x nil) x))
  :rule-classes :rewrite)
```

and that the command succeeded. This command instructs ACL2 to prove the formula `(implies (true-listp x) (equal (app x nil) x))`, which may be read "if x is a true-list then concatenating a `nil` on the right of x (with `app`) is the identity operation." If a proof is found, the command will store into the logical world a rewrite-rule derived from the formula, named `app-right-id`. Henceforth, whenever ACL2's simplifier encounters a term of the form (app α nil), then provided the user has not somehow instructed the system not to use the rule, it will replace the encountered term by α, provided it can prove (true-listp α). It will also report that it used `app-right-id`.

Other proof techniques used by ACL2 include induction, decision procedures for propositional calculus, equality, and linear arithmetic, elimination of "destructor" function symbols [e.g., in conventional notation we might eliminate p/d and $mod(p, d)$ by replacing p by $(q \times d) + r$, under suitable conditions on the variables, so that p/d becomes q and $mod(p, d)$ becomes r], heuristic use of equalities, generalization, and elimination of irrelevance. In addition, the user can introduce entirely new proof techniques via *metafunctions* [3, 23] and *clause-processors* [33] provided they can be proved sound by ACL2. As illustrated above with rewriting, almost all of ACL2's proof techniques are affected by the rules present in the logical world and the user's pragmatic advice given at the time an alleged theorem is presented.

Thus, the ACL2 user essentially *programs* the theorem prover by proving an appropriate collection of theorems. This is why we say that ACL2 is both interactive and automated.

ACL2 encourages users to share and reuse work by providing the "book" mechanism. A *book* is just a file of ACL2 commands, typically definitions and theorems. To admit the function definitions in a book or to prove all the theorems in a book, it is sometimes necessary to prove lemmas that are uninteresting to the ultimate user of the book. The book mechanism allows the author to mark certain commands within the book as *local*. When a book is *included* in a session, it is as though all the nonlocal commands of the book were executed in that session.

Books are hierarchical: a book may include other books in its construction. This raises a version control problem: What if the author of a subbook changes that book so that it no longer logically implies the results obtained from it in the construction of some superior book? To help prevent this, books may be *certified*, producing a "certificate" file which contains the "fingerprints" of all concerned books and other information meant both to detect version control problems as well as to make the loading of books faster by avoiding proofs.

The best designed books formalize a set of useful concepts and configure ACL2 to reason about those formal concepts effectively. ACL2's home page provides access to hundreds of user-developed books that attempt to provide convenient settings for reasoning about arithmetic in its many forms (integer, rational, bit-vector, floating point, etc.), list processing, and other more specialized domains. But virtually every project introduces important concepts that are idiosyncratic to that project or model and these become the standard books within that project.

The design of the ACL2 theorem prover – driven as it is by the available rules – puts a great deal of burden on the user in one sense and relieves a great burden in another. In the early- and mid-stages of a major project, ACL2 users are often less worried about proving the particular goal theorem than they are about discovering and codifying proof strategies in the form of books. This can make progress slow. But once a suitable collection of books has been created allowing the "automatic" proof of the main theorem, the investment pays off in the later stages of the project where the verified artifact is repeatedly modified, elaborated, and improved. Each such modification requires a proof of correctness. This is called *proof maintenance*. If each modification required a "hand-made" proof, even if it is produced from an explicitly described earlier proof, progress here would be much slower. But the ACL2 user frequently finds that minor modifications to the artifact can be verified automatically, and that when that verification fails the problem is just in the region changed, requiring the incremental formalization of the insight that justified the modification to the artifact.

3.4 *Efficiency*

To use ACL2 in industrial projects has required a great deal of engineering aside from the more obviously necessary attention to powerful proof techniques.

The ACL2 user highly values its capability as a programming language. This makes well-designed ACL2 models doubly useful: as simulation engines and as formal artifacts about which one can reason. But to be useful as a simulation platform, the user must have the means to make them efficiently executable without complicating their logical semantics. We give three examples.

It is possible to annotate Lisp code with declarations asserting the intended types of the values of the variables at runtime. The Common Lisp compiler merely assumes that these declarations are accurate and lay down suitably optimized code. To see how this can produce much more efficient execution, consider the representation of numbers in Lisp. To provide semantic cleanliness, every number is a first class "Object," a state of affairs that may be achieved by "boxing" every number, i.e., by representing every number as an instance of some class with the actual magnitude of the number somehow coded in the fields. Operationally, every number is then a pointer to one or more memory locations containing the binary representation of the number. But this would make common arithmetic exceedingly slow because, naively, one would have to allocate memory to add two numbers. Most Lisp implementations solve this by essentially preallocating all the "small" numbers, e.g., those representable in 30 bits, often by representing the small two's complement integers by the corresponding binary addresses. Thus, if two small numbers are to be added and their sum is known (or assumed) to be small, the compiler can generate the native add instruction on the host processor. If one of these conditions is not met, the compiler must lay down code that "unboxes," sums, and "boxes" appropriately.

But it is logically dangerous to assume that the declarations are accurate. Thus, ACL2 provides a mechanism (called *guard verification*) by which the user can not only annotate functions and formulas without affecting their logical meanings, but also prove mechanically the accuracy of those annotations. ACL2 will not execute optimized code unless the declarations have been verified; in the absence of verification, ACL2 arranges for Common Lisp to execute code as it would have no declarations present. Thus, the user can annotate code for efficiency; the annotations do not complicate the proof obligations when reasoning about the functional properties of the code, and if those annotations are subsequently verified, the user will observe his or her functions executing much faster.

A second example is the mbe ("must be equal") facility, described at length in [17]. This mechanism allows the user to provide, for example, two entirely different definitions for a function, one to use in logical reasoning and one to use in execution, but produces the obligation to prove the two definitions equivalent. For example, it may be easiest to reason about a function defined in a natural, recursive style but more efficient to compute it with an iterative (tail-recursive) scheme eliminating the need for a stack at the expense of some clever invariant among some auxiliary variables.

A third example is support for *single-threaded objects*, also called *stobjs* [5]. Semantically, these objects are just list structures. But "under the hood," they use destructive operations such as updating an array at a given index. Syntactic restrictions are imposed and enforced so that the user cannot detect the difference between the functional semantics alleged by the axioms and the imperative implementation. With stobjs one can attain execution that approaches the efficiency of C (e.g., 90% of the speed of C on a small microprocessor model is reported by Hardin et al. in [18]).

Other efficiency issues that have required careful engineering attention include manipulating and even printing large formulas and constants [25], implementing proof techniques efficient enough to deal with industrial-scale definitions and formulas, being able to load deep hierarchies of books into the session in a reasonable amount of time, and being able to recertify deep and broad hierarchies of books in parallel fast enough to allow overnight rebuilding of recently modified systems.

4 A Simple Microprocessor Model

What does a microprocessor model look like in ACL2? Below we show a complete description of a very simple machine akin to the Java Virtual Machine. To save space, we use a slightly small font in our displays. The state of the machine is a list of four items, a program counter (here called ipc because the name pc is already defined in ACL2), a list of local variable values (`locals`), a list (stack) of values pushed (`stack`), and a list of instructions (`code`). Below we define `make-state` to construct a state and the four accessor functions to return the corresponding components. A semicolon (`;`) delimits a comment to the end of the line.

```
(defun make-state (ipc locals stack code)
  (list ipc locals stack code))
(defun ipc     (s) (nth 0 s))
(defun locals (s) (nth 1 s))
(defun stack   (s) (nth 2 s))
(defun code    (s) (nth 3 s))
(defun next-inst (s)          ; fetch instruction at ipc in code
  (nth (ipc s) (code s)))
```

Macros can be written that make it easy to describe the shape of a state and have the appropriate functions defined automatically.

Instructions are represented by lists, where the 0th element of the list (i.e., the `car`, but below written (`nth 0 inst`)) is the symbolic name of the opcode and the remaining elements are the operands. For example, an `ICONST` instruction, which will cause the machine to push a literal constant onto the stack, will be represented (`ICONST` c), where c (i.e., (`nth 1 inst`)) is the literal constant to push. The list of local variable values is also accessed with `nth`; a new list of values may be obtained from an old one by `update-nth`, which "replaces" the item at a given location in a list by another item. Stacks will be represented here as lists, with the `car` being the top-most element and the `cdr` being the rest of the stack, i.e., the result of popping the stack. The code is the analog of an execute-only memory and is just a list of instructions.

Clearly we are describing this little machine at a very high level of abstraction. It will be possible to refine our description. For example, instructions and opcodes could be integers instead of lists and symbols; the opcode and operands could be obtained by arithmetic shifting and masking. The code could be refined into a list of integers representing the contents of successive machine addresses and would most likely become a read–write memory. The stack could be an integer address into a writable region of memory, etc. We return to this imagined lower level description below, but for now we continue to describe the machine at the abstract level.

For each opcode we define a function that executes the instructions with that opcode, i.e., the function takes an instruction (of the given opcode) and a state and returns the next state. Such functions are called *semantic functions* because they give the semantics of our instructions. The semantic function must define the program counter, locals, stack, and code of the next state.

```
(defun execute-ICONST (inst s) ; (ICONST c): push c onto stack
  (make-state (+ 1 (ipc s))
              (locals s)
              (cons (nth 1 inst) (stack s))
              (code s)))
```

```lisp
(defun execute-ILOAD (inst s)   ; (ILOAD n): push locals[n] onto stack
  (make-state (+ 1 (ipc s))
              (locals s)
              (cons (nth (nth 1 inst)
                         (locals s))
                    (stack s))
              (code s)))


(defun execute-IADD (inst s)    ; (IADD): pop first two items on stack
  (declare (ignore inst))       ;            and push their sum
  (make-state (+ 1 (ipc s))
              (locals s)
              (cons (+ (nth 1 (stack s))
                       (nth 0 (stack s)))
                    (cdr (cdr (stack s))))
              (code s)))


(defun execute-ISTORE (inst s) ; (ISTORE n): pop stack into local[n]
  (make-state (+ 1 (ipc s))
              (update-nth (nth 1 inst)
                          (nth 0 (stack s))
                          (locals s))
              (cdr (stack s))
              (code s)))

(defun execute-ISUB (inst s)    ; (ISUB): pop first two items on stack
  (declare (ignore inst))       ;           and push their difference
  (make-state (+ 1 (ipc s))
              (locals s)
              (cons (- (nth 1 (stack s))
                       (nth 0 (stack s)))
                    (cdr (cdr (stack s))))
              (code s)))


(defun execute-IMUL (inst s)     ; (IMUL): pop first two items on stack
  (declare (ignore inst))        ;           and push their product
  (make-state (+ 1 (ipc s))
              (locals s)
              (cons (* (nth 1 (stack s))
                       (nth 0 (stack s)))
                    (cdr (cdr (stack s))))
              (code s)))


(defun execute-GOTO (inst s) ; (GOTO k):  jump by k, i.e., add k to
  (make-state (+ (nth 1 inst) (ipc s)) ;  program counter
              (locals s)
              (stack s)
              (code s)))
```

```
(defun execute-IFLE (inst s)      ; (IFLE k):    if top item on stack
  (make-state (if (<= (nth 1 (stack s)) 0)  ; is less than or equal to 0
                  (+ (nth 1 inst) (ipc s))  ; jump by k
                (+ 1 (ipc s)))              ; else skip
              (locals s)
              (cdr (stack s))              ; always pop stack
              (code s)))
```

It should be obvious how to add other instructions and other state components to this machine. When the semantic functions for all instructions have been defined, we introduce, below, a single function which takes an arbitrary instruction and steps the state accordingly, by simply doing a "big switch" on the opcode of the instruction and invoking the appropriate semantic function. Note below that if an unknown instruction is encountered, its semantics is a no-op: the new state is the old state.

```
(defun do-inst (inst s)
  (case (nth 0 inst)
    (ICONST (execute-ICONST inst s))
    (ILOAD  (execute-ILOAD  inst s))
    (ISTORE (execute-ISTORE inst s))
    (IADD   (execute-IADD   inst s))
    (ISUB   (execute-ISUB   inst s))
    (IMUL   (execute-IMUL   inst s))
    (GOTO   (execute-GOTO   inst s))
    (IFLE   (execute-IFLE   inst s))
    (otherwise s)))
```

We finally define the single-step state transition function simply to fetch the next instruction and "do" it.

```
(defun istep (s)
   (do-inst (next-inst s) s))
```

We conclude by defining the function run that takes a "schedule" and a state and runs the state according to the schedule. Here, we just step the state once for every element of the schedule, but in general the schedule provides additional input to the istep function and may indicate signals received at that step on external pins, which "thread" in a multithreaded state to step, etc.

```
(defun run (sched s)
   (if (endp sched)
       s
     (run (cdr sched) (istep s))))
```

Recall our earlier hints of a lower level machine modeled mainly with integers. If we had such a lower level model we could proceed to formalize and prove the relation between it and this one, e.g., a commuting diagram or bisimulation between the two machines. This has been done many times in ACL2 (and its predecessor Nqthm [4]) for complex and realistic models, including some pipelined machines [1, 7, 20, 22, 38, 40, 50, 52].

The abstract model above may be executed on concrete data. For example, consider the code produced by a straightforward compilation of the pseudocode for factorial:

```
a = 1;
while (n > 0)
  a = n * a;
   n = n-1;;
return a;
```

If we allocate local variable n to `locals[0]` and local variable a to `locals[1]`, the resulting code is as shown below. We define the ACL2 constant `*fact-code*` to be this code snippet.

```
(defconst *fact-code*
  '((ICONST 1)        ;;;  0
    (ISTORE 1)        ;;;  1 a = 1;
    (ILOAD 0)         ;;;  2 while           ; loop: ipc=2
    (IFLE 10)         ;;;  3   (n > 0)
    (ILOAD 0)         ;;;  4
    (ILOAD 1)         ;;;  5
    (IMUL)            ;;;  6
    (ISTORE 1)        ;;;  7   a = n * a;
    (ILOAD 0)         ;;;  8
    (ICONST 1)        ;;;  9
    (ISUB)            ;;; 10
    (ISTORE 0)        ;;; 11   n = n-1;
    (GOTO -10)        ;;; 12                  ; jump to loop
    (ILOAD 1)         ;;; 13
    (HALT)            ;;; 14   return a;
    ))
```

Note that the unknown HALT instruction at program counter 14 halts the machine since stepping that instruction is a no-op.

The following expression evaluates a state by taking 100 steps. The term (repeat 'TICK 100) just returns a list of 100 repetitions of the symbol TICK and is used as a schedule here. The make-state below constructs the initial state: the program counter is 0; we have two locals, the 0th (called n in our pseudocode) having the value 5 and the 1st (called a) having the value 0; the stack is empty; and the code is our code constant. Note that this example illustrates running the code snippet with n= 5.

```
(run
 (repeat 'TICK 100)     ; 100 clock ticks
 (make-state
  0                     ; ipc
  '(5 0)                ; locals: n=5, a=0
  nil                   ; stack
  *ifact-code*          ; code
  ))
```

The answer computed by ACL2 is:

```
(14                      ; final ipc
 (0 120)                 ; final locals: n=0, a=120
 (120)                   ; final stack
 ((ICONST 1)             ; our code, again
  (ISTORE 1)
  ...
  (HALT)))
```

Note that the program counter points to the HALT instruction at location 14 and
the value 120 is on top of the stack. This aspect of ACL2, that models can serve
as simulation engines for the artifact, is highly valued and much engineering work
has been done, both by the implementors of ACL2 and by its user community, in
making these executions efficient [5, 13, 17, 18, 48].

Given an ACL2 semantics for the "pseudocode," e.g., an interpreter for that language, and a compiler modeled in ACL2, it is possible to state and prove (if true)
that the compiler is correct in the sense of preserving the semantics of the high-level
language when the object code is run on the machine. This has been done many
times in ACL2 and Nqthm [2, 8, 10, 40, 58]. It is also possible to prove general theorems about such machines. For example, in [35], Liu gives a realistic model of the
Java Virtual Machine and proves that class loading preserves certain invariants on
the heap and class table. At Rockwell Collins, security properties of microprocessor
and operating system models have been verified using this approach [12, 14–16].

It is possible to configure ACL2 to prove theorems about the behavior of code
when run on a given machine model. For example, here is an easily proved theorem
about the code above – "easy" when the appropriate books are loaded into ACL2.

```
(defthm ifact-correct
  (implies (natp n)
           (equal (run (ifact-sched n)
                       (make-state 0
                                   (cons n (cons a nil))
                                   stack
                                   *ifact-code*))
                  (make-state 14
                              (cons 0 (cons (! n) nil))
                              (cons (! n) stack)
                              *ifact-code*)))))
```

The two make-state expressions above are the initial and final states of a run of
our code. In the initial state, the program counter is set to 0 and the local variables
have the unknown symbolic values n and a. In the final state, obtained by running
the initial state some number of steps determined by the function ifact-sched,
the program counter is 14 (i.e., points to the HALT) and we find the factorial of
n pushed on the stack. This equivalence holds provided n is a natural number.
This theorem is proved by induction and establishes the correctness of the factorial snippet.

In [41], we present this same machine as well as a compiler for the pseudocode
used, and we describe the methodology used to configure ACL2 to do proofs about

code. Code is often verified with ACL2 and Nqthm this way, e.g., the binary code produced by gcc for the Berkeley C String Library is verified against a model of the Motorola 68020 in [6], a game of Nim is verified in [57] for the fabricated and verified microprocessor described in [22], and JVM bytecode is verified against a detailed model of the JVM in [35]. In [7], a commercially designed digital signal microprocessor is modeled and verified to implement a given ISA. The machine, the Motorola CAP, included a three-stage pipeline that exposed many programmer visible hazards. A tutorial on pipelined machine verification in ACL2 may be found in [50]. In [53], a pipelined microarchitecture with speculative execution, exceptions, and program modification capability is verified.

Because ACL2 is a general purpose mathematical logic, many different proof styles and strategies can be brought to bear on the problem of verifying properties of systems described in it [36, 43, 45]. These include commuting diagrams, bisimulation and stuttering bisimulation, direct proofs of functional equivalence, and a variety of methods related to inductive assertions. Different proof styles may be mixed.

5 Variations on the Theme

We have used a simple example to illustrate a methodology for using ACL2 to verify correctness properties for digital system models. But there are many ways to use ACL2 for digital system verification.

Another common approach to code verification in ACL2 skips the problem of formalizing the entire microprocessor and instead models the code as an ACL2 function. This is akin to modeling our factorial code with the expression `(ifact n 0)`, where

```
(defun ifact (n a)
  (if (zp n)              ; if n <= 0
      a                   ;   return a
      (ifact (- n 1)      ; else loop with a:=n*a, n:=n-1
             (* n a))))
```

The production of functions like `ifact` is sometimes done by hand and other times is automated by tools that embed the semantics of the code. For example, the correctness of the floating-point division algorithm on the AMD K5 microprocessor was proved using the former approach (in which the semantics of the relevant microcode was modeled as an ACL2 function) [42]. The correctness of divide and square root on the IBM Power4 was also proved that way [51]. At Rockwell Collins, the AAMP7G cryptoprocessor was modeled this way and a security-critical separation property was proved, allowing Rockwell to obtain NSA MILS certification[19]. Such certification requires a comparison of the model to the actual design (if they are different) and it behooves the modeler to produce a model with as much fidelity to the actual design as possible.

For this reason, or when code proofs are to be done repeatedly, it is more often worthwhile to build mechanical translators or to formalize the actual design language so that fidelity is assured – or at least has to be checked just once at the "metalevel." This methodology has been used many times in the verification of floating point designs at the register-transfer level (RTL) at AMD [9,48,49] and is similar to the use by Rockwell Collins [13] of "reader macros" that allow near-isomorphism between models written in ACL2 and in a very simple subset of C. Another approach is to formalize the hardware description language itself [21, 22, 46, 50].

Other chapters in this volume describe applications of ACL2 in more detail. In particular, the chapter by Hunt et al. describes research at Centaur Technology using an extension of ACL2 that supports a formalized hardware design language and efficient *symbolic simulation* to reason about RTL.

Traditional formal software verification often uses a *verification condition generator* (VCG) to take code annotated with assertions and produce proof obligations, so that the provability of those obligations implies that the assertions always hold. The "interpreter" approach of Sect. 4, by contrast, formalizes execution of the code so that properties can be proved directly against the semantics. However, the VCG approach can be emulated using the interpreter approach, see [39, 44].

In all ACL2 work, the problem arises of what to do when the theorem prover fails to find a proof. As noted, it is the responsibility of the user to determine whether the original goal formula was not a theorem (perhaps by constructing a counterexample from the failed proof) or to formulate lemmas and hints to lead ACL2 to a proof. Mechanizing the construction of counterexamples for ACL2 formulas is a topic of ongoing work in the ACL2 community, e.g., Reeber and Hunt [47] describe a system that unrolls a certain class of ACL2 formulas and attempts to prove them with a SAT solver, converting any counterexample produced by SAT into an ACL2 counterexample. Chamarthi, Dillinger, Kaufmann, and Manolios are working on an approach using failed proofs and testing (private communication 2009).

6 Summary

Because ACL2 is a functional programming language with a proof development environment, it can be used in a wide variety of ways to model digital systems and, once modeled in ACL2, system properties may be proved using a wide variety of proof styles and strategies. On the other hand, because ACL2 is a general-purpose system, albeit one with a computational orientation, much infrastructure often has to be built in it to analyze systems described in conventional design languages. Because the proof environment is rule-driven and heavily influenced by the database of previously proved or included results, ACL2 has a fairly steep learning curve. It is not enough to know what the language means and how to prove a given theorem: getting ACL2 to prove a theorem often requires learning how to program ACL2 effectively with lemmas. This upfront cost is often repaid in two ways. First, systems of large size can be tackled because the "automatic" strategy is carried out mechanically. Second, modifications to the system can often be verified with an incremental amount of effort on the part of the user. Nevertheless, in order to lead ACL2 to proofs about industrial-scale designs the user must be tenacious and must apply talent in mathematics, programming, and pattern recognition.

Acknowledgements We wish to thank the entire ACL2 community for helping push this work along. We especially thank Warren Hunt for his recognition that Nqthm was particularly well suited to modeling and verifying microprocessors and his decades of leadership in this area.

The preparation of this chapter was funded in part by NSF grants IIS-0417413 and EIA-0303609 and DARPA/NSF CyberTrust grant CNS-0429591. Kaufmann also thanks the Texas – United Kingdom Collaborative for travel support to Cambridge, England and the Computer Laboratory at the University of Cambridge for hosting him during preparation of this paper.

References

1. Bevier W, Hunt WA Jr, Moore JS, Young W (1989) Special issue on system verification. J Autom Reason 5(4):409–530
2. Boyer RS, Moore JS (1979) A computational logic. Academic, New York
3. Boyer RS, Moore JS (1981) Metafunctions: proving them correct and using them efficiently as new proof procedures. In: Boyer RS, Moore JS (eds) The correctness problem in computer science. Academic, London
4. Boyer RS, Moore JS (1997) A computational logic handbook, 2nd edn. Academic, New York
5. Boyer RS, Moore JS (2002) Single-threaded objects in ACL2. In: PADL 2002, LNCS 2257. Springer, Heidelberg, pp 9–27. http://www.cs.utexas.edu/users/moore/publications/stobj/main.ps.gz
6. Boyer RS, Yu Y (1996) Automated proofs of object code for a widely used microprocessor. J ACM 43(1):166–192
7. Brock B, Hunt WA Jr (1999) Formal analysis of the motorola CAP DSP. In: Hinchey M, Bowen J (eds) Industrial-strength formal methods. Springer, Heidelberg
8. Flatau AD (1992) A verified implementation of an applicative language with dynamic storage allocation. PhD thesis, University of Texas at Austin
9. Flatau A, Kaufmann M, Reed D, Russinoff D, Smith E, Sumners R (2002) Formal verification of microprocessors at AMD. In: Proceedings of designing correct circuits 2002. http://www.cs.chalmers.se/~ms/DCC02/Slides.html
10. Goerigk W, Hoffmann U (1998) Rigorous compiler implementation correctness: how to prove the real thing correct. In: Proceedings FM-TRENDS'98 international workshop on current trends in applied formal methods, Boppard, LNCS
11. Goodstein RL (1964) Recursive number theory. North-Holland, Amsterdam
12. Greve D, Wilding M (2002) Evaluatable, high-assurance microprocessors. In: NSA high-confidence systems and software conference (HCSS), Linthicum, MD. http://hokiepokie.org/docs/hcss02/proceedings.pdf
13. Greve D, Wilding M, Hardin D (2000) High-speed, analyzable simulators. In: Kaufmann M, Manolios P, Moore JS (eds) Computer-aided reasoning: ACL2 case studies. Kluwer, Boston, MA, pp 113–136
14. Greve D, Wilding M, Vanfleet WM (2003) A separation kernel formal security policy. In: ACL2 workshop 2003, Boulder, CO. http://www.cs.utexas.edu/users/moore/acl2/workshop-2003/
15. Greve D, Richards R, Wilding M (2004) A summary of intrinsic partitioning verification. In: ACL2 workshop 2004, Austin, TX. http://www.cs.utexas.edu/users/moore/acl2/workshop-2003/
16. Greve D, Wilding M, Richards R, Vanfleet M (2005) Formalizing security policies for dynamic and distributed systems. In: Proceedings of systems and software technology conference (SSTC) 2005, Salt Lake City, UT. http://hokiepokie.org/docs/sstc05.pdf

17. Greve D, Kaufmann M, Manolios P, Moore JS, Ray S, Ruiz-Reina JL, Sumners R, Vroon D, Wilding M (2008) Efficient execution in an automated reasoning environment. J Funct Program 18(01):15–46

18. Hardin D, Wilding M, Greve D (1998) Transforming the theorem prover into a digital design tool: from concept car to off-road vehicle. In: Hu AJ, Vardi MY (eds) Computer-aided verification – CAV '98, Lecture notes in computer science, vol 1427. Springer, Heidelberg. See http://pobox.com/users/hokie/docs/concept.ps

19. Hardin DS, Smith EW, Young WD (2006) A robust machine code proof framework for highly secure applications. In: ACL2 '06: proceedings of the sixth international workshop on the ACL2 theorem prover and its applications. ACM, New York, NY, pp 11–20. DOI http://doi.acm.org/10.1145/1217975.1217978

20. Hunt WA Jr (1994) FM8501: a verified microprocessor. LNAI 795. Springer, Heidelberg

21. Hunt WA Jr (2000) The DE language. In: Kaufmann M, Manolios P, Moore JS (eds) Computer-aided reasoning: ACL2 case studies. Kluwer, Boston, MA, pp 151–166

22. Hunt WA Jr, Brock B (1992) A formal HDL and its use in the FM9001 verification. Philosophical Transactions of the Royal Society: Physical and Engineering Sciences, 339(1652):35–47

23. Hunt WA Jr, Kaufmann M, Krug RB, Moore JS, Smith EW (2005) Meta reasoning in ACL2. In: Hurd J, Melham T (eds) 18th international conference on theorem proving in higher order logics: TPHOLs 2005, Lecture notes in computer science, vol 3603. Springer, Heidelberg, pp 163–178

24. Kaufmann M (2008) Aspects of ACL2 User interaction (Invited talk, 8th international workshop on user interfaces for theorem provers (UITP 2008), Montreal, Canada, August, 2008). See www.ags.uni-sb.de/~omega/workshops/UITP08/kaufmann-UITP08/talk.html

25. Kaufmann M (2009) Abbreviated output for input in ACL2: an implementation case study. In: Proceedings of ACL2 workshop 2009. http://www.cs.utexas.edu/users/sandip/acl2-09

26. Kaufmann M, Moore JS (1997) A precise description of the ACL2 logic. Technical report, Deparment of Computer Sciences, University of Texas at Austin. http://www.cs.utexas.edu/users/moore/publications/km97a.ps.gz

27. Kaufmann M, Moore JS (2001) Structured theory development for a mechanized logic. J Autom Reason 26(2):161–203

28. Kaufmann M, Moore JS (2008) An ACL2 tutorial. In: Proceedings of theorem proving in higher order logics, 21st international conference, TPHOLs 2008. Springer, Heidelberg. See http://dx.doi.org/10.1007/978-3-540-71067-7_4

29. Kaufmann M, Moore JS (2008) Proof search debugging tools in ACL2. In: A Festschrift in honour of Prof. Michael J. C. Gordon FRS. Royal Society, London

30. Kaufmann M, Moore JS (2009) The ACL2 home page. http://www.cs.utexas.edu/users/moore/acl2/

31. Kaufmann M, Manolios P, Moore JS (eds) (2000a) Computer-aided reasoning: ACL2 case studies. Kluwer, Boston, MA

32. Kaufmann M, Manolios P, Moore JS (2000b) Computer-aided reasoning: an approach. Kluwer, Boston, MA

33. Kaufmann M, Moore JS, Ray S, Reeber E (2009) Integrating external deduction tools with ACL2. J Appl Logic 7(1):3–25

34. Kaufmann M, Moore JS, Ray S (in press) Foundations of automated induction for a structured mechanized logic

35. Liu H (2006) Formal specification and verification of a jvm and its bytecode verifier. PhD thesis, University of Texas at Austin

36. Manolios P (2000) Correctness of pipelined machines. In: Formal methods in computer-aided design, FMCAD 2000, LNCS 1954. Springer, Heidelberg, pp 161–178

37. Manolios P, Vroon D (2003) Ordinal arithmetic in ACL2. In: ACL2 workshop 2003, Boulder, CO. http://www.cs.utexas.edu/users/moore/acl2/workshop-2003/

38. Manolios P, Namjoshi K, Sumners R (1999) Linking theorem proving and model-checking with well-founded bisimulation. In: Computed aided verification, CAV '99, LNCS 1633. Springer, Heidelberg, pp 369–379

39. Matthews J, Moore JS, Ray S, Vroon D (2006) Verification condition generation via theorem proving. In: Proceedings of 13th international conference on logic for programming, artificial intelligence, and reasoning (LPAR 2006), vol LNCS 4246, pp 362–376
40. Moore JS (1996) Piton: a mechanically verified assembly-level language. Automated reasoning series. Kluwer, Boston, MA
41. Moore JS (2008) Mechanized operational semantics: lectures and supplementary material. In: Marktoberdorf summer school 2008: engineering methods and tools for software safety and security. http://www.cs.utexas.edu/users/moore/publications/talks/marktoberdorf-08/index.html
42. Moore JS, Lynch T, Kaufmann M (1998) A mechanically checked proof of the correctness of the kernel of the AMD5K86 floating point division algorithm. IEEE Trans Comput 47(9): 913–926
43. Ray S, Hunt WA Jr (2004) Deductive verification of pipelined machines using first-order quantification. In: Proceedings of the 16th international conference on computer-aided verification (CAV 2004), vol LNCS 3117. Springer, Heidelberg, pp 31–43
44. Ray S, Moore JS (2004) Proof styles in operational semantics. In: Hu AJ, Martin AK (eds) Formal methods in computer-aided design (FMCAD-2004), Lecture notes in computer science, vol 3312. Springer, Heidelberg, pp 67–81
45. Ray S, Hunt WA Jr, Matthews J, Moore JS (2008) A mechanical analysis of program verification strategies. J Autom Reason 40(4):245–269
46. Reeber E, Hunt WA Jr (2005) Formalization of the DE2 language. In: Correct hardware design and verification methods (CHARME 2005), vol LNCS 3725. Springer, Heidelberg, pp 20–34
47. Reeber E, Hunt WA Jr (2006) A SAT-based decision procedure for the subclass of unrollable list functions in ACL2 (SULFA). In: Proceedings of 3rd international joint conference on automated reasoning (IJCAR 2006). Springer, Heidelberg, pp 453–467
48. Russinoff DM, Flatau A (2000) RTL verification: a floating-point multiplier. In: Kaufmann M, Manolios P, Moore JS (eds) Computer-aided reasoning: ACL2 case studies. Kluwer, Boston, MA, pp 201–232
49. Russinoff D, Kaufmann M, Smith E, Sumners R (2005) Formal verification of floating-point RTL at AMD using the ACL2 theorem prover. In: IMACS'2005 world congress
50. Sawada J (2000) Verification of a simple pipelined machine model. In: Kaufmann M, Manolios P, Moore JS (eds) Computer-aided reasoning: ACL2 case studies. Kluwer, Boston, MA, pp 137–150
51. Sawada J (2002) Formal verification of divide and square root algorithms using series calculation. In: Proceedings of the ACL2 workshop, 2002, Grenoble. http://www.cs.utexas.edu/users/moore/acl2/workshop-2002
52. Sawada J, Hunt WA Jr (1998) Processor verification with precise exceptions and speculative execution. In: Computer aided verification, CAV '98, LNCS 1427. Springer, Heidelberg, pp 135–146
53. Sawada J, Hunt WA Jr (2002) Verification of FM9801: an out-of-order microprocessor model with speculative execution, exceptions, and program modification capability. Formal Methods Syst Des 20(2):187–222
54. Shankar N (1994) Metamathematics, machines, and Godel's proof. Cambridge University Press, Cambridge
55. Shoenfield JR (1967) Mathematical logic. Addison-Wesley, Reading, MA
56. Steele GL Jr (1990) Common lisp the language, 2nd edn. Digital Press, Burlington, MA
57. Wilding M (1993) A mechanically verified application for a mechanically verified environment. In: Courcoubetis C (ed) Computer-aided verification – CAV '93, Lecture Notes in Computer Science, vol 697. Springer, Heidelberg. See ftp://ftp.cs.utexas.edu/pub/boyer/nqthm/wilding-cav93.ps
58. Young WD (1988) A verified code generator for a subset of Gypsy. Technical report 33. Computational Logic Inc., Austin, TX

A Mechanically Verified Commercial SRT Divider

David M. Russinoff

1 Introduction

The Sweeney–Robertson–Tocher (SRT) division algorithm [8, 13], owing to its susceptibility to efficient hardware implementation, is ubiquitous in contemporary microprocessor designs. It is also notoriously prone to implementation error. Analysis of the algorithm's most celebrated incarnation, the defective FDIV circuitry of the original Intel Pentium floating-point unit [7], suggests that thorough verification of an SRT divider by testing alone is a practical impossibility.

This development has been a boon to the enterprise of formal hardware verification. One early response to the 1994 revelation of the Pentium bug was Bryant's BDD-based analysis [2], which established a critical invariant of an SRT circuit but was limited by the practical constraints of the model-theoretic approach. More complete results were subsequently achieved with the use of mechanical theorem provers by Kapur and Subramanian [5], Ruess et al. [9], and Clarke et al. [3]. All of these efforts shared the goal of demonstrating the effectiveness of a particular prover in exposing a specific bug and consequently focused on the relevant aspects of the underlying algorithm. Moreover, although developed independently, all the three were coincidentally based on the same execution model, a high-level circuit design proposed in 1981 by Taylor [12].

A different objective is pursued in the work reported here: a comprehensive machine-checked proof of correctness of a commercial hardware design. The object of investigation is a register-transfer logic (RTL) model of a radix-4 SRT integer divider, to be implemented as a component of the AMD processor code-named "Llano." The theorem prover used in this project is ACL2 [1].

The required behavior of the module is concisely specified in terms of the integer values X and Y (divisor and dividend) represented by the primary data inputs and

D.M. Russinoff (✉)
Advanced Micro Devices, Inc., Austin, TX, USA
e-mail: david.russinoff@amd.com

D.S. Hardin (ed.), *Design and Verification of Microprocessor Systems for High-Assurance Applications*, DOI 10.1007/978-1-4419-1539-9_2,
© Springer Science+Business Media, LLC 2010

the corresponding values Q and R (quotient and remainder) of the data outputs. Under suitable input constraints, the following relations must be satisfied:

1. $Y = QX + R$
2. $|R| < |X|$
3. Either $R = 0$ or R and X have the same sign.

Regrettably (from a verification perspective), the simplicity of this behavioral specification is not reflected in the design. In contrast to Taylor's circuit, which uses only five state-holding registers, the divider of the Llano processor uses 56. In order to address this complexity, the proof is divided into four parts, which model the design at successively lower levels of abstraction.

At the highest level, as discussed in Sect. 2, we establish the essential properties of the underlying SRT algorithm. Our description of the algorithm is based on an unspecified radix, 2^r. In the case of interest, we have $r = 2$, which means that two quotient bits are generated per cycle. The main result of this section pertains to the iterative phase of the computation, which generates the sequences of partial remainders $p_0, \ldots, p_n$, quotient digits $m_1, \ldots, m_n$, and resulting partial quotients $Q_0, \ldots, Q_n$. We also address several relevant issues that are ignored in the proofs cited above (1) prescaling of the divisor and dividend and postscaling of the remainder; (2) determination of the required number n of iterations, which depend on the relative magnitudes of the operands; (3) incremental ("on-the-fly") computation of the quotient, which involves the integration of positive and negative quotient digits; and (4) derivation of the final remainder and quotient R and Q, as specified above, from the results R' and Q' of the iteration, which are characterized by $Y = Q'X + R'$ and $|R'| \leq |X|$.

In the radix-4 case, the quotient digits are confined to the range $-3 \leq m_k \leq 3$. Each m_k is read from a table of $4 \times 32 = 128$ entries according to indices derived from the normalized divisor d and the previous partial remainder p_{k-1} and is used to compute the next partial remainder by the recurrence formula $p_k = 4p_{k-1} - m_k d$. At the second level of abstraction, in Sect. 3, we present the actual table used in our implementation, which was adapted from the IBM z990 [4], and prove that it preserves the invariant $|p_k| \leq |d|$.

At the third level, the algorithm is implemented in XFL, a simple formal language developed at AMD for the specification of the AMD64 instruction set architecture. XFL is based on unbounded integer and arbitrary precision rational data types and combines the basic constructs of C with the logical bit vector operations of Verilog in which AMD RTL designs are coded. The XFL encodings of the lookup table and the divider are displayed in Appendices 1 and 2. Like most XFL programs, this code was automatically generated from a hand-coded C++ program, which has been subjected to testing for the purpose of validating the model.

The XFL model is significantly smaller than the RTL, which consists of some 150 kilobytes of Verilog code, but it is designed to perform the same sequence of register-transfer-level operations while avoiding low-level implementation concerns. Thus, much of the complexity of the design is captured at this third level, including several essential features that are absent from higher-level models such as Taylor's circuit specification:

1. A hardware implementation of Taylor's model, which computes an explicit representation of the partial remainder on each iteration, would require a time-consuming full-width carry-propagate adder, resulting in a prohibitively long cycle time. In contrast, a typical contemporary commercial implementation such as the Llano divider stores the remainder in a redundant form, which may be computed by a much faster carry-save adder. A single full-width addition is then performed at the end of the iterative phase.
2. The derivation of the final results R and Q from the intermediate values R' and Q' involves consideration of the special cases $R' = 0$ and $R' = \pm X$. Timing considerations dictate that these conditions be detected in advance of the full-width addition that produces R'. This requires special logic for predicting cancellation.
3. The module is also responsible for detecting overflow, i.e., a quotient that is too large to be represented in the target format. This involves an analysis that is performed concurrently with the final computation of the quotient.

Each of these complications introduces a possible source of design error that cannot be ignored. In Sect. 4, we present a complete proof of the claim that the algorithm is correctly implemented by the XFL model.

The lowest level of abstraction to be considered is that of the RTL itself. The proof of equivalence between the RTL and XFL models represents a significant portion of the overall effort, involving the analysis of a complex state machine, innumerable timing and scheduling issues, and various other implementation concerns. However, this part of the proof would be of relatively little interest to a general readership; moreover, neither space nor proprietary confidentiality allows its inclusion here.

Thus, the purpose of this paper is an exposition of the proof of correctness of the Llano divider as represented by the XFL model. The presentation is confined to standard mathematical notation, avoiding any obscure special-purpose formalism, but assumes familiarity with the general theory of bit vectors and logical operations, making implicit use of the results found in [10]. Otherwise, the proof is self-contained and surveyable, with one exception: Lemma 6, which provides a set of inequalities that are satisfied by the entries of the lookup table, involves machine-checked computation that is too extensive to be carried out by hand.

We emphasize, however, that a comprehensive statement of correctness of the RTL module itself has been formalized in the logic of ACL2 and its proof has been thoroughly checked with the ACL2 prover. This includes a formalization of the proof presented here, along with a detailed proof of equivalence between the XFL and RTL models. For this purpose, the XFL model was recoded directly in ACL2 and the RTL module was translated to ACL2 by a tool that was developed for this purpose [11]. Thus, the validity of the proof depends only on the semantic correctness of the Verilog–ACL2 translator and the soundness of the ACL2 prover, both of which have been widely tested.

While the ultimate objective of formal verification is a proof of correctness, its utility may best be demonstrated through the exposure of bugs that might otherwise have gone undetected. In the present case, three bugs that had already survived extensive directed testing were found through our analysis, at three distinct levels of abstraction. Once identified, all the three were readily corrected.

First, an entry of the lookup table (the one in the upper right corner of Fig. 1) was missing from the original design. This was strikingly similar to the original Pentium bug insofar as the entry was thought to be inaccessible. The error was not exposed by initial directed testing, which was designed to hit all accessible table entries on the first iteration. Subsequent analysis and testing revealed that the entry can be reached, but only after nine iterations.

A second problem was detected in connection with the same table entry. If the partial remainder is close to -2, it appears to be possible (although no test case has yet been constructed to confirm this possibility), as a result of the two's complement encoding scheme, for the approximation derived from the redundant representation to be close to $+2$ instead. The original design did not account for this occurrence, which would have resulted in a quotient digit with the wrong sign.

Finally, a timing problem was detected in the RTL implementation, related to the possible cancellation and retransmission of the divisor input. This issue, which of course is not reflected in the algorithm or the XFL encoding, illustrates the inadequacy of a correctness proof based solely on a high-level design model.

2 SRT Division

The description of a division algorithm is typically simplified by interpreting its parameters as fractions. Thus, our presentation begins with a normalized representation of the divisor X,

$$d = 2^{-expo(X)} X,$$

where $expo(X)$ is the integer defined by

$$2^{expo(X)} \leq |X| < 2^{expo(X)+1},$$

and consequently, $1 \leq |d| < 2$. The dividend Y is similarly shifted to produce the initial partial remainder,

$$p_0 = 2^{-expo(X)-rn} Y,$$

where 2^r is the underlying radix of the computation and n, the number of iterations to be performed, is chosen to ensure that $|p_0| \leq |d|$.

On each iteration, the current partial remainder p_{k-1} is shifted by r bits and an appropriate multiple of d is subtracted to form the next partial remainder,

$$p_k = 2^r p_{k-1} - m_k d,$$

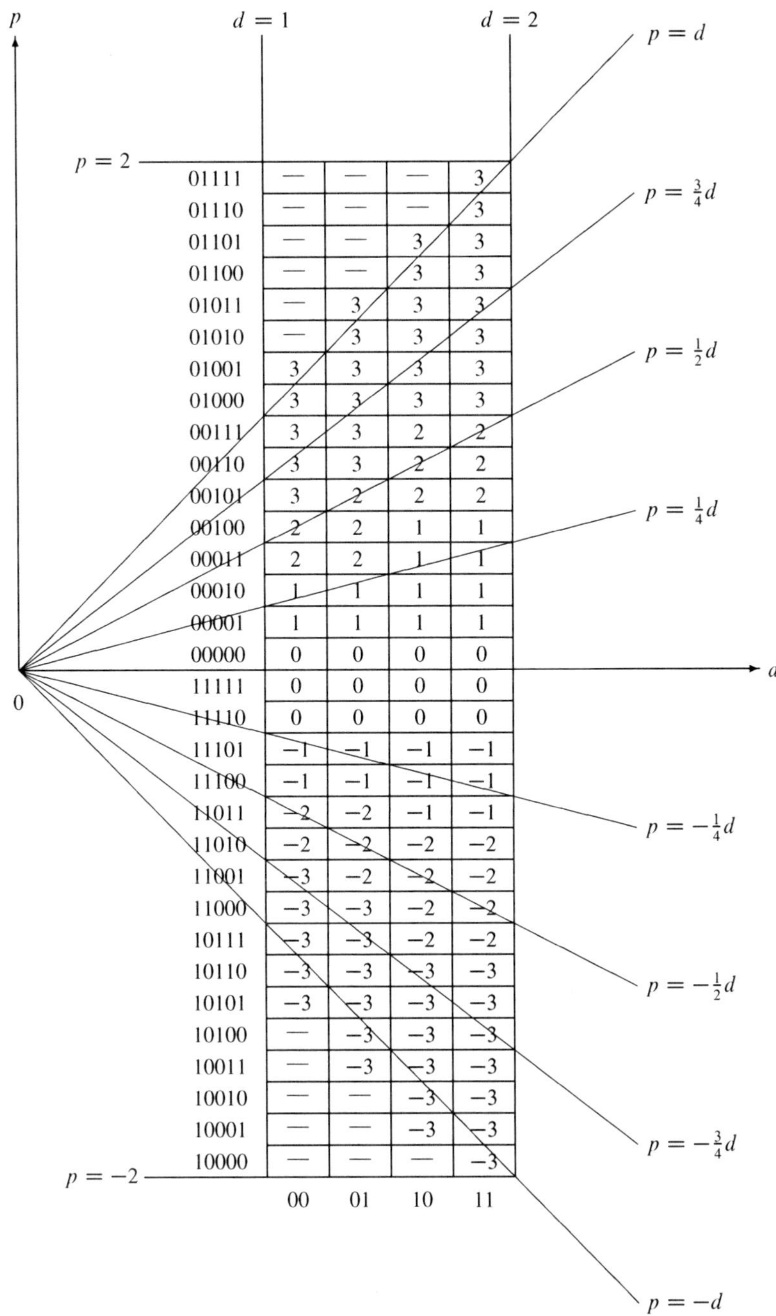

Fig. 1 SRT table

where the multiplier m_k contributes the accumulated quotient. The invariant $|p_k| \leq |d|$ is guaranteed by selecting m_k from the interval

$$\frac{2^r p_{k-1}}{d} - 1 \leq m_k \leq \frac{2^r p_{k-1}}{d} + 1.$$

For further motivation, refer to [6].

Lemma 1. *Let X, Y, r, and n be integers such that $X \neq 0$, $r > 0$, $n \geq 0$, and $|Y| \leq 2^{rn}|X|$. Let $d = 2^{-expo(X)}X$, $p_0 = 2^{-expo(X)-rn}Y$, and $Q_0 = 0$. For $k = 1,\ldots,n$, let*

$$p_k = 2^r p_{k-1} - m_k d$$

and

$$Q_k = 2^r Q_{k-1} + m_k,$$

where m_k is an integer such that if $|p_{k-1}| \leq |d|$, then $|p_k| \leq |d|$. Let $R = 2^{expo(X)}p_n$ and $Q = Q_n$. Then $Y = QX + R$ and $|R| \leq |X|$.

Proof. Clearly, $1 \leq |d| < 2$, and

$$|p_0| = 2^{-expo(X)-rn}|Y| \leq 2^{-expo(X)}|X| = |d|.$$

It follows by induction that $|p_k| \leq |d|$ for all $k \leq n$. We shall also show by induction that

$$p_k = 2^{rk} p_0 - Q_k d.$$

The claim clearly holds for $k = 0$, and for $0 < k \leq n$,

$$\begin{aligned}
p_k &= 2^r p_{k-1} - m_k d \\
&= 2^r \left(2^{r(k-1)} p_0 - Q_{k-1}d \right) - m_k d \\
&= 2^{rk} p_0 - (2^r Q_{k-1} + m_k)d \\
&= 2^{rk} p_0 - Q_k d.
\end{aligned}$$

In particular,

$$p_n = 2^{rn} p_0 - Q_n d$$

and

$$Y = 2^{expo(X)}2^{rn} p_0 = 2^{expo(X)}(Q_n d + p_n) = QX + R,$$

where $|R| = |2^{expo(X)}p_n| \leq |2^{expo(X)}d| = |X|$. $\qquad\square$

    A quotient and remainder that satisfy the conclusion of Lemma 1 may be easily adjusted to satisfy the specification stated in Sect. 1.

Lemma 2. *Let X, Y, Q', and R' be integers such that $X \neq 0$, $|R'| \leq |X|$, and $Y = Q'X + R'$. Let R and Q be defined as follows:*

(a) If $|R'| < |X|$ and either $R' = 0$ or $sgn(R') = sgn(Y)$, then $R = R'$ and $Q = Q'$;

(b) If (a) does not apply and $sgn(R') = sgn(X)$, then $R = R' - X$ and $Q = Q' + 1$;

(c) If (a) does not apply and $sgn(R') \neq sgn(X)$, then $R = R' + X$ and $Q = Q' - 1$.

Then $Y = QX + R$, $|R| < |X|$, and either $R = 0$ or $sgn(R) = sgn(Y)$.

Proof. We consider the three cases separately:

(a) In this case, the conclusion holds trivially.

(b) In this case,

$$Y = Q'X + R' = (Q' + 1)X + (R' - X) = QX + R$$

and $sgn(R') = sgn(X)$. If $|R'| = |X|$, then $R' = X$ and $R = 0$. Otherwise, we must have $Y \neq 0$ and $sgn(R') \neq sgn(Y)$. Since $|R'| < |X|$, $|R| = |R' - X| = |X| - |R'| < |X|$. Moreover, $sgn(R) \neq sgn(R')$, which implies $sgn(R) = sgn(Y)$.

(c) Here we have

$$Y = Q'X + R' = (Q' - 1)X + (R' + X) = QX + R$$

and $sgn(R') \neq sgn(X)$. If $|R'| = |X|$, then $R' = -X$ and $R = 0$. Otherwise, $Y \neq 0$ and $sgn(R') \neq sgn(Y)$. Since $|R'| < |X|$, $|R| = |R' + X| = |X| - |R'| < |X|$. Moreover, $sgn(R) \neq sgn(R')$, which implies $sgn(R) = sgn(Y)$. ☐

In an SRT implementation, the multiplier m_k of Lemma 1 represents a sequence of r bits that are appended to the quotient during the kth iteration. Although not required for the proof of the lemma, it may be assumed that in practice, $|m_k| < 2^r$. In particular, in our radix-4 implementation, we have $-3 \leq m_k \leq 3$. This provides a bound on the partial quotients.

Lemma 3. *Let $Q_0 = 0$ and for $k = 1, \ldots, n$, let $Q_k = 4Q_{k-1} + m_k$, where $-3 \leq m_k \leq 3$. Then $|Q_k| < 4^k$.*

Proof. By induction,

$$|Q_k| = |4Q_{k-1} + m_k| \leq |4Q_{k-1}| + |m_k| < 4\left(4^{k-1} - 1\right) + 4 = 4^k. \qquad ☐$$

If we could guarantee that $m_k \geq 0$, then we could maintain a bit vector encoding of the quotient simply by shifting in two bits at each step. In order to accommodate $m_k < 0$ without resorting to a full subtraction and simultaneously to provide an efficient implementation of Lemma 2, we adopt a scheme that involves three separate bit vectors representing the values Q_k, $Q_k - 1$, and $Q_k + 1$. The following lemma will be used in Sect. 4 to compute the final quotient. Note that each step in the computation may be implemented as a simple two-bit shift.

Lemma 4. *Let $Q_0 = 0$ and for $k = 1,\dots,n$, let $Q_k = 4Q_{k-1} + m_k$, where $-3 \le m_k \le 3$. Let $N > 0$. We define three sequences of bit vectors, E_k, E_k^-, and E_k^+, all of width N, as follows: $E_0 = 0$, $E_0^- = 2^N - 1$, $E_0^+ = 1$, and for $k = 1,\dots,n$,*

$$E_k = \begin{cases} (4E_{k-1} + m_k)[N-1:0] & \text{if } m_k \ge 0 \\ (4E_{k-1}^- + m_k + 4)[N-1:0] & \text{if } m_k < 0, \end{cases}$$

$$E_k^- = \begin{cases} (4E_{k-1} + m_k - 1)[N-1:0] & \text{if } m_k > 0 \\ (4E_{k-1}^- + m_k + 3)[N-1:0] & \text{if } m_k \le 0, \end{cases}$$

and

$$E_k^+ = \begin{cases} (4E_{k-1} + m_k + 1)[N-1:0] & \text{if } -1 \le m_k \le 2 \\ (4E_{k-1}^- + m_k + 5)[N-1:0] & \text{if } m_k < -1 \\ (4E_{k-1}^+)[N-1:0] & \text{if } m_k = 3. \end{cases}$$

Then for $k = 0,\dots,n$,

$$E_k = Q_k[N-1:0],$$

$$E_k^- = (Q_k - 1)[N-1:0],$$

and

$$E_k^+ = (Q_k + 1)[N-1:0].$$

Proof. The claim holds trivially for $k = 0$. In the inductive step, there are seven equations to consider. For example, if $m_k < -1$, then

$$\begin{aligned}
E_k^+ &= (4E_{k-1}^- + m_k + 5)[N-1:0] \\
&= (4(Q_{k-1} - 1)[N-1:0] + m_k + 5)[N-1:0] \\
&= (4(Q_{k-1} - 1) + m_k + 5)[N-1:0] \\
&= (4Q_{k-1} + m_k + 1)[N-1:0] \\
&= (Q_k + 1)[N-1:0].
\end{aligned}$$

The other six cases are handled similarly.   □

The implementation is also responsible for supplying the integer n of Lemma 1, which is required to satisfy $|Y| \le 2^{2n}|X|$ and represents the number of iterations to be performed. This may be accomplished by establishing an upper bound on the difference $expo(Y) - expo(X)$:

Lemma 5. *Let X, Y, and B be integers such that $X \ne 0$ and $expo(Y) - expo(X) \le B$. Let*

$$n = \begin{cases} \lfloor \frac{B}{2} \rfloor + 1 & \text{if } B \ge 0 \\ 0 & \text{if } B < 0. \end{cases}$$

Then $|Y| < 2^{2n}|X|$.

Proof. If $B \geq 0$, then

$$2n = 2\left(\left\lfloor \frac{B}{2} \right\rfloor + 1\right) \geq 2\left(\frac{B-1}{2} + 1\right) = B + 1,$$

so that $2n + expo(X) \geq B + 1 + expo(X) \geq expo(Y) + 1$ and

$$2^{2n}|X| \geq 2^{2n+expo(X)} \geq 2^{expo(Y)+1} > |Y|.$$

But if $B < 0$, then $expo(Y) < expo(X)$ and

$$|Y| < 2^{expo(Y)+1} \leq 2^{expo(X)} \leq |X| = 2^{2n}|X|. \qquad \square$$

The most intensive computation performed in the execution of the algorithm is that of the partial remainder, $p_k = 4p_{k-1} + m_k d$. In order for this to be completed in a single cycle, p_k is represented in a redundant form consisting of two bit vectors. Since $|m_k| \leq 3$, the term $m_k d$ is conveniently represented by up to two vectors corresponding to $\pm d$ and $\pm 2d$, depending on m_k. Thus, the computation of p_k is implemented as a two-bit shift (multiplication by 4) of p_{k-1} followed by a 4–2 compression. The details are deferred to Sect. 4.

The most challenging task is the computation of the quotient digit m_k. This is the subject of Sect. 3.

3 Quotient Digit Selection

In this section, we define a process for computing the quotient bits m_k of Lemma 1 and prove that the invariant $|p_k| \leq |d|$ is preserved. The problem may be formulated as follows:

> Given rational numbers d and p such that $1 \leq |d| < 2$ and $|p| \leq |d|$, find an integer m such that $-3 \leq m \leq 3$ and $|4p - dm| \leq |d|$.

We may restrict our attention to the case $d > 0$, since the inequalities in the above objective are unaffected by reversing the signs of both d and m. Thus, we have $1 \leq d < 2$ and $-2 < p < 2$. These constraints determine a rectangle in the dp-plane as displayed in Fig. 1, which is adapted from [4]. The rectangle is partitioned into an array of rectangles of width $\frac{1}{4}$ and height $\frac{1}{8}$. The columns and rows of the array are numbered with indices i and j, respectively, where $0 \leq i < 4$ and $0 \leq j < 32$. Let R_{ij} denote the rectangle in column i and row j, and let (δ_i, π_j) be its lower left vertex. Thus,

$$R_{ij} = \left\{(d, p) \mid \delta_i \leq d < \delta_i + \frac{1}{4} \text{ and } \pi_j \leq p < \pi_j + \frac{1}{8}\right\}.$$

The numbering scheme is designed so that if $(d, p) \in R_{ij}$, then i comprises the leading two bits of the fractional part of d and j comprises the leading 5 bits of the two's complement representation of p.

The contents of the rectangles of Fig. 1 represent a function

$$m = \phi(i, j),$$

which is defined formally in Appendix 1 and may be implemented as a table of $4 \times 32 = 128$ entries. For a given pair (d, p), we derive an approximation (δ_i, π_j), which determines the arguments of ϕ to be used to compute the corresponding value of m. Ideally, this approximation would be simply determined by the rectangle R_{ij} that contains (d, p), i.e., i and j would be derived by extracting the appropriate bits of d and p. Since our implementation generates the encoding of d explicitly, d may indeed be approximated in this manner. Thus, $i = d[-1 : -2] = \lfloor 4(d - 1) \rfloor$, which yields

$$\delta_i \leq d < \delta_i + \frac{1}{4}.$$

On the other hand, as noted in Sect. 2, p is represented redundantly as a sum of two vectors. The index j may be derived by adding the high-order bits of these vectors, but as a consequence of this scheme, as we shall see in Sect. 4, instead of the optimal range of $\frac{1}{8}$, the accuracy of π_j is given by

$$\pi_j \leq p < \pi_j + \frac{1}{4}.$$

Thus, in geometric terms, we may assume that (d, p) is known to lie within the square S_{ij} formed as the union of the rectangle R_{ij} and the rectangle directly above it:

$$S_{ij} = \left\{ (d, p) \mid \delta_i \leq d < \delta_i + \frac{1}{4} \text{ and } \pi_j \leq p < \pi_j + \frac{1}{4} \right\}.$$

We would like to show that if $(d, p) \in S_{ij}$ and $m = \phi(i, j)$, then $|4p - dm| \leq d$, or, equivalently,

$$\frac{m - 1}{4} \leq \frac{p}{d} \leq \frac{m + 1}{4}.$$

We first present an informal argument, which will then be formalized and proved analytically.

The definition of ϕ is driven by the following observations:

1. Since $|p| \leq d$, (d, p) lies between the lines $p = d$ and $p = -d$. Therefore, if S_{ij} lies entirely above the line $p = d$ or entirely below the line $p = -d$, then m is inconsequential and left undefined. In all other cases, m is defined.
2. Since $p \leq d$, the upper bound

$$\frac{p}{d} \leq \frac{m + 1}{4}$$

is satisfied trivially if $m = 3$. In order to guarantee that this bound holds generally, it suffices to ensure that if $m \neq 3$, then S_{ij} lies below the line $p = \frac{(m+1)d}{4}$.

3. Similarly, since $p \geq -d$, the lower bound

$$\frac{p}{d} \geq \frac{m-1}{4}$$

is satisfied trivially if $m = -3$. In order to guarantee that this bound holds generally, it suffices to ensure that if $m \neq -3$, then S_{ij} lies above the line $p = \frac{(m-1)d}{4}$.

It is easily verified by inspection of Fig. 1 that in all cases in which m is defined, the conditions specified by (2) and (3) are satisfied and, consequently, the desired inequality holds. It should also be noted that in some cases, there is a choice between two acceptable values of m. If S_{ij} lies within the region bounded by $p = \frac{m}{4}d$ and $p = \frac{m+1}{4}d$, where $-3 \leq m \leq 2$, then the inequality is satisfied by both m and $m + 1$. For example, although we have assigned 3 as the value of $\phi(11b, 01000b)$, since $S_{11,01000}$ lies between $p = \frac{1}{2}d$ and $p = \frac{3}{4}d$, we could have chosen 2 instead.

The first step toward formalization is to express the conditions listed above in precise analytical terms:

1. S_{ij} lies entirely above the line $p = d$ if and only if its lower right vertex, $(\delta_i + \frac{1}{4}, \pi_j)$, lies on or above that line, a condition expressed by the inequality

$$\pi_j \geq \delta_i + \frac{1}{4}.$$

The condition that S_{ij} lies entirely below the line $p = -d$ is similarly determined by the location of its upper right vertex, $(\delta_i + \frac{1}{4}, \pi_j + \frac{1}{4})$, and is expressed by the inequality

$$\pi_j \leq -\left(\delta_i + \frac{1}{4}\right) - \frac{1}{4} = -\delta_i - \frac{1}{2}.$$

Thus, $m = \phi(i, j)$ is defined if and only if neither of these inequalities holds, i.e.,

$$-\delta_i - \frac{1}{2} < \pi_j < \delta_i + \frac{1}{4}.$$

2. The maximum value of the quotient $\frac{d}{p}$ in S_{ij} occurs at either the upper left or the upper right vertex, depending on the sign of their common p-coordinate, $\pi_j + \frac{1}{4}$. Thus, S_{ij} lies below the line $p = \frac{(m+1)d}{4}$ if and only if both vertices lie on or below the line, i.e,

$$\frac{\pi_j + \frac{1}{4}}{\delta_i} = \frac{4\pi_j + 1}{4\delta_i} \leq \frac{m+1}{4}$$

and

$$\frac{\pi_j + \frac{1}{4}}{\delta_i + \frac{1}{4}} = \frac{4\pi_j + 1}{4\delta_i + 1} \leq \frac{m+1}{4}.$$

3. The minimum value of $\frac{d}{p}$ in S_{ij} occurs at either the lower left or the lower right vertex, depending on the sign of π_j. Thus, S_{ij} lies above the line $p = \frac{(m-1)d}{4}$ if and only if both vertices lie on or above the line, i.e.,

$$\frac{\pi_j}{\delta_i} \geq \frac{m-1}{4}$$

and

$$\frac{\pi_j}{\delta_i + \frac{1}{4}} = \frac{4\pi_j}{4\delta_i + 1} \geq \frac{m-1}{4}.$$

We shall also require analytical expressions for δ_i and π_j as functions of i and j. The definition of δ_i is trivial.

Definition 1. For each integer i such that $0 \leq i < 4$,

$$\delta_i = 1 + \frac{i}{4}.$$

Since j is the five-bit two's complement representation of the signed integer $8\pi_j$, we have the following definition, in which the function $SgndIntVal(w, x)$ computes the value represented by a bit vector x with respect to a signed integer format of width w:

Definition 2. For each integer j such that $0 \leq j < 32$,

$$\pi_j = SgndIntVal(5, j) = \begin{cases} \dfrac{j}{8} & \text{if } j < 16 \\ \dfrac{j}{8} - 32 & \text{if } j \geq 16. \end{cases}$$

The formal statement of correctness of ϕ appears below as Lemma 7. The constraints on ϕ that were derived above are required in the proof. These are summarized in Lemma 6, which is proved by straightforward exhaustive computation.

Lemma 6. *Let i and j be integers, $0 \leq i < 4$ and $0 \leq j < 32$. Assume that $-\delta_i - \frac{1}{2} < \pi_j < \delta_i + \frac{1}{4}$ and let $m = \phi(i, j)$.*

(a) If $m \neq 3$, then $\max\left(\frac{4\pi_j + 1}{4\delta_i}, \frac{4\pi_j + 1}{4\delta_i + 1}\right) \leq \frac{m+1}{4}$;

(b) If $m \neq 3$, then $\min\left(\frac{\pi_j}{\delta_i}, \frac{4\pi_j}{4\delta_i + 1}\right) \geq \frac{m-1}{4}$.

Lemma 7. *Let d and p be rational numbers, $1 \leq d < 2$ and $|p| \leq d$. Let i and j be integers, $0 \leq i < 4$ and $0 \leq j < 32$, such that $\delta_i \leq d < \delta_i + \frac{1}{4}$ and $\pi_j \leq p < \pi_j + \frac{1}{4}$. Let $m = \phi(i, j)$. Then $|4p - dm| \leq d$.*

Proof. First note that since

$$\pi_j \leq p \leq d < \delta_i + \frac{1}{4}$$

and

$$\pi_j > p - \frac{1}{4} \geq -d - \frac{1}{4} > -\left(\delta_i + \frac{1}{4}\right) - \frac{1}{4} = -\delta_i - \frac{1}{2},$$

we may apply Lemma 6.

We must show that $-d \leq 4p - dm \leq d$, i.e.,

$$\frac{m-1}{4} \leq \frac{p}{d} \leq \frac{m+1}{4}.$$

First we establish the upper bound. Since

$$\frac{p}{d} \leq 1 = \frac{3+1}{4},$$

we may assume $m \neq 3$. If $\pi_j \geq -\frac{1}{4}$, then

$$\frac{p}{d} < \frac{\pi_j + \frac{1}{4}}{d} \leq \frac{\pi_j + \frac{1}{4}}{\delta_i} = \frac{4\pi_j + 1}{4\delta_i} \leq \frac{m+1}{4}.$$

On the other hand, if $\pi_j < -\frac{1}{4}$, then

$$\frac{p}{d} < \frac{\pi_j + \frac{1}{4}}{d} < \frac{\pi_j + \frac{1}{4}}{\delta_i + \frac{1}{4}} = \frac{4\pi_j + 1}{4\delta_i + 1} \leq \frac{m+1}{4}.$$

As for the lower bound, since

$$\frac{p}{d} \geq -1 = \frac{-3-1}{4},$$

we may assume $m \neq -3$. If $\pi_j \geq 0$, then

$$\frac{p}{d} \geq \frac{\pi_j}{d} \geq \frac{\pi_j}{\delta_i + \frac{1}{4}} = \frac{4\pi_j}{4\delta_i + 1} \geq \frac{m-1}{4}.$$

But if $\pi_j < 0$, then

$$\frac{p}{d} \geq \frac{\pi_j}{d} \geq \frac{\pi_j}{d} \geq \frac{m-1}{4}. \qquad \square$$

4 Implementation

The results of this section refer to the values assumed by variables during a hypothetical execution of the XFL function SRT, defined in Appendix 2. With the exception of the loop variables b and k, each variable of SRT belongs to one of the two classes:

- Some variables assume at most one value during an execution. The value of such a variable will be denoted by the name of the variable in italics, e.g., X, $dEnc$, and YNB.
- Variables that are assigned inside the main for loop may assume only one value during each iteration and may or may not be assigned an initial value before the loop is entered. The value assigned to such a variable during the kth iteration will be denoted with the subscript k, e.g., p_k, $mAbs_k$, and $addA_k$. If such a variable is assigned to an initial value outside of the loop, it will be denoted with the subscript 0, e.g., p_0 and $QPart_0$. When convenient, the subscript may be omitted and understood to have the value k. When replaced with an accent ($'$), it will be understood to have the value $k - 1$. For example, in the statement of Lemma 14, m and p' represent m_k and p_{k-1}, respectively.

SRT has four input parameters:

- $isSigned$ is a boolean indication of a signed or unsigned integer format
- w is the format width, which is assumed to be 8, 16, 32, or 64
- $XEnc$ is the signed or unsigned w-bit encoding of the divisor
- $YEnc$ is the signed or unsigned $2w$-bit encoding of the dividend

Three values are returned:

- A boolean indication of whether the computation completed successfully
- The signed or unsigned w-bit encoding of the quotient
- The signed or unsigned w-bit encoding of the remainder

The last two values are of interest only when the first is true, in which case they are the values of the variables $QOut$ and $ROut$, respectively.

Some of the variables of SRT do not contribute to the outputs, but are used only in our analysis and in embedded assertions. Of these (listed in a preamble to the function), X and Y are the integer values represented by $XEnc$ and $YEnc$, and Q and R are the quotient and remainder, which, unless $X = 0$, satisfy $Y = QX + R$, $|R| < |X|$, and either $R = 0$ or $sgn(R) = sgn(Y)$.

Our objective is to show that success is indicated if and only if $X \neq 0$ and Q is representable with respect to the indicated format, in which case Q and R are the integer values of $QOut$ and $ROut$. Since this obviously holds when $X = 0$, we shall assume $X \neq 0$ in the following. The main result is the theorem at the end of this section.

The computation is naturally partitioned into three phases, which are described in the following three subsections.

4.1 Analysis of Operands

In the first phase, the operands are analyzed and normalized in preparation for the iterative computation of the quotient and remainder, and the number n of iterations is established.

The variable XNB represents the "number of bits" of X, derived by counting the leading zeroes or ones:

Lemma 8. $XNB = expo(X) + 1$.

Proof. If $X > 0$, then $X = XEnc$ and $XNB - 1$ is the index of the leading 1 of X, which implies $2^{XNB-1} \leq X < 2^{XNB}$, and the claim follows.

If $XNegPower2 = 1$, then $XEnc[b] = 1$ if and only if $w > b \geq XNB - 1$. It follows that $XEnc = 2^w - 2^{XNB-1}$ and

$$X = XEnc - 2^w = 2^w - 2^{XNB-1} - 2^w = -2^{XNB-1}.$$

In the remaining case, $X < 0$, $XNB - 1$ is the index of the leading 0 of $XEnc$, and $XEnc[XNB - 2 : 0] \neq 0$. It follows that

$$2^w - 2^{XNB} < XEnc < 2^w - 2^{XNB-1},$$

which implies $-2^{XNB} < X < -2^{XNB-1}$, i.e., $2^{XNB-1} < |X| < 2^{XNB}$. $\square$

The variable $dEnc$ is an encoding of $d = 2^{-expo(X)}X$:

Lemma 9. $dEnc = (2^{65}d)[67 : 0] = d[2 : -65]$.

Proof. Since $d = 2^{-expo(X)}X$ and $expo(X) \leq 63$, $2^{63}d = 2^{63-expo(X)}X$ is an integer and

$$X = 2^{expo(X)}d = 2^{XNB-1}d.$$

Clearly,

$$\begin{aligned}
dEnc[65 : 66 - XNB] &= XEnc[XNB - 1 : 0] \\
&= X[w - 1 : 0][XNB - 1 : 0] \\
&= X[XNB - 1 : 0]
\end{aligned}$$

and since $|X| < 2^{XNB}$,

$$dEnc[66] = dEnc[67] = XSign = X[XNB] = X[XNB + 1].$$

Thus, $dEnc[67 : 66 - XNB] = X[XNB + 1 : 0]$, and hence

$$
\begin{aligned}
dEnc &= 2^{66-XNB} X[XNB + 1 : 0] \\
&= (2^{66-XNB} X)[67 : 0] \\
&= (2^{XNB-1+66-XNB} d)[67 : 0] \\
&= (2^{65} d)[67 : 0] \\
&= d[2 : -65].
\end{aligned}
$$

□

Lemma 10 gives an expression for i, the first argument of the table access function ϕ:

Lemma 10. $i = \lfloor 4(|d| - 1) \rfloor$.

Proof. If $X > 0$, then since $4 \le 4d < 8$,

$$
i = dEnc[64 : 63] = d[-1 : -2] = mod(\lfloor 4d \rfloor, 4) = \lfloor 4d \rfloor - 4 = \lfloor 4(d - 1) \rfloor.
$$

If $XNegPower2 = 1$, then $X = -2^{expo(X)}$, $d = -1$, and

$$
i = 0 = \lfloor 4(|d| - 1) \rfloor.
$$

In the remaining case, $X < 0$, $dEnc[66] = 1$, $dEnc[65] = 0$, and $dEnc[64 : 0] \neq 0$. Since $2^{65} d = 2^{65-expo(X)} X$ is an integer and $2^{65} d < 2^{66}$,

$$
\begin{aligned}
2^{65} d &= SgndIntVal(67, (2^{65} d)[66 : 0]) \\
&= SgndIntVal(67, dEnc[66 : 0]) \\
&= dEnc[64 : 0] - 2^{66}.
\end{aligned}
$$

Thus, $|d| = -d = 2 - 2^{-65} dEnc[64 : 0]$ and

$$
\lfloor 4(|d| - 1) \rfloor = \lfloor 4 - 2^{-63} dEnc[64 : 0] \rfloor = \lfloor 4 - dEnc[64 : 63] - 2^{-63} dEnc[62 : 0] \rfloor.
$$

Suppose that $dEnc[62 : 0] = 0$. Then $dEnc[64 : 63] \neq 0$ and

$$
\lfloor 4(|d| - 1) \rfloor = 4 - dEnc[64 : 63].
$$

An exhaustive case analysis ($dEnc[64 : 63] = 1, 2$, or 3) shows that

$$
4 - dEnc[64 : 63] = ((\verb|~|dEnc[64] \mid \verb|~|dEnc[63]) << 1) \mid dEnc[63] = i.
$$

Finally, suppose that $dEnc[62 : 0] \neq 0$. Then

$$
\lfloor 4(|d| - 1) \rfloor = 3 - dEnc[64 : 63] = \verb|~|dEnc[64 : 63] = i.
$$

□

YNB is the "number of bits" of Y, including, in the negative case, the final trailing sign bit.

Lemma 11. *If $Y > 0$, then*

$$2^{YNB-1} \leq Y < 2^{YNB},$$

and if $Y < 0$, then

$$2^{YNB-2} < |Y| \leq 2^{YNB-1}.$$

Consequently, in either case, $YNB \geq expo(Y) + 1$.

Proof. If $Y > 0$, then $YNB - 1$ is the index of the leading 1 of $YEnc = Y$, i.e., $expo(Y) = YNB - 1$.

If $Y = -1$, then $YNB = 1$ and

$$2^{YNB-2} = \frac{1}{2} < |Y| = 1 = 2^{YNB-1}.$$

In the remaining case, $Y < -1$, $YNB - 2$ is the index of the leading 0 of $YEnc$, which implies

$$2^w - 2^{YNB-1} \leq YEnc < 2^w - 2^{YNB-2}.$$

But since $Y = YEnc - 2^w$,

$$-2^{YNB-1} \leq Y < -2^{YNB-2}$$

and

$$2^{YNB-2} < |Y| \leq 2^{YNB-1}. \qquad \square$$

The number of iterations, n, satisfies the requirement of Lemma 1.

Lemma 12. $|Y| < 2^{2n}|X|$.

Proof. This is an immediate consequence of Lemmas 8, 11, and 5. $\qquad \square$

The bit vector $pEnc$ is an encoding of $p_0 = 2^{-expo(X)-2n}Y$:

Lemma 13.

(a) If $n = 0$, then

$$pEncHi_0 = \begin{cases} (2^{64-YNB}Y)[67:0] & \text{if } YNB[0] = XNB[0] \\ (2^{65-YNB}Y)[67:0] & \text{if } YNB[0] \neq XNB[0]. \end{cases}$$

(b) If $n > 0$, then $2^{129}p_0$ is an integer and

$$pEnc = (2^{129}p_0)[131:0] = p_0[2:-129].$$

Proof. First consider the case $YNB[0] = XNB[0]$. We may assume $YNB > 0$; otherwise, $Y = 0$ and the lemma is trivial.

Note that $YNB \leq 128$ and

$$pEnc[127 : 128 - YNB] = YEnc[YNB - 1 : 0] = Y[YNB - 1 : 0].$$

Therefore,

$$pEnc[127 : 0] = 2^{128-YNB}Y[YNB - 1 : 0] = (2^{128-YNB}Y)[127 : 0].$$

Since $Y < 2^{YNB}$, for $\ell = 128, \ldots, 131$,

$$pEnc[\ell] = YSign = (2^{128-YNB}Y)[\ell].$$

Thus,

$$pEnc = pEnc[131 : 0] = (2^{128-YNB}Y)[131 : 0].$$

If $n = 0$, then $YNB < XNB \leq 64$ and

$$pEnc = (2^{128-YNB}Y)[131 : 0] = 2^{64}(2^{64-YNB}Y)[67 : 0],$$

which implies $pEncHi_0 = (2^{64-YNB}Y)[67 : 0]$.

On the other hand, if $n > 0$, then

$$2n = 2\left(\left\lfloor \frac{YNB - XNB}{2} \right\rfloor + 1\right) = YNB - XNB + 2 = YNB - expo(X) + 1$$

and

$$p_0 = 2^{-expo(X)-2n}Y = 2^{-YNB-1}Y.$$

Thus, $2^{129}p_0 = 2^{128-YNB}Y$ is an integer and

$$pEnc = (2^{128-YNB}Y)[131 : 0] = (2^{129}p_0)[131 : 0].$$

The proof for the case $YNB[0] \neq XNB[0]$ is similar, with every occurrence of 127 or 128 replaced by 128 or 129. Thus, we have

$$pEnc = (2^{129-YNB}Y)[131 : 0],$$

which, in the $n = 0$ case, leads to

$$pEncHi_0 = (2^{65-YNB}Y)[67 : 0]. \qquad \qquad \square$$

4.2 *Iteration*

The second phase is the iteration loop in which the quotient digits are selected and the partial remainder and quotient are updated accordingly. The main results pertaining to the iterative computation of the partial remainder are given by Lemmas 14 and 16:

1. The quotient digit m is correctly computed as the value of $\phi(i, j)$, as stated in Lemma 14.
2. The partial remainder $p_k = 4p_{k-1} - m_k d$ is encoded by *pEncHi*, *carryHi*, and *pEncLo*, as stated in Lemma 16.

The proof of (2) depends on (1), and that of (1) requires the assumption that (2) holds on the preceding iteration.

Lemma 14. *Let $0 < k \le n$. Suppose that $|p'| \le |d| < 2$, $2^{129}p'$ is an integer, and*

$$(2^{129}p')[131:0] = 2^{64}(pEncHi' + carryHi')[67:0] + pEncLo'.$$

Then

$(a)\ m = \begin{cases} \phi(i, j) & \text{if } X \ge 0 \\ -\phi(i, j) & \text{if } X < 0; \end{cases}$

$(b)\ \pi_j \le p' < \pi_j + \frac{1}{4}.$

Proof. First suppose $pEncHi' + carryHi' \ge 2^{68}$. Then $pTop = 63$; otherwise,

$$\begin{aligned} pEncHi' + carryHi' &= 2^{62}pTop + pEncHi'[61:0] + carryHi' \\ &\le 2^{62}62 + 2^{62} - 1 + 2^{62} - 1 \\ &< 2^{68}. \end{aligned}$$

Consequently, $j = pIndex = pTop[4:0] = 31$, which implies $m = \phi(i, j) = 0$, and (a) follows. To prove (b), we note that

$$\begin{aligned} (2^{129}p')[131:64] &= (pEncHi' + carryHi')[67:0] \\ &= pEncHi' + carryHi' - 2^{68} \\ &< 2^{68} + 2^{62} - 2^{68} \\ &= 2^{62} \end{aligned}$$

and therefore,

$$(2^{129}p')[131:0] < 2^{64}(2^{62} - 1) + 2^{62} < 2^{126}.$$

Since $|2^{129}p'| < 2^{131}$,

$$2^{129}p' = SgndIntVal(132, (2^{129}p')[131:0]) = (2^{129}p')[131:0]$$

and thus, $0 \leq 2^{129} p' < 2^{126}$ and

$$\pi_j = -\frac{1}{8} < 0 \leq p' < \frac{1}{8} = \pi_j + \frac{1}{4}.$$

We may assume, therefore, that $pEncHi' + carryHi' < 2^{68}$ and hence

$$\begin{aligned}
(2^{129} p')[131:0] &= 2^{64}(pEncHi' + carryHi')[67:0] + pEncLo' \\
&= 2^{64}(pEncHi' + carryHi') + pEncLo' \\
&= 2^{126} pTop + 2^{64}(pEncHi'[61:0] + carryHi') + pEncLo' \\
&< 2^{126} pTop + 2^{64}(2^{62} - 1 + 2^{62} - 1) + 2^{64} \\
&< 2^{126}(pTop + 2).
\end{aligned}$$

Suppose $p' \geq 0$. Then $2^{129} p' = (2^{129} p')[131:0]$ and

$$\frac{1}{8} pTop \leq p' < \frac{1}{8} pTop + \frac{1}{4}.$$

Since $pTop \leq 8p' < 16$, $j = pIndex = pTop$ and $pSign = 0$. Thus,

$$|m| = mAbs = SRTLookup(i, j) = |\phi(i, j)| = \phi(i, j)$$

and $mSign = XSign$, which implies (a). To prove (b), we need only observe that

$$SgndIntVal(5, j) = SgndIntVal(5, pTop) = pTop.$$

Now suppose $p' < 0$. Then $2^{129} p' = (2^{129} p')[131:0] - 2^{132}$ and the above estimate yields

$$\frac{1}{8}(pTop - 64) \leq p' < \frac{1}{8}(pTop - 64) + \frac{1}{4}.$$

Thus, $pTop > 8p' + 62 > -16 + 62 = 46$, so $pTop \geq 47$. Let us assume that $pTop \geq 48$. Then $j = pIndex$ and $|m| = |\phi(i, j)| = -\phi(i, j)$. Thus, to establish (a), we need only show that m and X have opposite signs. But this follows from $mSign = XSign \; \hat{} \; pSign$ and $pSign = 1$. To prove (b), it suffices to show that $pTop = SgndIntVal(5, j) + 64$. But in this case, $j = pTop[4:0] = pTop - 32 \geq 16$, so $SgndIntVal(5, j) = j - 32 = pTop - 64$.

There remains the special case $pTop = 47$. Since

$$p' < \frac{1}{8}(pTop - 64) + \frac{1}{4} = -\frac{17}{8} + \frac{1}{4} = -\frac{15}{8},$$

$2 > |d| \geq |p'| > \frac{15}{8}$, which implies

$$dIndex = i = \lfloor 4(|d| - 1) \rfloor = 3.$$

Thus,

$$|m| = mAbs = SRTLookup(3, 15) = 3.$$

On the other hand, $\phi(i, j) = \phi(3, 16) = -3$. But again, since $pSign = 1$, m and X have opposite signs and (a) follows. To prove (b), note that $SgndIntVal(5, j) = -16$; hence,

$$\pi_j = -2 < p' < -\frac{17}{8} + \frac{1}{4} < \pi_j + \frac{1}{4}. \qquad \square$$

The computation of the partial remainder, as described in Lemma 16, involves a "compression" that reduces four addends to two. This is performed by the serial operation of two carry-save adders, as described by the following basic result, taken from [10]:

Lemma 15. *Given n-bit vectors x, y, and z, let*

$$a = x \; \hat{} \; y \; \hat{} \; z$$

and

$$b = 2(x \; \& \; y \; | \; x \; \& \; z \; | \; y \; \& \; z).$$

Then

$$x + y + z = a + b.$$

Lemma 16. *If $n > 0$ and $0 \leq k \leq n$, then $|p| \leq |d| < 2$, $2^{129} p$ is an integer, and*

$$(2^{129} p)[131 : 0] = p[2 : -129] = 2^{64}(pEncHi + carryHi)[67 : 0] + pEncLo.$$

Proof. The proof is by induction on k.

For $k = 0$, we have

$$|p| = |2^{-expo(X)-2n} Y| < |2^{-expo(X)} X| = |d|$$

by Lemma 12, and since

$$2^{64}(pEncHi + carryHi)[67 : 0] + pEncLo = pEnc,$$

the other two claims follow from Lemma 13.

In the inductive case, we shall derive the bound on $|p|$ from Lemma 7. By Lemma 10, since $|d| \leq 1, 0 \leq i < 4$ and $\delta_i \leq |d| < \delta_i + \frac{1}{4}$. Clearly, $j < 32$, and by Lemma 13, $\pi_j \leq p < \pi_j + \frac{1}{4}$ and

$$m = \begin{cases} \phi(i, j) & \text{if } X \geq 0 \\ -\phi(i, j) & \text{if } X < 0. \end{cases}$$

Thus, applying Lemma 7, with the signs of d and m reversed if $d < 0$, we have $|p| = |4p' - md| \leq d$.

By induction,

$$2^{129}p = 2^{131}p' - 2^{129}md$$

is an integer. The computation of $(2^{129}p)[131 : 0]$ involves a 4–2 compressor with inputs $addA$, $addB$, $addC$, $addD$. We shall show that

$$(addA + addB + addC + addD)[67 : 0] = (2^{129}p)[131 : 64].$$

The first two terms, $addA$ and $addB$, if not 0, represent $\pm 2d$ and $\pm d$, respectively, depending on the value of m. However, in the negative case, in order to avoid a full 67-bit addition, the simple complement of $2d$ or d is used in place of its negation, and the missing 1 is recorded in the variable $inject$, which is more conveniently combined later with $addD$. Thus, our first goal is to prove that

$$(addA + addB + inject)[67 : 0] = (-2^{65}dm)[67 : 0].$$

If $mSign = 1$, then

$$addA = mAbs[1] \cdot (2 \cdot dEnc)[67 : 0] = (2 \cdot mAbs[1] \cdot dEnc)[67 : 0],$$

$$addB = mAbs[0] \cdot dEnc,$$

and $inject = 0$. Hence,

$$\begin{aligned}
(addA + addB + inject)[67 : 0] &= (2 \cdot mAbs[1] \cdot dEnc + mAbs[0] \cdot dEnc)[67 : 0] \\
&= ((2 \cdot mAbs[1] + mAbs[0]) \cdot dEnc)[67 : 0] \\
&= (mAbs \cdot dEnc)[67 : 0] \\
&= (-dEnc \cdot m)[67 : 0]) \\
&= (-2^{65}dm)[67 : 0]).
\end{aligned}$$

On the other hand, if $mSign = 0$, then

$$\begin{aligned}
addA &= addA[67 : 0] \\
&= mAbs[1] \cdot (2(\tilde{}dEnc[66 : 0]) + 1)[67 : 0] \\
&= mAbs[1] \cdot (2(-dEnc - 1)[66 : 0]) + 1)[67 : 0] \\
&= mAbs[1] \cdot (-2dEnc - 2)[67 : 0]) + 1)[67 : 0] \\
&= mAbs[1] \cdot (-2 \cdot dEnc - 1)[67 : 0] \\
&= (-2 \cdot mAbs[1] \cdot dEnc - mAbs[1])[67 : 0],
\end{aligned}$$

$$\begin{aligned}
addB &= mAbs[0] \cdot \tilde{}dEnc[67 : 0] \\
&= mAbs[0] \cdot (-dEnc - 1)[67 : 0] \\
&= (-mAbs[0] \cdot dEnc - mAbs[0])[67 : 0],
\end{aligned}$$

and $inject = mAbs[0] + mAbs[1]$, so that

$$(addA + addB + inject)[67:0]$$
$$= (-2 \cdot mAbs[1] \cdot dEnc - mAbs[1] - mAbs[0] \cdot dEnc$$
$$-mAbs[0] + mAbs[0] + mAbs[1])[67:0]$$
$$= (-(2 \cdot mAbs[1] + mAbs[0]) \cdot dEnc)[67:0]$$
$$= (-m \cdot dEnc)[67:0]$$
$$= (-2^{65}dm)[67:0])$$
$$= (-2^{129}dm)[131:64]).$$

The remaining two terms, $addC$ and $addD$, represent the shifted result of the previous iteration, $4p'$. Thus,

$$addC = 4 \cdot pEncHi + pEncLo[63:62],$$

$$addD = 2 \cdot carryHi + inject,$$

and

$$(addC + addD - inject)[67:0]$$
$$= (4 \cdot pEncHi + pEncLo[63:62] + 4 \cdot carryHi + inject - inject)[67:0]$$
$$= (4(pEncHi + carryHi)[67:0] + pEncLo[63:62])[67:0]$$
$$= (4(2^{64}(pEncHi + carryHi)[67:0] + 2^{62} \cdot pEncLo[63:62]))[131:64]$$
$$= 4(2^{64}(pEncHi + carryHi)[67:0] + 2^{62} \cdot pEncLo[63:62])[129:62]$$
$$= 4(2^{64}(pEncHi + carryHi)[67:0] + pEncLo)[129:62]$$
$$= 4(2^{129}p')[131:0][129:62]$$
$$= 4(2^{129}p')[129:62]$$
$$= (4 \cdot 2^{129}p')[131:64].$$

Combining these last two results, we have

$$(addA + addB + addC + addD)[67:0]$$
$$= ((addA + addB + inject)[67:0] + (addC + addD - inject)[67:0])[67:0]$$
$$= ((2^{129}4p')[131:64] + (-2^{129}dm)[131:64])[67:0].$$

Since $(-2^{129}dm)[63:0] = 0$, this may be reduced to

$$((2^{129}(4p' - dm))[131:64] = (2^{129}p)[131:64].$$

Two applications of Lemma 15 yield

$$addA + addB + addC + addD = sum1 + carry1 + addD = sum2 + carry2,$$

and therefore,

$$
\begin{aligned}
(2^{129}p)&[131:64] \\
&= (sum2 + carry2)[67:0] \\
&= ((2^{62}(sum2[67:62] + carry2[67:62]))[67:0] \\
&\quad + sum2[61:0] + carry2[61:0])[67:0] \\
&= (2^{62}(sum2[67:62] + carry2[67:62])[5:0] \\
&\quad + sum2[61:0] + carry2[61:0])[67:0] \\
&= (pEncHi + carryHi)[67:0].
\end{aligned}
$$

But by Lemma 9,

$$
\begin{aligned}
(2^{129}p)[63:0] &= (2^{129}(4p' - md))[63:0] \\
&= ((2^{129}4p')[63:0] + (-2^{64}m2^{65}d)[63:0])[63:0] \\
&= (2^{129}4p')[63:0] \\
&= (4 \cdot pEncLo')[63:0] \\
&= pEncLo,
\end{aligned}
$$

and thus,

$$
\begin{aligned}
(2^{129}p)[131:0] &= 2^{64}(2^{129}p)[131:64] + (2^{129}p)[63:0] \\
&= 2^{64}(pEncHi + carryHi)[67:0] + pEncLo. \qquad \square
\end{aligned}
$$

As a result of the iterative shifting of the partial remainder, $pEncLo = 0$ upon exiting the loop. This is proved recursively:

Lemma 17. *If $n > 0$ and $0 \leq k \leq n$, then $pEncLo[63 - 2(n - k):0] = 0$).*

Proof. The proof is by induction on k. For $k = 0$, since $pEncLo[127 - YNB:0] = 0$, we need only show that $127 - YNB \geq 63 - 2n$, or $2n \geq YNB - 64$. But

$$2n = 2\left(\left\lfloor \frac{YNB - XNB}{2} \right\rfloor + 1\right) \geq YNB - XNB + 1 \geq YNB - 63.$$

For $k > 0$,

$$
\begin{aligned}
pEncLo[63 - 2(n - k):0] &= (4 \cdot pEncLo')[63 - 2(n - k):0] \\
&= 4 \cdot pEncLo'[63 - 2(n - k + 1):0] \\
&= 0. \qquad \square
\end{aligned}
$$

The partial quotient *QPart* is encoded by *Q0Enc*. Its computation, as described in Lemma 4, is facilitated by simultaneously maintaining encodings of $QPart \pm 1$:

Lemma 18. *For* $0 \leq k \leq n,$

$$Q0Enc = QPart[66:0],$$

$$QMEnc = (QPart - 1)[66:0],$$

and

$$QPEnc = (QPart + 1)[66:0].$$

Proof. We shall invoke Lemma 4 with $N = 67$ and $Q_k = QPart$. We need only show that $Q0Enc = E_k$, $QMEnc = E_k^-$, and $QPEnc = E_k^+$. The claim is trivial for $k = 0$. For $k > 0$, it may be readily verified by examining each value of m, $-3 \leq m \leq 3$. For example, if $m = -1$, then $mSign = 1$, $mAbs = 1$,

$$Q0Enc = (4 \cdot Q0Enc')[66:0] = (4E_{k-1})[66:0] = (4E_{k-1} + m_k - 1)[66:0] = E_k^+,$$

$$\begin{aligned}
QMEnc &= (4 \cdot Q0Enc')[66:0] \mid 2 \\
&= (4E_{k-1}^-)[66:0] \mid 2 \\
&= (4E_{k-1}^- \mid 2)[66:0] \\
&= (4E_{k-1}^- + 2)[66:0] \\
&= (4E_{k-1}^- + m_k + 3)[66:0] \\
&= E_k^-,
\end{aligned}$$

and

$$\begin{aligned}
QPEnc &= (4 \cdot Q0Enc')[66:0] \mid 3 \\
&= (4E_{k-1}^-)[66:0] \mid 3 \\
&= (4E_{k-1}^- \mid 3)[66:0] \\
&= (4E_{k-1}^- + 3)[66:0] \\
&= (4E_{k-1}^- + m_k + 4)[66:0] \\
&= E_k^+.
\end{aligned}$$
$\square$

4.3 *Final Computation*

In the final phase of the computation, a full addition is performed to generate an explicit (nonredundant) representation of the remainder. This result is then adjusted, along with the quotient, to produce the final results as specified by Lemma 25.

Lemma 19 refers to the quotient and remainder before the correction step:

Lemma 19. $Y = QPre \cdot X + RPre$ and $|RPre| \leq |X|$.

Proof. This is an immediate consequence of Lemma 1, with $r = 2$, $Q_k = QPart_k$, $R = RPre$, and $Q = QPre$. We need only note that the condition $|p_k| \leq |d|$ is ensured by Lemma 16. $\square$

RPre is encoded by *REncPre*:

Lemma 20.

$$
REncPre = \begin{cases} (2^{66-XNB}RPre)[66:0] & \text{if } n > 0 \\ (2^{64-YNB}RPre)[66:0] & \text{if } n = 0 \text{ and } YNB[0] = XNB[0] \\ (2^{65-YNB}RPre)[66:0] & \text{if } n = 0 \text{ and } YNB[0] \neq XNB[0]. \end{cases}
$$

Proof. If $n > 0$, then by Lemmas 16 and 8,

$$
\begin{aligned}
REncPre &= (pEncHi_n + carryHi_n)[66:0] \\
&= (2^{129}p_n)[130:64] \\
&= (2^{64}2^{65-expo(X)}RPre)[130:64] \\
&= (2^{66-expo(X)}RPre)[66:0] \\
&= (2^{66-XNB}RPre)[66:0].
\end{aligned}
$$

On the other hand, if $n = 0$, then $REncPre = pEncHi_0[66:0]$ and the lemma follows from Lemma 13. $\square$

The encoding *REnc* of the final remainder, which is derived from *REncPre*, depends on the signs of *RPre* and *Y* and the special cases *RPre* is 0 or $\pm X$. Timing considerations dictate that these conditions must be detected before the full addition that produces *REncPre* is actually performed. This requires a technique for predicting cancellation, which is provided by the following result, found in [10]:

Lemma 21. *Given n-bit vectors a and b and a one-bit vector c, let*

$$
\tau = a \mathbin{\char`^} b \mathbin{\char`^} (2(a \mid b) + c).
$$

If $0 \leq k < n$, then

$$
(a + b + c)[k:0] = 0 \Leftrightarrow \tau[k:0] = 0.
$$

Lemma 21 is used in the proofs of the following three lemmas:

Lemma 22. *RIs0 is true if and only if RPre = 0.*

Proof. By Lemma 21, *RIs0* is true if and only if *REncPre* $= 0$. If $n > 0$, then by Lemma 20, *REncPre* $= (2^{65-expo(X)}RPre)[66:0]$. But by Lemma 19,

$$|2^{65-expo(X)}RPre| \leq |2^{65-expo(X)}X| < 2^{66},$$

and it follows that *REncPre* $= 0$ if and only if *RPre* $= 0$.

Now suppose $n = 0$. Then *RPre* $= 2^{expo(X)}p_0 = Y$, and $|Y| < 2^{YNB} \leq 2^{XNB-1} \leq |X|$. By Lemma 20, *REncPre* $= (2^{e-YNB}RPre)[66:0]$, where $e = 64$ or 65. Thus,

$$|2^{e-YNB}RPre| \leq |2^{65-YNB}Y| \leq |2^{64-expo(Y)}Y| < 2^{65},$$

and again, *REncPre* $= 0$ if and only if *RPre* $= 0$. $\square$

Lemma 23. *RNegX is true if and only if RPre* $= -X$.

Proof. First note that by Lemma 9,

$$dEnc[66:0] = (2^{65}d)[66:0] = (2^{65-expo(X)}X)[66:0].$$

Now by Lemma 15,

$$(RNegXSum + RNegXCarry)[66:0] = (pEncHi + CarryHi + dEnc)[66:0]$$
$$= (REncPre + dEnc)[66:0];$$

hence, by Lemma 21,

$$RNegX = 1 \Leftrightarrow (RNegXSum + RNegXCarry)[66:0] = 0$$
$$\Leftrightarrow (REncPre + dEnc))[66:0] = 0$$
$$\Leftrightarrow (REncPre + 2^{65-expo(X)}X)[66:0] = 0.$$

If $n > 0$, then we have

$$RNegX = 1 \Leftrightarrow (2^{65-expo(X)}(RPre + X))[66:0] = 0,$$

where

$$|2^{65-expo(X)}(RPre+X)| < |2^{65-expo(X)}(2X)| \leq |2^{65-expo(X)}2^{expo(X)+2}| = 2^{67},$$

and the result follows.

If $n = 0$, then since $|RPre| = |Y| < |X|$, we must show that *RNegX* $= 0$. By Lemma 22, *REncPre* $= (2^{e-YNB}Y)[66:0]$, where $e \leq 65$ and as noted above, $|2^{e-YNB}Y| < 2^{65}$. If *RNegX*$=1$, then $(2^{e-YNB}Y + 2^{65-expo(X)}X)[66:0]=0$. But since

$$|2^{e-YNB}Y + 2^{65-expo(X)}X| < 2|2^{65-expo(X)}X| < 2|2^{65-expo(X)}2^{expo(X)+1}| = 2^{67},$$

this implies $2^{e-YNB}Y + 2^{65-expo(X)}X = 0$, which is impossible. $\square$

Lemma 24. *RPosX is true if and only if RPre $= X$.*

Proof. By Lemma 15,

$$\begin{aligned}
(RPosXSum + RPosXCarry)[66:0] &= (pEncHi + CarryHi + \tilde{\ }dEnc + 1)[66:0]\\
&= (REncPre - dEnc - 1 + 1)[66:0]\\
&= (REncPre - dEnc)[66:0],
\end{aligned}$$

and hence, by Lemma 21,

$$\begin{aligned}
RPosX = 1 &\Leftrightarrow (RPosXSum + RPosXCarry)[66:0] = 0\\
&\Leftrightarrow (REncPre - dEnc)[66:0] = 0.
\end{aligned}$$

The rest of the proof is similar to that of Lemma 23. $\square$

Lemma 25. $Y = QX + R$, *where* $|R| < |X|$ *and either* $R = 0$ *or* $sgn(R) = sgn(Y)$.

Proof. This is an immediate consequence of Lemmas 2, 19, and 22. $\square$

The final remainder is encoded by *REnc*.

Lemma 26.

$$REnc = \begin{cases} (2^{64-XNB}R)[63:0] & \text{if } n > 0\\ (2^{62-YNB}R)[63:0] & \text{if } n = 0 \text{ and } YNB[0] = XNB[0]\\ (2^{63-YNB}R)[63:0] & \text{if } n = 0 \text{ and } YNB[0] \neq XNB[0]. \end{cases}$$

Proof. If *fixupNeeded* is false, then $R = RPre$, $REnc = REncPre[66:2]$, and the lemma follows from Lemma 20. If $n = 0$, then as noted in the proof of Lemma 22, $RPre = Y$ and $|Y| < |X|$, from which it follows that *fixupNeeded* is false. Thus, we may assume that *fixupNeeded* is true and $n > 0$. We may further assume that $RIsX = 0$; otherwise, $REnc = R = 0$. If $RSign = XSign$, then

$$\begin{aligned}
REnc &= (REncPre + \tilde{\ }dEnc[66:0] + 1)[65:2]\\
&= (2^{66-XNB}RPre + 2^{67} - 2^{65-expo(X)}X - 1 + 1)[65:2]\\
&= (2^{66-XNB}(RPre - X))[65:2]\\
&= (2^{66-XNB}R)[65:2]\\
&= (2^{64-XNB}R)[63:0].
\end{aligned}$$

The case $RSign \neq XSign$ is similar. $\square$

Our main result follows:

Theorem 1. *If Q is representable in the integer format determined by isSigned and w, then QTooLarge = false and QOut and ROut are the encodings of Q and R, respectively. Otherwise, QTooLarge = true.*

Proof. We shall first prove that *QTooLarge* is false if and only if Q is representable. We begin with the case $YNB - XNB > w$ in which we must show that Q is not representable. If $Y > 0$, then by Lemmas 8 and 11, $|X| < 2^{XNB}$ and $Y \geq 2^{YNB-1}$, and hence,

$$\left|\frac{Y}{X}\right| > 2^{YNB-1-XNB} \geq 2^{w},$$

which implies

$$|Q| = \left|\frac{Y-R}{X}\right| \geq \left|\frac{Y}{X}\right| - \left|\frac{R}{X}\right| > \left|\frac{Y}{X}\right| - 1 > 2^{w} - 1$$

and $|Q| \geq 2^{w}$.

Now suppose $Y < 0$. Then $|X| < 2^{XNB}$ and $|Y| > 2^{YNB-2}$. Since the format is signed, it will suffice to show that $|Q| > 2^{w-1}$ or $\left|\frac{Y}{X}\right| \geq 2^{w-1} + 1$. If $XNB = w$, then $|X| \geq 2^{w-1}$ and we must have $X = -2^{w-1}$ and

$$\left|\frac{Y}{X}\right| > \frac{2^{YNB-2}}{2^{XNB-1}} = 2^{YNB-XNB-1} \geq 2^{w}.$$

We may assume, therefore, that $XNB < w$. Since $|X| \leq 2^{XNB} - 1$,

$$\left|\frac{Y}{X}\right| > \frac{2^{YNB-2}}{2^{XNB} - 1} \geq \frac{2^{XNB+w-1}}{2^{XNB} - 1},$$

and we need only show that

$$\frac{2^{XNB+w-1}}{2^{XNB} - 1} \geq 2^{w-1} + 1,$$

or, equivalently,

$$2^{XNB+w-1} \geq (2^{XNB} - 1)(2^{w-1} + 1) = 2^{XNB+w-1} + 2^{XNB} - 2^{w-1} - 1,$$

which follows from $XNB < w$.

In the case $n \leq 1$, *QTooLarge* is false and we must show that Q is representable. But this is trivially true, since $YNB - XNB \leq 1$, $|Y| < 2^{YNB}$, and $X \geq 2^{XNB-1}$ imply

$$|Q| \le \left| \frac{Y}{X} \right| < 2^{YNB-XNB+1} \le 4.$$

In the remaining case, $YNB - XNB \le w$ and $n > 1$. The first of these conditions implies that

$$2n \le 2\left(\left\lfloor \frac{w}{2} \right\rfloor + 1\right) \le w + 2;$$

thus, by Lemma 3,

$$|QPre| = |QPart_n| < 2^{w+2},$$

from which we conclude that $|Q| \le 2^{w+2}$.

The second condition implies that $YNB - XNB \ge 2$, so

$$|Y| > 2^{YNB-2} \ge 2^{XNB} > |X|,$$

from which we conclude that $Q \ne 0$. Thus, if $YSign = XSign$, then $Q > 0$, and if $YSign \ne XSign$, then $Q < 0$.

Suppose $Q > 0$. Then since $Q \le 2^{w+2}$, $Q = Q[w + 2 : 0] = QEnc[w + 2 : 0]$. If the format is unsigned, then

$$Q \text{ is representable} \Leftrightarrow Q < 2^w \Leftrightarrow QEnc[w + 2 : w] = 0 \Leftrightarrow QTooLarge = 0,$$

while if the format is signed, then

$$Q \text{ is representable} \Leftrightarrow Q < 2^{w-1} \Leftrightarrow QEnc[w + 2 : w - 1] = 0 \Leftrightarrow QTooLarge = 0.$$

Finally, suppose $Q < 0$. Then $Q \ge -2^{w+2}$ and

$$QEnc[w + 1 : 0] = Q[w + 1 : 0] = mod(Q, 2^{w+2}) = Q + 2^{w+2}.$$

Since the format must be signed,

$$\begin{aligned}
Q \text{ is representable} &\Leftrightarrow Q \ge -2^{w-1} \\
&\Leftrightarrow QEnc[w + 1 : 0] \ge 2^{w+2} - 2^{w-1} \\
&\Leftrightarrow QEnc[w + 1 : w - 1] = 7 \\
&\Leftrightarrow QTooLarge = 0. \qquad\qquad\qquad \square
\end{aligned}$$

Next, we show that if Q is representable, then $QOut$ and $ROut$ are the encodings of Q and R. Clearly, $QOut = QEnc[w - 1 : 0] = Q[w - 1 : 0]$, which is the encoding of Q. We must also show that $ROut = R[w - 1 : 0]$.

Consider the case $n > 0$. By Lemma 26,

$$REnc[63 : 64 - XNB] = (2^{64-XNB}R)[63 : 0][63 : 64 - XNB] = R[XNB - 1 : 0].$$

Since $|R| < |X| < 2^{XNB}$,

$$R[w - 1 : XNB] = \begin{cases} 0 & \text{if } R \geq 0 \\ 2^{w-XNB} - 1 & \text{if } R < 0 \end{cases}$$

and in either case, $R[w - 1 : XNB] = ROut[w - 1 : XNB]$.

Suppose $n = 0$ and $YNB[0] = XNB[0]$. By Lemma 26,

$$\begin{aligned} REnc[63 : 62 - YNB] &= (2^{62-YNB}R)[63 : 0][63 : 62 - YNB] \\ &= (2^{62-YNB}R)[63 : 62 - YNB] \\ &= R[YNB + 1 : 0]. \end{aligned}$$

Since $|R| = |Y| < 2^{YNB}$,

$$R[w - 1 : YNB + 1] = \begin{cases} 0 & \text{if } R \geq 0 \\ 2^{w-YNB-1} - 1 & \text{if } R < 0 \end{cases}$$

and in either case, $R[w - 1 : YNB + 1] = ROut[w - 1 : YNB + 1]$.

The case $YNB[0] \neq XNB[0]$ is similar. $\qquad\qquad\square$

Acknowledgments The author is grateful to Mike Achenbach, the principal designer of the Llano divider, for facilitating this work and especially for his patience in explaining the design and answering endless questions about it.

References

1. ACL2 Web site. http://www.cs.utexas.edu/users/moore/acl2/
2. Bryant RE, Chen YA (1996) Verification of arithmetic circuits with binary moment diagrams. In: Proceedings of the 32nd design automation conference, San Francisco, CA, June 1996
3. Clarke EM, German SM, Zhou X (1999) Verifying the SRT division algorithm using theorem proving techniques. Formal Methods Syst Des 14(1):7–44. http://www-2.cs.cmu.edu/~modelcheck/ed-papers/VtSRTDAU.pdf
4. Gerwig G, Wetter H, Schwarz EM, Haess J, Krygowski CA, Fleischer BM, Kroener M (2004) The IBM eServer z990 floating-point unit. IBM J Res Dev 48(3/4):311–322. http://www.research.ibm.com/journal/rd/483/gerwig.html
5. Kapur D, Subramaniam M (1997) Mechanizing verification of arithmetic circuits: SRT division. In: Invited Talk, Proceedings of FSTTCS-17, Kharagpur, India, LNCS 1346. Springer, New York, pp 103–122. http://www.cs.unm.edu/~kapur/myabstracts/fsttcs97.html
6. Parhami B (2000) Computer arithmetic: algorithms and hardware designs. Oxford University Press, Oxford
7. Pratt V (1995) Anatomy of the pentium bug. In: TAPSOFT '95: theory and practice of software development, LNCS 915. Springer, Heidelberg. https://eprints.kfupm.edu.sa/25851/1/25851.pdf
8. Robertson JE (1958) A new class of digital division methods. IRE Trans Electron Comput EC-7:218–222
9. Ruess H, Shankar N (1999) Modular verification of SRT division. Formal Methods Syst Des 14(1):45–73. http://www.csl.sri.com/papers/srt-long/srt-long.ps.gz
10. Russinoff DM (2007) A formal theory of register-transfer logic and computer arithmetic. http://www.russinoff.com/libman/

11. Russinoff DM (2005) Formal verification of floating-point RTL at AMD using the ACL2 theorem prover, IMACS World Congress, Paris, 2005. http://www.russinoff.com/papers/paris.html
12. Taylor GS (1981) Compatible hardware for division and square root. In: Proceedings of the 5th symposium on computer arithmetic. IEEE Computer Society, Washington, DC
13. Tocher KD (1958) Techniques of multiplication and division for automatic binary computers. Q J Mech Appl Math 11(3):364–384

Appendix 1: XFL Definition of ϕ

```
int phi(nat i, nat j) {
 switch (i) {
 case 0:
   switch (j) {
   case 0x09: case 0x08: case 0x07: case 0x06: case 0x05:
     return 3;
   case 0x04: case 0x03:
     return 2;
   case 0x02: case 0x01:
     return 1;
   case 0x00: case 0x1F: case 0x1E:
     return 0;
   case 0x1D: case 0x1C:
     return -1;
   case 0x1B: case 0x1A:
     return -2;
   case 0x19: case 0x18: case 0x17: case 0x16: case 0x15:
     return -3;
   default: assert(false);
   }
 case 1:
   switch (j) {
   case 0x0B: case 0x0A: case 0x09: case 0x08: case 0x07:  case 0x06:
     return 3;
   case 0x05: case 0x04: case 0x03:
     return 2;
   case 0x02: case 0x01:
     return 1;
   case 0x00: case 0x1F: case 0x1E:
     return 0;
   case 0x1D: case 0x1C:
     return -1;
   case 0x1B: case 0x1A: case 0x19:
     return -2;
   case 0x18: case 0x17: case 0x16: case 0x15:  case 0x14: case 0x13:
     return -3;
   default: assert(false);
   }
```

```
 case 2:
   switch (j) {
   case 0x0D: case 0x0C: case 0x0B: case 0x0A: case 0x09: case 0x08:
     return 3;
   case 0x07: case 0x06: case 0x05:
     return 2;
   case 0x04: case 0x03: case 0x02: case 0x01:
     return 1;
   case 0x00: case 0x1F: case 0x1E:
     return 0;
   case 0x1D: case 0x1C: case 0x1B:
     return -1;
   case 0x1A: case 0x19: case 0x18: case 0x17:
     return -2;
   case 0x16: case 0x15: case 0x14: case 0x13:  case 0x12: case 0x11:
     return -3;
   default: assert(false);
   }
 case 3:
   switch (j) {
   case 0x0F: case 0x0E: case 0x0D: case 0x0C:
   case 0x0B: case 0x0A: case 0x09: case 0x08:
     return 3;
   case 0x07: case 0x06: case 0x05:
     return 2;
   case 0x04: case 0x03: case 0x02: case 0x01:
     return 1;
   case 0x00: case 0x1F: case 0x1E:
     return 0;
   case 0x1D: case 0x1C: case 0x1B:
     return -1;
   case 0x1A: case 0x19: case 0x18: case 0x17:
     return -2;
   case 0x16: case 0x15: case 0x14: case 0x13:
   case 0x12: case 0x11: case 0x10:
     return -3;
   default: assert(false);
   }
 default: assert(false);
 }
}

// The table that is actually used by the implementation contains
// only non-negative entries; the sign is computed separately:
nat SRTLookup(nat i, nat j) {
 return abs(phi(i, j));
}
```

Appendix 2: XFL Model of the Implementation

```
// The function SRT is an XFL model of the Llano integer divider.
// It has four input parameters:
//    (1) isSigned: a boolean indication of whether the dividend,
//        divisor, quotient, and remainder are represented as
//        signed or unsigned integers.
//    (2) w: the width of the divisor, quotient, and remainder,
//        which may be 8, 16, 32, or 64; the width of the
//        dividend is 2*w.
//    (3) XEnc: the encoding of the divisor.
//    (4) YEnc: the encoding of the dividend.
// Three values are returned:
//    (1) A boolean indication of successful completion, which is
//        false if either the divisor is zero or the quotient is
//        too large to be represented in the indicated format.
//        The other two values are invalid in this event.
//    (2) The encoding of the quotient.
//    (3) The encoding of the remainder.

<bool, nat, nat> SRT(nat YEnc, nat XEnc, nat w, bool isSigned) {
  assert((w == 8) || (w == 16) || (w == 32) || (w == 64));

  // Division by 0 signals an error:
  if (XEnc == 0) {
    return <false, 0, 0>;
  }


  // The following variables appear in assertions but are not
  // involved in the computation of the function values:
  int Y;   // value of dividend
  int X;    // value of divisor
  int QPart;  // value of quotient during iteration
  int QPre;  // value of quotient before fix-up
  int RPre; // value of remainder before fix-up
  int Q;  // value of quotient after fix-up
  int R; // value of remainder after fix-up
  int m;  // value derived from table, -3 <= m <= 3
   rat d;  // shifted divisor, 1 <= abs(d) < 2
  rat p;  // partial remainder, abs(p) <= abs(d)
  nat i;  // first argument of phi
  nat j;  // second argument of phi

  // Decode operands:
  if (isSigned) {
    Y = SgndIntVal(2*w, YEnc[2*w-1:0]);
    X = SgndIntVal(w, XEnc[w-1:0]);
  }
  else {
    Y = YEnc[2*w-1:0];
    X = XEnc[w-1:0];
  }
```

```
// Compute the number of divisor bits that follow the leading sign
// bits. In the case of the negative of a power of 2, the trailing
// sign bit is included as a divisor bit:
bool XSign = isSigned ? XEnc[w-1] : false;
nat b = w;
while ((b > 0) && (XEnc[b-1] == XSign)) {
  b--;
}
bool XNegPower2 = XSign && ((b == 0) || (XEnc[b-1:0] == 0));
nat XNB = XNegPower2 ? b+1 : b;
assert(XNB == expo(X) + 1);


// Compute dEnc, a bit vector encoding of d = X >> expo(X):
nat dEnc = 0;
dEnc[67] = XSign;
dEnc[66] = XSign;
dEnc[65:66-XNB] = XEnc[XNB-1:0];
d = X >> expo(X);
assert(dEnc == d[2:-65]);
// Compute leading 2 bits of fractional part of d:
nat dIndex;
if (XSign == 0) {
  dIndex = dEnc[64:63];
}
else if (XNegPower2) {
  dIndex = 0;
}
else if (dEnc[62:0] == 0) {
  dIndex = ((~dEnc[64] | ~dEnc[63]) << 1) | dEnc[63];
}
else {
  dIndex = ~dEnc[64:63];
}
i = dIndex; // first argument of phi
assert(i == fl(4*(abs(d) - 1)));


// Compute the number of dividend bits that follow the leading sign
// bits. In the negative case, the trailing sign bit is
// included as a dividend bit.
bool YSign = isSigned ? YEnc[2*w-1] : false;
b = 2*w;
while ((b > 0) && (YEnc[b-1] == YSign)) {
  b--;
}
nat YNB = YSign ? b + 1 : b;
if (Y > 0) {
  assert(1 << (YNB - 1) <= Y && Y < 1 << YNB);
}
else if (Y < 0) {
  assert(1 << (YNB - 2) < abs(Y) && abs(Y) <= 1 << (YNB - 1));
};
assert(Y == 0 || YNB >= expo(Y)+1);
```

```
// Compute number of iterations:
nat n;
if (YNB >= XNB) {
  n = fl((YNB - XNB)/2) + 1;
}
else {
  n = 0;
}
assert(abs(Y) <= abs(X) << 2*n);

// Initialize pEncHi, pEncLo, and carryHi, which form a
// redundant representation of the partial remainder:
nat pEnc = 0;
pEnc[131] = YSign;
pEnc[130] = YSign;
pEnc[129] = YSign;
if (YNB != 0) {
  if (YNB[0] == XNB[0]) {
    pEnc[128] = YSign;
    pEnc[127:128-YNB] = YEnc[YNB-1:0];
  }
  else {
    pEnc[128:129-YNB] = YEnc[YNB-1:0];
  }
}
nat pEncHi = pEnc[131:64];
nat pEncLo = pEnc[63:0];
nat carryHi = 0;
assert(n >= 32 || pEncLo[63-2*n:0] == 0);
p = Y >> (expo(X) + 2*n); // initial partial remainder
if (YNB >= XNB) {
  assert(((pEncHi << 64) | pEncLo) == p[2:-129]);
}
else if (YNB[0] == XNB[0]) {
  assert(pEncHi == (Y << (64 - YNB))[67:0]);
}
else {
  assert(pEncHi == (Y << (65 - YNB))[67:0]);
}

// Initialize the quotient:
QPart = 0;  // partial quotient
nat Q0Enc = 0;  // encoding of QPart
nat QPEnc = 1;  // encoding of QPart+1
nat QMEnc = 0x7FFFFFFFFFFFFFFF;  // encoding of QPart-1

// On each iteration, the next partial remainder is computed
// and the quotient is updated:
for (nat k=1; k<=n; k++) {

  // Table lookup:
  nat pTop = pEncHi[67:62];
  bool pSign = pTop[5];
  nat pIndex = pTop[4:0]; // second argument of SRTLookup
```

```
nat mAbs = SRTLookup(dIndex, pIndex);
bool mSign = XSign ^ pSign;
m = mSign ? -mAbs : mAbs;
if (pTop == 0x2F) {
  j = 0x10;
}
else {
  j = pIndex;
}
assert(m == (XSign ? -phi(i, j) : phi(i, j)));
assert(SgndIntVal(5, j)/8 <= p && p < SgndIntVal
(5, j)/8 + 1/4);

// 4*p - dm is computed as a sum of four terms.
// The first two, addA and addB, represent -d*m:
nat addA;
if (mAbs[1] == 0) {
  addA = 0;
}
else if (mSign) {
  addA = dEnc[66:0] << 1;
}
else {
  addA = (~dEnc[66:0] << 1) | 1;
}
nat addB;
if (mAbs[0] == 0) {
  addB = 0;
}
else if (mSign) {
  addB = dEnc;
}
else {
  addB = ~dEnc[67:0];
}
// A correction term is required to complete the 2's
// complement in case m > 0:
nat inject = 0;
if (!mSign) {
  if (mAbs[0] ^ mAbs[1]) {
    inject = 1;
  }
  else if (mAbs[0] & mAbs[1]) {
    inject = 2;
  }
}
assert((addA + addB + inject)[67:0] ==  (-d*m)[2:-65]);
// addC and addD represent the upper bits of 4*p:
nat addC = (pEncHi << 2) | pEncLo[63:62];
nat addD = (carryHi << 2) | inject;
assert((addC + addD - inject)[67:0] == (4*p)[2:-65]);
// The next partial remainder:
p = 4*p - m*d;
assert(abs(p) <= abs(d));
```

```
assert((addA + addB + addC + addD)[67:0] == p[2:-65]);

// 4-2 compression:
nat sum1 = addA ^ addB ^ addC;
nat carry1 = (addA & addB | addB & addC | addA & addC) << 1;
nat sum2 = sum1 ^ carry1 ^ addD;
nat carry2 = (sum1 & carry1 | carry1 & addD | sum1 & addD) << 1;
assert((sum2 + carry2)[67:0] == p[2:-65]);

// Update the redundant representation of p:
pEncHi = ((sum2[67:62] + carry2[67:62])[5:0] << 62) | sum2[61:0];
pEncLo = (pEncLo << 2)[63:0];
carryHi = carry2[61:0];
assert((((pEncHi + carryHi)[67:0] << 64) | pEncLo) == p[2:-129]);
assert(n >= k + 32 || pEncLo[63-2*(n-k):0] == 0);

// Update quotient:
QPart = 4*QPart + m;
assert(abs(QPart) < (1 << 2*k));
if (mAbs == 0) {
  QPEnc = (Q0Enc << 2)[66:0] | 1;
  QMEnc = (QMEnc << 2)[66:0] | 3;
  Q0Enc = (Q0Enc << 2)[66:0];
}
else if (mSign == 0) {
  switch (mAbs) {
  case 1:
    QPEnc = (Q0Enc << 2)[66:0] | 2;
    QMEnc = (Q0Enc << 2)[66:0];
    Q0Enc = (Q0Enc << 2)[66:0] | 1;
    break;
  case 2:
    QPEnc = (Q0Enc << 2)[66:0] | 3;
    QMEnc = (Q0Enc << 2)[66:0] | 1;
    Q0Enc = (Q0Enc << 2)[66:0] | 2;
    break;
  case 3:
    QPEnc = (QPEnc << 2)[66:0];
    QMEnc = (Q0Enc << 2)[66:0] | 2;
    Q0Enc = (Q0Enc << 2)[66:0] | 3;
    break;
  default: assert(false);
  }
}
else { // mSign == 1
  switch (mAbs) {
  case 1:
    QPEnc = (Q0Enc << 2)[66:0];
    Q0Enc = (QMEnc << 2)[66:0] | 3;
    QMEnc = (QMEnc << 2)[66:0] | 2;
    break;
  case 2:
    QPEnc = (QMEnc << 2)[66:0] | 3;
    Q0Enc = (QMEnc << 2)[66:0] | 2;
```

```
      QMEnc = (QMEnc << 2)[66:0] | 1;
      break;
    case 3:
      QPEnc = (QMEnc << 2)[66:0] | 2;
      Q0Enc = (QMEnc << 2)[66:0] | 1;
      QMEnc = (QMEnc << 2)[66:0];
      break;
    default: assert(false);
    }
  }
  assert(Q0Enc == QPart[66:0]);
  assert(QMEnc == (QPart - 1)[66:0]);
  assert(QPEnc == (QPart + 1)[66:0]);
}

// Remainder and quotient before fix-up:
RPre = p << expo(X);
QPre = QPart;
assert(Y == RPre + QPre*X);
assert(abs(RPre) <= abs(X));

// Encoding of remainder:
nat REncPre = (pEncHi + carryHi)[66:0];
if (YNB >= XNB) {
  assert(REncPre == (RPre << (66 - XNB))[66:0]);
}
else if (YNB[0] == XNB[0]) {
  assert(REncPre == (RPre << (64 - YNB))[66:0]);
}
else {
  assert(REncPre == (RPre << (65 - YNB))[66:0]);
}

// Fix-up is required if either the remainder and the dividend have
// opposite signs or the absolute value of the remainder is the same
// as that of the divisor. The signals RIs0, RPosX, and RNegX, which
// indicate whether the remainder is 0, X, or -X, are computed in
// parallel with the addition and may not refer to the sum:

bool RSignPre = REncPre[66];
bool RIs0 = (pEncHi[66:0] ^ carryHi ^ ((pEncHi[65:0] | carryHi)
<< 1)) == 0;
assert(RIs0 == (RPre == 0));

nat RPosXSum = pEncHi[66:0] ^ carryHi ^ ~dEnc[66:0];
nat RPosXCarry = (pEncHi[66:0] & carryHi |
                  pEncHi[66:0] & ~dEnc[66:0] |
                  carryHi & ~dEnc[66:0]) << 1;
bool RPosX = (RPosXSum ^ RPosXCarry ^
             (((RPosXSum[65:0] | RPosXCarry[65:0]) << 1) | 1))
             == 0;;
assert(RPosX == (RPre == X));

nat RNegXSum = pEncHi[66:0] ^ carryHi ^ dEnc[66:0];
```

```
nat RNegXCarry = (pEncHi[66:0] & carryHi |
                       pEncHi[66:0] & dEnc[66:0] |
                       carryHi & dEnc[66:0]) << 1;
bool RNegX = (RNegXSum ^ RNegXCarry ^
                  ((RNegXSum[65:0] | RNegXCarry[65:0]) << 1)) == 0;;
assert(RNegX == (RPre == -X));

bool RIsX = RPosX | RNegX;
assert(RIsX == (abs(RPre) == abs(X)));

bool fixupNeeded = RIsX || (!RIs0) && (RSignPre != YSign);

nat REnc; // final encoding of remainder
if (!fixupNeeded) {
  REnc = REncPre[65:2];
  R = RPre;
}
else if (RIsX) {
  REnc = 0;
  R = 0;
}
else if (RSignPre == XSign) {
  REnc = ((REncPre[65:2] << 2) + ~dEnc[65:0] + 1)[65:2];
  R = RPre - X;
}
else {
  REnc = ((REncPre[65:2] << 2) + dEnc[65:0])[65:2];
  R = RPre + X;
}
bool RSign = YSign & ~RIs0[0] & ~RIsX[0];
if (YNB >= XNB) {
   assert(REnc == (R << (64 - XNB))[63:0]);
}
else if (YNB[0] == XNB[0]) {
   assert(REnc == (R << (62 - YNB))[63:0]);
}
else {
   assert(REnc == (R << (63 - YNB))[63:0]);
}
nat QEnc; // final encoding of quotient
if (!fixupNeeded) {
  QEnc = Q0Enc;
  Q = QPre;
}
else if (RSignPre == XSign) {
  QEnc = QPEnc;
  Q = QPre + 1;
}
else {
  QEnc = QMEnc;
  Q = QPre - 1;
}
assert(Y == R + Q*X);
assert(abs(R) < abs(X));
```

```
assert((R == 0) || ((R < 0) == (Y < 0)));
assert(n > 33 || QEnc == Q[66:0]);


// Determine whether the quotient is representable:
bool QTooLarge;
if (YNB > XNB + w) {
  QTooLarge = true;
}
else if (n <= 1) {
  QTooLarge = false;
}
else if (YSign == XSign) {
  QTooLarge = (QEnc[w+2:w] != 0) || isSigned && (QEnc[w-1]
  != 0);
}
else {
  QTooLarge = (QEnc[w+1:w-1] != 7);
}
if (isSigned) {
  assert(QTooLarge == ((Q > MaxSgndIntVal(w)) ||
                       (Q < MinSgndIntVal(w))));
}
else {
  assert(QTooLarge == (Q >= 1 << w));
}

if (QTooLarge) {
  return <false, 0, 0>;
}


// Compute the final results:
nat QOut = QEnc[w-1:0];
nat ROut;
if (YNB >= XNB) {
  ROut = ((RSign << w) - (RSign << XNB)) | REnc[63:64-XNB];
}
else if (YNB[0] == XNB[0]) {
  ROut = ((RSign << w) - (RSign << (YNB+2))) | REnc[63:62-YNB];
}


  else {
  ROut = ((RSign << w) - (RSign << (YNB+1))) | REnc[63:63-YNB];
}
assert(QOut == Q[w-1:0]);
assert(ROut == R[w-1:0]);
return <true, QOut, ROut>;
}
```

Use of Formal Verification at Centaur Technology

Warren A. Hunt, Jr., Sol Swords, Jared Davis, and Anna Slobodova

1 Introduction

We have developed a formal-methods-based hardware verification toolflow to help ensure the correctness of our x86-compatible microprocessors. Our toolflow uses the ACL2 theorem-proving system as a design database and a verification engine. We verify Verilog designs by first translating them into a formally defined hardware description language and then using a variety of automated verification algorithms controlled by theorem-proving scripts.

In this chapter, we describe our approach to verifying components of VIA Centaur's 64-bit Nano x86-compatible microprocessor (referred to herein as CN). We have successfully verified a number of media-unit operations, such as the packed addition/subtraction instructions. We have verified the integer multiplication unit, and we are in the process of verifying microcode sequences that perform arithmetic operations.

1.1 Overview of Verification Methodology

In our verification process, we first translate the Verilog RTL source code of Centaur's design into EMOD, a formally defined HDL. This process captures a design as an ACL2 object that can be interpreted by an ACL2-based HDL simulator. The HDL simulator is used both to run concrete test cases and to extract symbolic representations of the circuit logic of blocks of interest. We then use a combination of theorem proving and equivalence checking to prove that the functionality of the circuit in question is equivalent to a higher-level specification. A completed verification yields an ACL2 theorem that precisely states what we have proven.

We have developed a deep embedding of our hardware description language, EMOD [12], in the ACL2 logic. We describe the EMOD language in Sect. 4.1. Our

W.A. Hunt, Jr. (✉)
Centaur Technology, Austin, TX, USA
and
University of Texas at Austin, Austin, TX, USA
e-mail: hunt@centtech.com; hunt@cs.utexas.edu

D.S. Hardin (ed.), *Design and Verification of Microprocessor Systems for High-Assurance Applications*, DOI 10.1007/978-1-4419-1539-9_3,

implementation includes a syntax checker for well-formed EMOD modules and an interpreter that gives meaning to such modules. The EMOD interpreter can operate in several different modes to perform concrete or symbolic simulations, analyze dependencies, and estimate delays. Simulations may use either a Boolean or four-valued logic mode, and symbolic simulations may use either binary-decision diagrams (BDDs) [6] or and-inverter graphs (AIGs) as the representation for symbolic bits. We believe our approach to representing the hardware design reduces the risk of translation errors, since we may perform cosimulation between Verilog and EMOD to ensure the veracity of the translation. We can also translate the design as represented in the EMOD language back to Verilog.

Our Verilog translator consists of a parser and a series of code transformations that simplify the design until it can be easily translated into the EMOD language, which lacks features such as continuous assignments and always blocks. We describe the translator in Sect. 2.

To prove that output from a hardware simulation is equivalent to that produced by a specification function, we produce BDDs representing both the hardware and the specification outputs and compare them for equivalence. We use case splitting to avoid certain BDD-size explosions. We sometimes use AIGs as an intermediate form before creating the BDDs to avoid size explosions. To produce ACL2 theorems using these methods, we have created a verified symbolic execution framework that uses these procedures. We describe our proof methodology in Sect. 3.

1.2 Timeline

The integration of formal methods into Centaur's design methodology has been ongoing for several years. Hunt first met with Centaur representatives in April 2007. This led Hunt and Swords to join Centaur in June 2007 to see if our existing (ACL2-based) tools could be usefully deployed on Centaur verification problems. Our use of formal methods is not new, and AMD [19] has been using ACL2 for many years for floating-point hardware verification. However, there are several things that differentiate our effort from all others: the Centaur design is converted into our EMOD-formalized hardware description language (described later), our verification (BDD and AIG) algorithms are themselves verified, and all of our claims are all checked as ACL2 theorems.

The use of formal methods to aid hardware design has been ongoing for many years. Possibly the earliest adopter was IBM with equivalence checking mechanisms that they developed in the early 1980s; IBM protected these mechanisms as trade secrets. With the development of simple microprocessor verification examples, such as the FM8501 [9] and the VIPER [5], and introduction of BDDs [6], commercial organizations started integrating some use of formal methods into their design flow. A big impetus for the use of formal methods came from the Intel FDIV bug [18].

Work that allowed us to get an immediate start was just being finished when our Centaur-based effort began. Boyer and Hunt had implemented BDDs [3, 4] with an extended version of ACL2 that included unique object representation and function memoization [3]. Separately, Hunt and Reeber had previously embedded the DE2 HDL into ACL2 [10], and this greatly influenced the development of the EMOD HDL.

Our initial efforts were directed along two fronts: analyzing microcode for integer division and verifying the floating-point addition/subtraction hardware. Our analysis of the microcode for the integer divide algorithm involved creating an abstraction of the microcode with ACL2 functions and then using the ACL2 theorem-prover to mechanically check that our model of the divide microcode computes the correct answer. This effort discovered an anomaly that was subsequently corrected.

Our work on the verification of the floating-point addition/subtraction hardware was much more involved. Because of the size of the design – some 34,000 lines of Verilog – it was necessary for us to create a translator from Verilog into our EMOD hardware description language. We enhanced a Verilog parser, written by Terry Parks (of Centaur), so that it emitted an EMOD-language version of the floating-point hardware design; this translator created an EMOD-language representation of the entire module hierarchy, including all interface and wire names. The semantics of the EMOD language are given by the EMOD simulator which allows an EMOD-language-based design to be simulated or symbolically simulated with a variety (e.g., BDDs and AIGs) of mechanisms. Simultaneously, we developed an extension to ACL2 that provides a symbolic simulator for the entire ACL2 logic; this system was called **G**. Given these components, we were able to attempt the verification of Centaur's designs; this was done by comparing the symbolic equations produced by the EMOD HDL symbolic simulator to the equations produced by the **G**-based symbolic simulation of our ACL2 floating-point specifications.

Our verification of Centaur's floating-point addition/subtraction instructions led to the discovery of two design flaws: for two of the four floating-point adders, the floating-point control flag inputs arrived one cycle early and for one pair of 80-bit numbers (described more fully later), the sum/difference was incorrect. Both of these very subtle problems were fixed. This work was completed within the first year of our efforts at Centaur. This effort strained our Verilog translator and illuminated areas where we wanted to better integrate symbolic simulation into the ACL2 system.

In the summer of 2008, Davis arrived and began developing a more capable Verilog translator named VL. The new translator was itself written in ACL2, and it was designed with simplicity and assurance in mind. The translator has provisions for translating Verilog annotations and property specifications into the EMOD language.

Starting in the summer of 2008, Swords began an effort to build a verified version of the ACL2 **G** symbolic simulator, called **GL** (for **G** in the **L**ogic). This new system represents symbolic ACL2 expressions as ACL2 data objects, which allows proofs to be carried out which show that such objects are manipulated correctly.

In the fall of 2008, Slobodova joined Centaur as manager of the formal verification team and began using these tools to verify a number of different(integer and floating-point) multiplier implementations. These multipliers are actually quite complicated as they can be reconfigured on a clock-by-clock basis to create different (e.g., four 32×32-bit or one 64×64-bit) multipliers. The verification of the multipliers has stressed the capacity of our tools in a variety of ways and this effort has led to many improvements in capacity and speed.

By the spring of 2009, the outcome of the two efforts mentioned above resulted in the replacement of our original prototype Verilog translator with VL, and the replacement of the **G** system with **GL**, a verified symbolic simulator for ACL2 functions. All of our proofs are now carried out using these new tools.

1.3 Centaur Media Unit

As an example to illustrate our methodology, we will discuss our verification of the floating-point addition instructions implemented in the media unit of Centaur's CPU design. The part of the media unit that handles floating-point addition and subtraction is called the **fadd** unit; this unit is highly optimized for low latency arithmetic operations and implements SIMD 32- and 64-bit floating-point additions as well as scalar x87 80-bit floating point additions. All floating-point addition operations are performed with a two-cycle latency; the **fadd** unit can also forward results internally so that operations may be chained.

The **fadd** RTL-level design is composed of 680 modules, which we convert from Verilog into our EMOD hardware description language; it is this EMOD form of Centaur's design that we subject to analysis. The physical implementation is composed of 432,322 transistors, almost evenly split between PMOS and NMOS devices. This represents less than 5% of the total transistors in the implementation, but its 33,700 line Verilog description represents more than 6% of the CN design specification. The **fadd** unit has 374 output signals and 1,074 inputs including 26 clock inputs. Multiple clock inputs are used to manage power usage.

The **fadd** unit is composed of four adders: two 32-bit units, one 64-bit unit, and one 80-bit unit (see Fig. 1). When a 32-bit packed addition is requested, all four units are used, and the 64-bit and 80-bit adders each takes 32-bit operands and produces a 32-bit result. When a 64-bit packed addition is requested, the 64-bit and 80-bit adders each takes 64-bit operands and produces a 64-bit result. The **fadd** unit can only add one pair of 80-bit operands per clock cycle. Other combinations are possible when a memory-resident operand is added to a register-resident, x87-style, 80-bit operand; the **fadd** unit also manages such x87-mode, mixed-size addition requests.

There are multiple paths through the addition logic that operate in parallel. The relevant path for a particular pair of operands is determined by characteristics such as the operand types (NaN, zero, denormal, etc.) and their sign and exponent bits, and the result from that path is selected as the result of the addition.

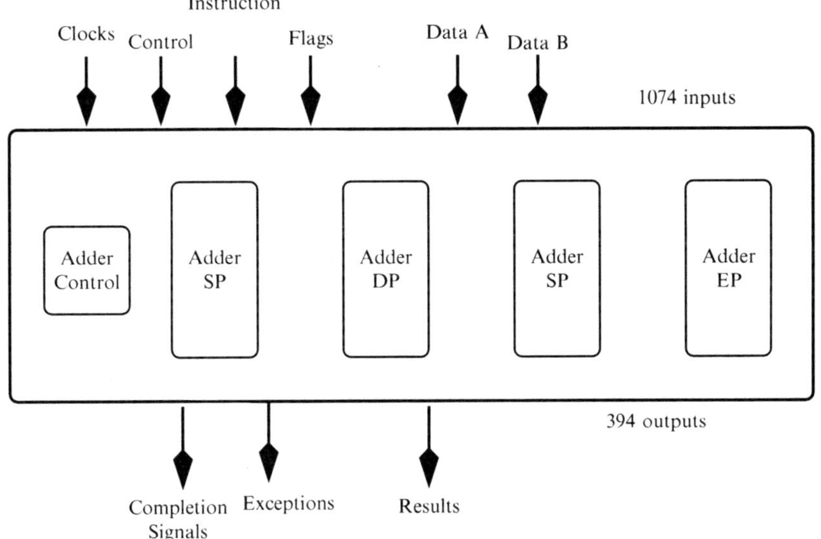

Fig. 1 Adder units inside of **fadd**

2 Modeling Effort

The specification of the CN processor consists of over half a million lines of Verilog; this Verilog is frequently updated by the logic designers. To bring this design into our EMOD HDL, we have developed a translator named VL. This is a challenge since Verilog is such a large language with no formal semantics. Our work is based on the IEEE Verilog 1364-2005 standard [13], and we do not yet support the SystemVerilog extensions. This standard usually explains things well, but sometimes it is vague; in these cases, we have carried out thousands of tests and attempted to emulate the behavior of Cadence's Verilog simulator.

VL needs to produce a "sound" translation or our verification results may be meaningless. Because of this, we have written VL in the purely functional programming language of the ACL2 theorem prover, and our emphasis from the start has been on correctness rather than performance. For instance, our parser is written in a particularly naive way: to begin, each source file is read, in its entirety, into a simple list of *extended characters*, which associate each character with its filename and position. This makes the remaining steps in the parsing process ordinary list-transforming functions:

- read : filename → echar list
- preprocess : echar list → echar list
- lex : echar list → token list
- eat-comments : token list → token list × comment map
- parse : token list → module list

The parser itself is written in a conventional, recursive-descent style, which is a good match for Verilog since keywords like module, assign, etc. usually say what comes next. Since the entire list of tokens has been computed before parsing begins, we can take advantage of arbitrary look-ahead, and backtracking is completely straightforward.

This simple-minded approach lends itself well to informal validation. For instance, since we actually construct each intermediate list, we can add assertions relating them to one another, e.g., we can test that flattening the parsed input is equal to the original input. Since our functions operate on lists, instead of files, it is very easy to write unit tests directly in our source code, and we have developed a number of these tests. Furthermore, since these routines are written in the ACL2 theorem prover, we can actually prove some theorems about the parser, e.g., on success it produces a list of syntactically well-formed modules.

2.1 *Conversion to the* EMOD *Language*

To implement the translation into EMOD, we adopt a program-transformation-like [23] style: to begin with, the entire parse tree for the Verilog sources is constructed; we then apply a number of rewriting passes to the tree which result in simpler Verilog versions of each module. The final conversion into EMOD is really almost incidental, with the resulting EMOD modules differing from our most-simplified Verilog modules only in syntax.

Each transformation tends to be fairly short and easy to understand and can be studied in isolation, either informally or with the theorem prover. Since each rewriting pass produces well-formed Verilog modules, we can simulate the original and simplified Verilog modules against each other, either at the end of the simplification process or at some intermediate point.

We can also run a number of common sanity checks after each rewrite to catch any gross errors. These sorts of checks serve to answer questions such as:

- Are only supported constructs used?
- Are only defined modules instanced?
- Is each module's namespace free of collisions?
- Are the ports compatible with the port declarations?
- Are the port and wire declarations compatible?
- Have we determined the size of every declaration?
- Have the widths and signs of all expressions been determined?
- Are the widths of the arguments to every submodule correct?
- Are the indices for every bit- and part-select in bounds?

These sorts of checks are often useful as "guards" (preconditions) for our transformation steps. In few cases, we also prove, using ACL2, that some of these properties will be satisfied after a certain transformation is run. But usually we do not try to do this because it is easier to just run the checks after each transformation.

We now present an overview of our transformation sequence.

2.1.1   Unparameterization

Verilog modules can have parameters, e.g., an `adder` module might take input wires of some arbitrary `width`, and other modules can then instantiate `adder` with different `widths`, say 8, 16, and 32. Our first transformation is to eliminate parametrized modules, e.g., we would introduce three new modules, `adder$width=8`, `adder$width=16`, and `adder$width=32`, and change the instances of `adder` to point to these new modules as appropriate.

2.1.2   Declaring Implicit and Port Wires

Verilog permits undeclared identifiers to be used as one-bit wires. We would like to prohibit this to reduce the chance of typos (a la Perl's *use strict*), but this idea is unpopular so we only issue warnings. In this transformation, we add a `wire` declaration for each undeclared wire and each port, which has been declared to be an `input` or `output` but which has not also been declared as a `wire`.

2.1.3   Standardizing Argument Lists

Modules may be instantiated using either positional or named argument lists. For instance, given a module M with ports a, b, and c, the following instances of M are equivalent:

```
M my_instance(1, 2, 3);
M my_instance(.b(2), .c(3), .a(1));
```

In this transformation, we convert all instances to the positional style and annotate the arguments as inputs or outputs.

2.1.4   Resolving Ranges

Wires and registers in Verilog can have widths. For instance,

```
wire [3:0] w;
```

declares a four-bit wire, w, whose bits are w[3] through w[0]. Unparameterization sometimes leaves us with expressions here, e.g., in the `adder` module, we might have

```
wire [width-1:0] a;
```

which, in `adder$width=8`, will become

```
wire [8-1:0] a;
```

We now resolve these expressions to constants. The specification seems vague about how these expressions are to be evaluated (e.g., with respect to widths and signedness), so we are quite careful and only allow signed, 32-bit, overflow-free computations of +, -, and *.

2.1.5 Operator Rewriting

We can reduce the variety of operators we need to deal with by simply rewriting
some operators away. In particular, we perform rewrites such as

$$a \ \&\& \ b \rightarrow (|a) \ \& \ (|b),$$
$$a \ != \ b \rightarrow |(a \ \hat{} \ b), \text{and}$$
$$a \ < \ b \rightarrow \sim(a \ >= \ b).$$

This process eliminates all logical operators (&&, ||, and !), equality comparisons
(== and !=), negated reduction operators ($\sim$&, $\sim$|, and $\sim\hat{}$), and standardizes all
inequality comparisons (<, >, <=, and >=) to the >= format. We have a considerable
simulation test suite to validate these rewrites.

2.1.6 Sign and Width Computation

We now annotate every expression with its type (sign) and width. This is tricky.
The rules for determining widths are quite complicated, and if they are not properly
implemented then, for instance, carries might be inappropriately kept or dropped. It
took a lot of experimenting with Cadence and many readings of the standard to be
sure that we had it right.

2.1.7 Expression Splitting

After the widths have been computed, we introduce explicit wires to hold the intermediate values in expressions,

```
assign w = (a + b) - c;
  →
wire [width:0] newname;
assign newname = a + b;
assign w = newname - c;
```

We also split inputs to module and gate instances

```
my_mod my_inst(a + b, ...);
  →
wire [width:0] newname;
assign newname = a + b;
my_mod my_inst(newname, ...);
```

2.1.8 Making Truncation Explicit

Verilog allows for implicit truncations in assignment statements; for instance,
one can assign the result of a five-bit addition a + b to a three-bit bus

(collection of wires), w. We now make these truncations explicit by introducing a new wire for the intermediate result, for example,

```
wire [4:0] newname;
assign newname = a + b;
assign w = newname[2:0];
```

We print warnings about such truncations since they are not good form and may point to problems.

2.1.9   Eliminating Assignments

We now replace all assignments with module instances. First, we develop a way to generate modules to perform each operation at a given width, and we write these modules using only gates and submodule instances. Next, we replace each assignment with an instance of the appropriate module, e.g.,

```
assign w = a + b;
    →
VL_13_BIT_PLUS newname(w, a, b);
```

This is one of our more complicated transformations, so we have developed a test suite which, for instance, uses Cadence to exhaustively test VL_4_BIT_PLUS against an ordinary addition operation. We are careful to handle the X and Z behavior appropriately. We go out of our way so that all of w's bits become X if any bit of a or b is X or Z, even though this makes our generated adders more complex.

2.1.10   Eliminating Instance Arrays

Gate and module instances can be put into arrays,

```
and foo [13:0] (o, a, b);
```

declares 14 and-gates. We now convert such arrays into explicit instances, such as, `foo0, ..., foo13`. The rules for partitioning the bits of the arguments are not too difficult.

2.1.11   Eliminating Higher-Arity Gates

Primitive gate instances in Verilog can use variable-length argument lists;

```
not multinot(o_1, ..., o_n, i);
```

represents a `not` gate with one input, i, and n outputs, $o_1, \ldots, o_n$. We now split these into lists of gates,

```
not multinot_1(o_1, i);

...

not multinot_n(o_n, i);
```

Afterward, each `not` and `buf` gate has one input and output, and each `and`, `or`, `nand`, `nor`, `xor`, and `xnor` gate has two inputs and one output.

We have left out a few other rewrites like naming any unnamed instances, eliminating supply wires, and some minor optimizations. But the basic idea is that, taken all together, our simplifications leave us with a new list of modules where only simple gate and module instances are used. This design lets us focus on each task separately instead of needing to consider all of Verilog at once.

2.2 Modeling Flow

It takes around 20 min to run our full translation process on the whole of CN. A lot of memory is needed, and we ordinarily use a machine with 64 GB of physical memory to do the translation. Not all modules can be translated successfully (e.g., because they use constructs which are not yet supported). However, a large portion of the chip is fully supported.

The translator is run against multiple versions of the chip each night, and the resulting EMOD modules are stored on disk into files that can be loaded into an ACL2 executable in seconds. This process also results in internal Web pages that allow the original source code, translated source code, and warnings about each module to be easily viewed and some other Lint-like reports for the benefit of the logic designers and verification engineers.

3 Verification Method

Our verification efforts so far have concentrated on proving the functional correctness of instructions running on certain execution units; that is, showing that they operate equivalently to a high-level specification. However, we believe our methodology would also be useful for proving nonfunctional properties of the design.

Our specifications are functions written in ACL2. They are executable and can therefore be used to run tests against the hardware model or a known implementation. In most cases, we write specifications that operate at the integer level on vectors of signals. Often these specifications are simple enough that we are satisfied that they are correct by examination; by comparison with the RTL designs of the corresponding hardware units, they are very small indeed. For floating-point addition, we use a low-level integer-based specification that is somewhat optimized for symbolic execution performance and is relatively complicated compared to our other specifications. However, this specification has been separately proven equivalent to a high-level, rational-number-based specification. Before this proof was completed, we had also tested the specification by running it on millions of inputs and comparing the results to those produced by running the same floating-point operations directly on the local CPU.

Figure 2 shows the verification methodology we used in proving the correctness of the **fadd** unit's floating-point addition instructions. We compare the result of symbolic simulations of an instruction specification and our model of the **fadd**

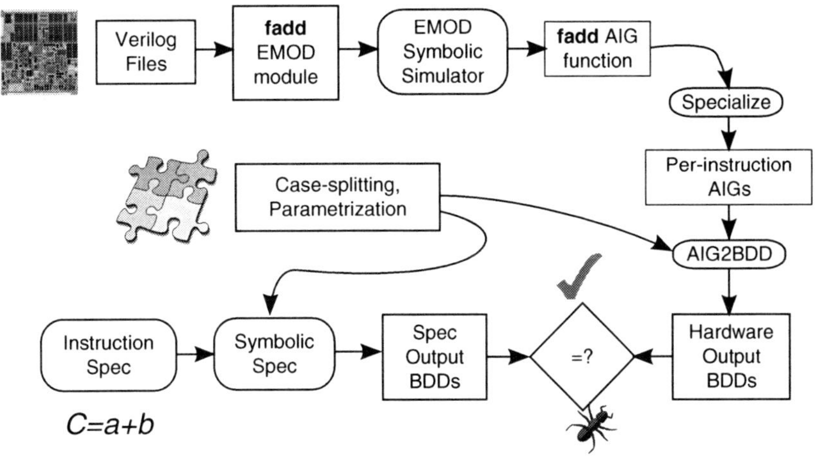

Fig. 2 Verification method

hardware. To obtain our model of the hardware, we translate the **fadd** unit's Verilog design into our EMOD hardware description language. We then run an AIG-based symbolic simulation of the **fadd** model using the EMOD symbolic simulator; the results of this simulation describe the outputs of the **fadd** unit as four-valued functions of the inputs, and we represent these functions with AIGs. We then specialize these functions by setting input control bits to values appropriate for the desired instruction. To compare these functions with those produced by the specification, we then convert these AIGs into BDDs.

For many instructions, it is feasible to simply construct BDDs representing the outputs as functions of the inputs, and we therefore may verify these instructions directly using symbolic simulation. For the case of floating-point addition, however, there is a capacity problem due to the shifted addition of mantissas. We therefore use case splitting via BDD parametrization [1, 16] to restrict the analysis to subsets of the input space. This allows us to choose a BDD variable ordering specially for each input subset, which is essential to avoid this blowup. For each case split, we run a symbolic simulation of the instruction specification and an AIG-to-BDD conversion of the specialized AIGs for the instruction. If corresponding BDDs from these results are equal, this shows that the **fadd** unit operates identically to the specification function on the subset of the input space covered by the case split; otherwise, we can generate counterexamples by analyzing the differences in the outputs.

For each instruction, we produce a theorem stating that evaluation of the instruction-specialized AIGs yields the same result as the instruction's specification function. This theorem is proven using the GL symbolic simulation framework [4], which automates the process of proving theorems by BDD-based symbolic execution, optionally with parametrized case splitting. Much of the complexity of the flow is hidden from the user by the automation provided by GL; the user provides the statement of the desired theorem and high-level descriptions of the case split, symbolic simulation inputs, and suitable BDD variable orderings. BDD parametrization and the AIG to BDD conversion algorithm are used automatically based on these parameters. The statement of the theorem is independent of the symbolic execution mechanism; it is stated in terms of universally quantified variables which collectively represent a (concrete) input vector for the design.

In the following subsections, we will describe in more detail the case-splitting mechanism, the process of translating the Verilog design into an EMOD description, and the methods of symbolic simulation used for the **fadd** unit model and the instruction specification.

3.1   Case-Splitting and Parametrization

For verifying the floating-point addition instructions, we use case splitting to avoid BDD blowup that occurs due to a nonconstant shift of the operand mantissas based on the difference in their exponents. By choosing case-splitting boundaries appropriately, the shift amount can be reduced to a constant. The strategy for choosing these boundaries is documented by others [1, 7, 15, 20], and we believe it to be reusable for new designs.

In total, we split into 138 cases for single, 298 for double, and 858 for extended precision. Most of these cases cover input subsets over which the exponent difference of the two operands is constant and either all input vectors are effective additions or all are effective subtractions. Exponent differences greater than the maximum shift amount are considered as a block. Special inputs such as NaNs and infinities are considered separately. For performance reasons, we use a finer-grained case-split for extended precision than for single or double precision.

For each case split, we restrict the simulation coverage to the chosen subset of the input space using BDD parametrization. This generates a symbolic input vector (a BDD for each input bit) that covers exactly and only the appropriate set of inputs; we describe BDD parametrization in more detail in Sect. 4.3. Each such symbolic input vector is used in both an AIG-to-BDD conversion and a symbolic simulation of the specification. The BDD variable ordering is chosen specifically for each case split, thereby reducing the overall size of the intermediate BDDs. No knowledge of the design was used to determine the case-splitting approach.

3.2   Symbolic Simulation of the Hardware Model

We use the EMOD symbolic simulator to obtain Boolean formulas (AIGs) representing the outputs of a unit in terms of its inputs. In such simulations, we use a four-valued logic in which each signal may take values 1 (true), 0 (false), X (unknown), or Z (floating). This is encoded using two AIGs (onset and offset) per signal. The Boolean values taken by each AIG determine the value taken by the signal as in Fig. 3.

Fig. 3 Four-valued signal
logic

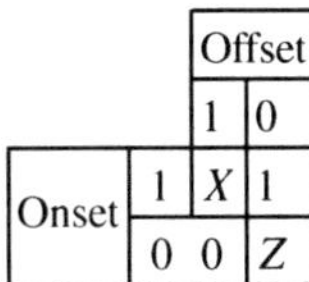

The **fadd** unit is mainly a pipeline, where each instruction is bounded by a fixed latency. To verify its instructions, we set all bits of the initial state to unknown (X) values – the onsets and offsets of all nonclock inputs are set to free Boolean variables at each cycle, so that every input signal but the clocks can take any of the four values. We then symbolically simulate it for a fixed number of cycles. This results in a fully general formula for each output in terms of the inputs at each clock cycle.

To obtain symbolic outputs for a particular instruction, we restrict the fully general output formulas by setting control signals to the values required for performing the given instruction and any signals we know to be irrelevant to unknown (X) input values. This reduces the number of variables present in these functions and keeps our result as general as possible. Constant propagation with these specified values restricts the AIGs to formulas in terms of only the inputs relevant to the instruction we are considering. For the floating-point addition instructions of the **fadd** unit, the remaining inputs are the operands and the status register, which are the same as the inputs to the specification function.

The theorems produced by our verifications typically say that for any well-formed input vector, the evaluation of the instruction-specialized AIGs using the variable assignment generated from the input vector is equivalent to the output of the specification function on that input vector. Such a theorem may often be proven automatically, given appropriate BDD ordering and case splitting, by the GL symbolic execution framework. GL has built in the notion of symbolically evaluating an AIG using BDDs, effectively converting the Boolean function representation from one form to the other. It uses the procedure `AIG2BDD` described in Sect. 4.4 for this process; this algorithm avoids computing certain intermediate-value BDDs that are irrelevant to the final outputs, which helps to solve some BDD size explosions.

3.3 Symbolic Simulation of Specification

The specification for an instruction is generally an ACL2 function that takes integers or Booleans representing some of the inputs to a block and produces integers or Booleans representing the relevant outputs. Such functions are usually defined in terms of word-level primitives such as shifts, bit-wise logical operations, plus, and minus. For the floating-point addition instructions, the function takes integers representing the operands and the control register and produces integers representing the result and the flag register. It is optimized for symbolic simulation performance rather than referential clarity; however, it has separately been proven equivalent to a high-level, rational arithmetic-based specification of the IEEE floating-point standard [14]. Additionally, it has been tested against floating-point instructions running on Intel and AMD CPUs on many millions of input operand pairs, including a test suite designed to detect floating-point corner-cases [22] as well as random tests.

To support symbolic simulation of our specifications, we developed the GL symbolic execution framework for ACL2 [4]. The GL framework allows user-provided ACL2 code to be symbolically executed using a BDD-based symbolic object representation. The symbolic execution engine is itself verified in ACL2 so that its results provably reflect the behavior of the function that was symbolically executed. GL also provides automation for proving theorems based on such symbolic executions. Since these theorems do not depend on any unverified routines, they offer the same degree of assurance as any proof in ACL2: that is, they can be trusted if ACL2 itself can be trusted.

GL automates several of the steps in our verification methodology. For a theorem in which we show that the evaluation of an AIG representation of the circuit produces results equivalent to a specification function, the GL symbolic execution encompasses the AIG-to-BDD transformation and the comparison of the results, as well as the counterexample generation if there is a bug. If the proof requires case splitting, the parametrization mechanism is also handled by GL. The user specifies the BDD variable ordering used to construct the symbolic input vectors, as well as the case split. To specify the case split, the user provides a predicate which determines whether an input vector is covered by a given case; like the theorem itself, this predicate is written at the level of concrete objects. Typically, all computations at the symbolic (BDD) level are performed by GL; the user programs only at the concrete level.

3.4 *Comparison of Specification to Hardware Model*

For each case split in which the results from the symbolic simulations of the specification and the hardware model are equal, this serves to prove that for any concrete input vector drawn from the coverage set of the case, a simulation of the **fadd** model will produce the same result as the instruction specification. If the results are not equal, we can generate a counterexample by finding a satisfying assignment for the XOR of two corresponding output BDDs.

To prove the top-level theorem that the **fadd** unit produces the same result as the specification for all legal concrete inputs, we must also prove that the union of all such input subsets covers the entire set of legal inputs. This is handled automatically by the GL framework. For each case, GL produces a BDD representing the indicator function of the coverage set (the function which is true on inputs that are elements of the set and false on inputs that are not.) As in [7], the OR of all such BDDs is shown to be implied by the indicator function BDD of the set of legal inputs; therefore, if an input vector is legal then it is in one or more of the coverage sets of the case split.

4 Mechanisms Used to Achieve the Verification

4.1 EMOD *Symbolic Simulator*

The EMOD interpreter is capable of running various simulations and analyses on a hardware model; examples include concrete-value simulations in two- or four-value mode, symbolic simulations in two- or four-value mode using AIGs or BDDs as the Boolean function representations, and delay and dependency analyses. The interpreter can also easily be extended with new analyses. The language supports multiple clocks with different timing behavior, clock gating, and both latch- and flip-flop-based sequential designs as well as implicitly clocked finite state machines (FSMs). Its language for representing hardware models is a hierarchical, gate-level HDL. A hardware model in the EMOD language is either a primitive module (such as basic logic gates, latches, and flip-flops), or a hierarchically defined module, containing a list of submodules and a description of their interconnections. The semantics of primitive modules are built into the EMOD interpreter, whereas hierarchical modules are simulated by recursively simulating submodules.

A pair of small example modules, ***half-adder-module*** and ***one-bit-cntr***, are shown in Fig. 4. Both are hierarchically defined since they each have a list of occurrences labeled :occs. Connectivity between submodules, inputs, and outputs is defined by the :i (input) and :o (output) fields of the modules and the occurrences. We translate the Verilog RTL design unit into this format for our analysis.

A novel feature of our approach is that we can actually print the theorem we are checking; thus, we have an explicit, circuit-model representation that includes all of the original hierarchy, annotations, and wire names. This is different than all other approaches of which we are aware; for instance, the Forte tool reads Intel design descriptions and builds a FSM in its memory image. Our representation allows us to search the design using database-like commands to inspect our representation of Centaur's design; this explicit representation also enables easy tool construction for users as they can write ACL2 programs to investigate the design in a manner of their choosing.

```
(defm *half-adder-module*
  '(:i (a b)
    :o (sum carry)
    :occs
    ((:u o0 :o (sum)   :op ,*xor2* :i (a b))
     (:u o1 :o (carry) :op ,*and2* :i (a b)))))

(defm *one-bit-cntr*
  '(:i (c-in reset-)
    :o (out c)
    :occs
    ((:u o2 :o out        :op ,*ff* :i (sum-reset))
     (:u o0 :o (sum c)    :op ,*half-adder-module* :i (c-in out))
     (:u o1 :o (sum-reset) :op ,*and2* :i (sum reset-)))))
```

Fig. 4 EMOD examples

4.2 BDDs and AIGs

BDDs and AIGs both are data objects that represent Boolean-valued functions of Boolean variables. We have defined evaluators for both BDDs and AIGs in ACL2. The BDD (resp. AIG) evaluator, given a BDD (AIG) and an assignment of Boolean values to the relevant variables, produces the Boolean value of the function it represents at that variable assignment. Here, for brevity, we use the notation $\langle x \rangle_{\mathrm{bdd}}\,(env)$ or $\langle x \rangle_{\mathrm{aig}}\,(env)$ for the evaluation of x with variable assignment env. We use the same notation when x is a list to denote the mapping of $\langle \_ \rangle_{\mathrm{bdd}}\,(env)$ over the elements of x.

The BDD and AIG logical operators are defined in the ACL2 logic and proven correct relative to the evaluator functions. For example, the following theorem shows the correctness of the BDD AND operator (written $\wedge_{\mathrm{bdd}}$); similar theorems are proven for every basic BDD and AIG operator such as NOT, OR, XOR, and ITE:

$$\langle a \wedge_{\mathrm{bdd}} b \rangle_{\mathrm{bdd}}\,(env) = \langle a \rangle_{\mathrm{bdd}}\,(env) \wedge \langle b \rangle_{\mathrm{bdd}}\,(env)$$

4.3 Parametrization

BDD parametrization is also implemented in ACL2. The parametrization algorithm is described in [1]; we describe its interface here. Assume that a hardware model has n input bits. To run a symbolic simulation over all 2^n possible input vectors, one possible set of symbolic inputs is n distinct BDD variables – say, $\mathbf{v} = [v_0, \ldots, v_{n-1}]$. This provides complete coverage because $\langle \mathbf{v} \rangle_{\mathrm{bdd}}\,(env)$ may equal any list of n Booleans. (In fact, if env has length n, then $\langle \mathbf{v} \rangle_{\mathrm{bdd}}\,(env) = env$.) However, to avoid BDD blowups, we sometimes run symbolic simulations that each cover only a subset of the well-formed inputs. For each such case, we first represent the desired coverage set as a BDD p, so that an input vector $\mathbf{w}$ is in the coverage set if and only if $\langle p \rangle_{\mathrm{bdd}}\,(\mathbf{w})$. We then parametrize $\mathbf{v}$ by predicate p and use the resulting BDDs $\mathbf{v}_p$ as the symbolic inputs. These parametrized BDDs have the following important properties, which have been proven in ACL2:

- The parametrized BDDs $\mathbf{v}_p$ evaluate under every environment to a list of Booleans satisfying the parametrization predicate p:

$$\forall \mathbf{w} \,.\, \langle p \rangle_{\mathrm{bdd}}\left(\langle \mathbf{v}_p \rangle_{\mathrm{bdd}}\,(\mathbf{w})\right)$$

  Therefore, a concrete input vector is only covered by a symbolic simulation of $\mathbf{v}_p$ if it satisfies p.
- Any input vector $\mathbf{u}$ that does satisfy p is covered by $\mathbf{v}_p$; that is, there is some environment under which $\mathbf{v}_p$ evaluates to $\mathbf{u}$:

$$\langle p \rangle_{\mathrm{bdd}}\,(\mathbf{u}) \Rightarrow \exists \mathbf{u}' . \langle \mathbf{v}_p \rangle_{\mathrm{bdd}}\,(\mathbf{u}') = \mathbf{u}.$$

Therefore, any concrete input vector satisfying p will be covered by a symbolic simulation of $\mathbf{v}_p$.

It can be nontrivial to produce "by hand" a BDD p that correctly represents a particular subset of the input space. Instead, this is handled by the GL symbolic execution framework. The user defines an ACL2 function that takes an input vector and determines whether or not that input vector is in a particular desired subset; GL then symbolically executes this function on (unparametrized) symbolic inputs, yielding a symbolic Boolean value (represented as a BDD) that exactly represents the accepted subset of the inputs.

4.4 AIG-to-BDD Translation

In the symbolic simulation process for the **fadd** unit, we obtain AIGs representing the outputs as a function of the primary inputs and subsequently assign parametrized input BDDs to each primary input, computing BDDs representing the function composition of the AIG with the input BDDs. A straightforward (but inefficient) method to obtain this composition is an algorithm that recursively computes the BDD corresponding to each AIG node: at a primary input, look up the assigned BDD; at an AND node, compute the BDD AND of the BDDs corresponding to the child nodes; and at a NOT node, compute the BDD NOT of the BDD corresponding to the negated node. This method proves to be impractical for our purpose; we describe here the algorithm AIG2BDD that we use instead.

To improve the efficiency of the straightforward recursive algorithm, one necessary modification is to memoize it so as to traverse the AIG as a DAG (without examining the same node twice) rather than as a tree: due to multiple fanouts in the hardware model, most AIGs produced would take time exponential in the logic depth if traversed as a tree. The second important improvement is to attempt to avoid computing the full BDD translation of nodes that are not relevant to the primary outputs. For example, if there is a multiplexer present in the circuit and its selector is set to 1 for all settings of the inputs possible under the current parametrization, then the value of the unselected input is irrelevant unless it has another fanout that is relevant. In AIGs, such irrelevant branches appear as fanins to ANDs in which the other fanin is unconditionally false. More generally, an AND of two child AIGs a and b can be reduced to a if it can be shown that $a \Rightarrow b$ (though the most common occurrence of this is when a is unconditionally false.) The AIG2BDD algorithm applies in iterative stages of two methods that can each detect certain of these situations without fully translating b to a BDD. In both methods, we calculate exact BDD translations for nodes, beginning at the leaves and moving toward the root, until some node's translation exceeds a BDD size limit. We replace the over-sized BDD with a new representation that loses some information but allows the computation to continue while avoiding blowup. When the primary outputs are computed, we check to see whether or not they are exact BDD translations. If so, we are done; if not, we increase the size limit and try again. During each iteration of the translation, we check each AND node for an irrelevant branch; if a branch is irrelevant it is removed from the AIG so that it will be ignored in subsequent iterations. We use the weaker of the two methods first with small size limits, then switch to the stronger method at a larger size limit.

In the weaker method, the translated value of each AIG node is two BDDs that are upper and lower bounds for its exact BDD translation, in the sense that the lower-bound BDD implies the exact BDD and the exact BDD implies the upper-bound BDD. If the upper and lower bound BDDs for a node are equal, then they both represent the exact BDD translation for the node. When a BDD larger than the size limit is produced, it is thrown away and the constant-*true* and constant-*false* BDDs are instead used for its upper and lower bounds. If an AND node $a \wedge b$ is encountered for which the upper bound for a implies the lower bound for b, then we have $a \Rightarrow b$; therefore we may replace the AND node with a. Thus using the weak method we can, for example, replace an AIG representing $a \wedge (a \vee b)$ with a whenever the BDD translation of a is known exactly, without computing the exact translation for b.

In the stronger method, instead of approximating BDDs by an upper and lower bound, fresh BDD variables are introduced to replace over-sized BDDs. (We necessarily take care that these variables are not reused.) The BDD associated with a node is its exact translation if it references only the variables used in the primary input assignments. This catches certain additional pruning opportunities that the weaker method might miss, such as $b \neq (a \neq b) \rightarrow a$.

These two AIG-to-BDD translation methods, as well as the combined method `AIG2BDD` that uses both in stages, have been proven in ACL2 to be equivalent, when they produce an exact result, to the naive AIG-to-BDD translation algorithm described above.

When symbolically simulating the **fadd** unit, using input parametrization in conjunction with the `AIG2BDD` procedure works around the problem that BDD variable orderings that are efficient for one execution path are inefficient for another. Input parametrization allows cases where one execution path is selected to be analyzed separately from cases where others are used. However, a naive method of building BDDs from the hardware model might still construct the BDDs of the intermediate signals produced by multiple paths, leading to blowups. The `AIG2BDD` procedure ensures that unused paths do not cause a blowup.

4.5 GL Symbolic Execution Framework

The GL framework is designed to allow proof by symbolic execution within ACL2, particularly targeted at hardware verification. A typical theorem to be proven by GL consists of some hypotheses, which restrict our consideration to a finite (but often large) set of input vectors, and a conclusion, which states some desired property that should hold on every input vector within this set. For example, to prove the correctness of a 16-bit adder circuit, we could hypothesize that inputs x and y are both 16-bit natural numbers, and we conclude that when x and y are given as inputs to the circuit, the result produced is $x + y$. To prove this, the user specifies what shape of input objects should be used for symbolic execution (in this case, 16-bit natural numbers). This shape specification also gives the BDD ordering for the bits of x and y. From the shape specification, GL constructs symbolic objects representing x and y. It then symbolically executes the conclusion. Ideally, the result of this symbolic execution will be a symbolic object that can syntactically be determined to always represent *true*. If not, GL will extract counterexamples from the resulting symbolic object, giving concrete values of x and y that falsify the conjecture. When the symbolic execution produces a true result, the final step in proving this theorem is to show that the symbolic objects used as inputs to the simulation cover the finite set of concrete inputs recognized by the hypothesis. In this example, 16-bit symbolic natural numbers suffice to cover the input space provided all the bits are free, independent variables; smaller symbolic naturals would not be adequate.

Symbolic objects are structures that describe functions over Booleans. Depending on the shape of such objects, they may take as their values any object in the ACL2 universe. For example, we represent symbolic integers as the pairing of a tag, which distinguishes such an object from other symbolic types such as Booleans and ordered pairs, and a list of BDDs, which represents the two's-complement digits of the integer. We define an evaluator function for symbolic objects, which gives the concrete value represented by an object under an assignment of Booleans to each variable. For the integer example, the evaluator recognizes the tag and evaluates each BDD in the representation under the given assignment. Then it produces the integer whose two's-complement representation matches the resulting list of bits.

To perform a symbolic execution, we employ two methods. We may create a *symbolic counterpart* f_{sym} for a user-specified function f. f_{sym} is an executable ACL2 function that operates on symbolic objects in the same way as f operates on concrete objects. It is defined by examining the definition of f, creating symbolic counterparts recursively for all its subfunctions and nesting them in the same manner as in the definition. Alternatively, we may symbolically interpret an ACL2 term under an assignment of symbolic objects to that term's free variables. In this case, we walk over the given term. At each function call, we either call that function's symbolic counterpart if it exists or else look up the function's definition and recursively symbolically interpret it under an assignment that pairs its formals with the corresponding symbolic values produced by the given actual parameters.

In both methods of symbolic execution, it is necessary for any ACL2 primitives to have predefined symbolic counterparts, since they do not have definitions. We have defined many of these functions manually and proven the correctness of their symbolic counterparts. For example, the symbolic counterpart of $+$ is defined such that on symbolic integers, it performs a BDD-level ripple-carry algorithm, producing a new symbolic integer that provably always evaluates to the sum of the evaluations of the inputs. We have also manually defined symbolic counterparts for certain functions for which symbolic interpretation of the ACL2 definitions would be inefficient. For example, the bit-wise negation function *lognot* is defined as $lognot(x) = (-x) - 1$, but for symbolic execution it is more efficient to perform the bit-wise negation directly by negating the BDDs in the symbolic integer representation; in fact, we define the negation operator in terms of *lognot*, rather than the reverse.

The correctness condition for a symbolic counterpart f_{sym} states a correspondence between the operation of f_{sym} on symbolic objects and the operation of f on concrete objects. Namely, the evaluation of the (symbolic) result of f_{sym} on some symbolic inputs is the same as the (concrete) result of running f on the evaluations of those inputs:

$$\mathbf{ev}\left(f_{\text{sym}}\left(s\right),a\right) = f\left(\mathbf{ev}\left(s,a\right)\right).$$

The following diagram illustrates the correspondence:

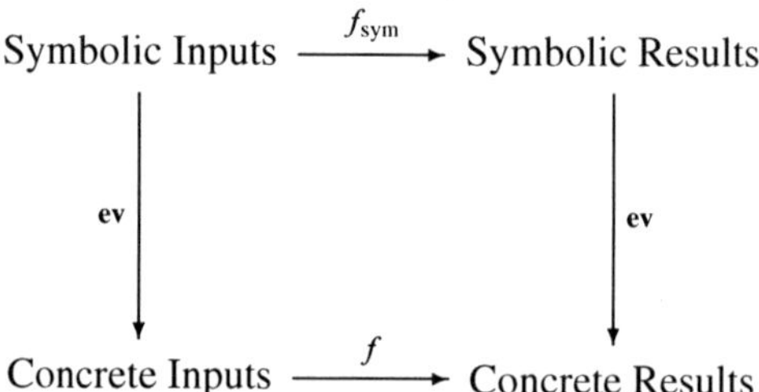

Each primitive symbolic counterpart we have defined is proven (using standard ACL2 proof methodology) to provide this correctness condition. The correctness of symbolic counterparts of functions defined in terms of these primitives follows from this; the correctness proofs are automated in the routine that creates symbolic counterparts. The symbolic interpreter is also itself verified; its correctness condition is similar. Suppose we symbolically interpret a term x with a binding of its variables v_i to symbolic objects s_i, yielding a symbolic result. We have proven that the evaluation of this result under assignment a equals the result of running the term x with its variables v_i each bound to $\mathbf{ev}(s_i, a)$.

These correctness conditions allow theorems to be proven using symbolic execution. Consider our previous example of the 16-bit adder. Suppose we symbolically execute the conclusion of our theorem on symbolic inputs s_x, s_y and the result is an object that evaluates to *true* under every variable assignment:

$$\forall a . \mathbf{ev}\left(conc_{\text{sym}}\left(s_x, s_y\right), a\right).$$

By the symbolic execution correctness condition, this commutes to:

$$\forall a . conc\left(\mathbf{ev}\left(s_x, a\right), \mathbf{ev}\left(s_y, a\right)\right).$$

That is, the conclusion holds of any pair of values x and y such that s_x and s_y evaluate to that pair under some assignment. The coverage side condition then requires us to show that s_x and s_y are general enough to cover any pair that satisfies the theorem's hypothesis. Once this is proven, the proof of the theorem (hypotheses imply conclusion) is complete.

5   Verification Results and Observations

We have used our ACL2-based verification methodology to prove the correctness
of several instructions in Centaur's execution cluster, including packed floating-
point addition/subtractions and comparisons, format conversions, logical operations, shuffles, and integer and packed-integer multiplication. We have also verified
the one-cycle invariant for the divider and proven the correctness of several microcode routines.

Our floating-point addition verification was performed using a machine with four
Intel Xeon X7350 processors running at 2.93 GHz with 128 GB of system memory.
However, each of our verification runs is a single-threaded procedure, and we limit
our memory usage to 35 GB for each process so that we can run single, double,
and extended-precision verifications concurrently on this machine without swapping. The symbolic simulations run in 8 min 40 s for single precision, 36 min for
double precision, and 48 min for extended precision. Proof scripts required to complete the three theorems take an additional 10 min of real time when using multiple
processors, totaling 25 min of CPU time. The process of reading the Verilog design
into ACL2, which is done as part of a process that additionally reads in a number of
other units, takes about 17 min. In total it takes about 75 min of real time (125 min
of CPU time) to reread the design from Verilog sources and complete verifications
of all the three instructions.

We found two bugs with our verification process, which began after the floating-
point addition instructions had been thoroughly checked using a testing-based
methodology. The first bug was a timing anomaly affecting SSE addition instructions, which we found during our initial investigation of the media unit. Later, a bug
in the extended precision instruction was detected by symbolic simulation. This bug
affected a total of four pairs of input operands out of the 2^{160} possible, producing
a denormal result of twice the correct magnitude. Because of the small number of
inputs affected, it is unlikely that random testing would have encountered the bug;
directed testing had also not detected it. Both bugs have been fixed in the current
design.

Working in an industrial environment forced us to be able to respond to design
changes quickly. Every night, we run our Verilog translator on the entire 570,000
lines of Verilog that comprise the Centaur CN and produce output for all of the
Verilog that we can translate. We build a new copy of ACL2 with our EMOD representation of the design already included so when we sit down in the morning,
we are ready to work with the current version of the design. Also, each night, we
rerun many of the verifications that have been done previously to make sure that
recent changes are safe. Each week, we attempt to rerun our entire regression suite
of previously proven results.

Our major challenges involved getting our toolsuite to be sufficiently robust,
getting the specification correct, dealing with the complicated clocking and power-
saving schemes employed, and creating a suitable circuit input environment. It
is difficult for us to provide a meaningful labor estimate for this verification because we were developing the translator, flow, our tools, our understanding of

floating-point arithmetic, and our specification style simultaneously. Now, we could likely check another IEEE-compatible floating-point design in the time it would take us to understand the clocking and input requirements. Centaur will certainly be using this methodology in the future; it is much faster, cheaper, and more thorough than nonexhaustive simulation. Although our verification approach is currently dependent on BDDs, we have considered what would be required to have an AIG-and-SAT flow.

The improvements in ACL2 that permitted this verification will be included in future ACL2 releases. The specifics of Centaur's two-cycle, floating-point design are considered proprietary. We plan to publish our ACL2-checked proof that our integer-level specification is equal to our IEEE floating-point specification; this level of proof is similar to work by Harrison[8].

6 Related Work

Several groups have completed floating-point addition and other related verifications. Notably, Intel has largely supplanted testing-based validation of the execution unit, instead using full formal verification based on symbolic trajectory evaluation in the Forte/reFLect system [17]. Also, IBM has integrated formal verification based on model checking and equivalence checking into their mainstream verification flow [2]. Our formal verification has only covered a small fraction of the instructions that run on Centaur's execution unit; we hope to expand this coverage in the future. However, we differ from previous verifications in that we obtained our result using verified automated methods within a general-purpose theorem prover and in that we base our verification on a formally defined HDL operating on a data representation mechanically translated from the RTL design.

An AMD floating-point addition design was verified using ACL2. It was proven to comply with the primary requirement of the IEEE 754 floating-point addition specification, namely that the result of the addition operation must equal the result obtained by performing the addition at infinite precision and subsequently rounding to the required precision [19]. The design was represented in ACL2 by mechanically translating the RTL design into ACL2 functions. A top-level function representing the full addition unit was proven to always compute a result satisfying the specification. This theorem was proved in ACL2 by the usual method of mechanical theorem proving, wherein numerous human-crafted lemmas are proven until they suffice to prove the final theorem. A drawback to this method is that even small changes to the RTL design may require the proof script to be updated. We avoid this pitfall by using a symbolic simulation-based methodology. Our method also differs in that we use a deep-embedding scheme, translating the RTL design to be verified into a data object in an HDL rather than a set of special-purpose functions.

We described our floating-point addition verification previously [11]. Among bit-level symbolic simulation-based floating-point addition verifications, many have used a similar case splitting and BDD parametrization scheme as ours [1, 15,

20, 21]. The symbolic simulation frameworks used in all of these verifications, including the symbolic trajectory evaluation implementation in Intel's Forte prover, are themselves unverified programs. Similarly, the floating-point verification described in [7] uses the SMV model checker and a separate argument that its case split provides full coverage. To obtain more confidence in our results, we construct our symbolic simulation mechanisms within the theorem prover and prove that they yield sound results. Combining tool verifications with the results of our symbolic simulations yields a theorem showing that the instruction implementation equals its specification.

7 Conclusion

In the verification methodology used at Centaur, we use a combination of symbolic simulation and conventional theorem proving to verify equivalences between hardware models and specifications written as ACL2 functions. Because our toolflow consists, to a large extent, of programs that have been verified by the ACL2 theorem prover, we are able to obtain ACL2 theorems reflecting our verification results even though the proofs are done in large part through symbolic simulation.

We model the design using a deep embedding in the EMOD formal HDL, which we obtain by automatic translation of the Verilog RTL design. We run a new translation nightly so as to keep current on the design effort, because we primarily use "black box" verification methods on the design. We rarely need to update proof scripts in response to design changes.

Our verification efforts have yielded correctness proofs for several instructions including floating-point addition, subtraction, and integer multiplication, conversions between integer and float formats, and comparisons. These proof efforts resulted in the discovery of two flaws in floating-point addition instructions that had escaped extensive simulation; these flaws have been corrected in Centaur's current design.

Acknowledgments We would like to acknowledge the support of Centaur Technology, Inc. and ForrestHunt, Inc. We would also like to thank Bob Boyer for development of much of the technology behind EMOD and the ACL2 BDD package, Terry Parks for developing a very detailed floating-point addition specification, and Robert Krug for his proof that our integer-level, floating-point addition specification performs the rational arithmetic and rounding specified by the IEEE floating-point standard. Portions of this chapter originally appeared in [11].

References

1. Aagaard MD, Jones RB, Seger CJH (1999) Formal verification using parametric representations of boolean constraints. In: Proceedings of the 36th design automation conference, pp 402–407
2. Baumgartner J (2006) Integrating FV into main-stream verification: the IBM experience. Tutorial given at FMCAD. Available at http://domino.research.ibm.com/comm/research_projects.nsf/pages/sixthsense.presentations.html

3. Boyer RS, Hunt WA Jr (2006) Function memoization and unique object representation for ACL2 functions. In: ACL2 '06: Proceedings of the sixth international workshop on the ACL2 theorem prover and its applications. ACM, New York, NY, pp 81–89

4. Boyer RS, Hunt WA Jr (2009) Symbolic simulation in ACL2. In: Proceedings of the eighth international workshop on the ACL2 theorem prover and its applications

5. Brock B, Hunt WA Jr (1991) Report on the formal specification and partial verification of the VIPER microprocessor. In: NASA contractor report, 187540

6. Bryant RE (1986) Graph-based algorithms for boolean function manipulation. IEEE Trans Comput C-35(8):677–691

7. Chen Y-A, Bryant RE (1998) Verification of floating-point adders. In: Hu AJ, Vardi MY (eds) Computer aided verification, Lecture notes in computer science, vol 1427. Springer, Heidelberg

8. Harrison J (2006) Floating-point verification using theorem proving. In: Bernardo M, Cimatti A (eds) Formal methods for hardware verification, Sixth international school on formal methods for the design of computer, communication, and software systems, SFM 2006, Lecture notes in computer science, Bertinoro, Italy, vol 3965. Springer, New York, pp 211–242

9. Hunt WA Jr (1994) FM8501: a verified microprocessor. Springer, London

10. Hunt WA Jr, Reeber E (2005) Formalization of the DE2 language. In: Proceedings of the 13th conference on correct hardware design and verification methods (CHARME 2005), pp 20–34

11. Hunt WA Jr, Swords S (2009a) Centaur technology media unit verification: case study: floating-point addition. In: Computer aided verification, Lecture notes in computer science 5643. Springer, Berlin, pp 353–367

12. Hunt WA Jr, Swords S (2009b) Use of the E language. In: Martin A, O'Leary J (eds) Hardware design and functional languages ETAPS 2009 Workshop, York, UK

13. IEEE (2005) IEEE standard (1364-2005) for verilog hardware description language

14. IEEE Computer Society (2008) IEEE standard for floating-point arithmetic, IEEE std 754^{TM}-2008 edn

15. Jacobi C, Weber K, Paruthi V, Baumgartner J (2005) Automatic formal verification of fused-multiply-add FPUs. In: Proceedings of design, automation and test in Europe, vol 2, pp 1298–1303

16. Jones RB (2002) Symbolic simulation methods for industrial formal verification. Kluwer, Dordrecht

17. Kaivola R, Ghughal R, Narasimhan N, Telfer A, Whittemore J, Pandav S, Slobodová A, Taylor C, Frolov V, Reeber E, Naik A (2009) Replacing testing with formal verification in Intel$^®$ Core^{TM}i7 processor execution engine validation. In: Computer aided verification, Lecture notes in computer science. Springer, Berlin, pp 414–429

18. Price D (1995) Pentium FDIV flaw – lessons learned. IEEE Micro 15(2):88–87

19. Russinoff D (2000) A case study in formal verification of register-transfer logic with ACL2: the floating point adder of the AMD Athlon (TM) processor. In: Hunt WA Jr, Johnson SD (eds) Formal methods in computer-aided design, LNCS 1954. Springer, Berlin, pp 22–55

20. Seger CJH, Jones RB, O'Leary JW, Melham T, Aagaard MD, Barrett C, Syme D (2005) An industrially effective environment for formal hardware verification. IEEE Trans Comput Aided Des Integr Circuits Syst 24(9):1381

21. Slobodová A (2006) Challenges for formal verification in industrial setting. In: Brim L, Haverkort B, Leucker M, van de Pol J (eds) Formal methods: applications and technology, LNCS 4346. Springer, Berlin, pp 1–22

22. University of California at Berkeley, Department of Electrical Engineering and Computer Science, Industrial Liaison Program. A compact test suite for P754 arithmetic – version 2.0.

23. Visser E (2005) A survey of strategies in rule-based program transformation systems. J Symbolic Comput 40:831–873

Designing Tunable, Verifiable Cryptographic Hardware Using Cryptol

Sally Browning and Philip Weaver

1 Introduction

Cryptographic functional units are increasingly being incorporated into microprocessor designs. Since producing a flawed implementation can have grave financial, as well as security, implications, it is highly desirable to be able to design efficient, high-performance hardware implementations of cryptographic algorithms that are provably equivalent to their high-level specifications.

Cryptol®[1] is a domain-specific language that is well suited for expressing cryptographic algorithms, as well as other data-flow algorithms. Cryptol's types and language constructs also provide an appropriate abstraction of hardware. As we shall see in the course of this chapter, the Cryptol toolchain provides a framework for producing verified hardware implementations that can be "tuned" for performance or for area.

Cryptol is a pure, declarative, functional language. A pure function depends only on its inputs, not I/O or any internal state, and does not produce any side effects. Cryptol functions naturally model combinatorial circuits in hardware, because any combinatorial circuit is just a pure mapping from inputs to outputs. Cryptol supports homogeneous fixed-length sequences, which naturally model bit-vectors in hardware description languages, and supports mapping sequences over time in order to model sequential circuits.

This chapter is intended to teach the reader how to design hardware circuits in Cryptol and uses the Cryptol toolchain to produce efficient, pipelined, high-assurance circuits (for more detailed documentation on this topic, see [3]). This document is not an appropriate resource for learning the Cryptol language itself; such a resource is available in [2]. It is intended for Computer Scientists, Engineers,

---

[1] Cryptol is a registered trademark of Galois, Inc. in USA and other countries. This chapter is derived from materials copyrighted by Galois, Inc., and used by permission.

S. Browning (✉)
Galois, Inc., Portland, OR, USA
e-mail: sally@galois.com

D.S. Hardin (ed.), *Design and Verification of Microprocessor Systems
for High-Assurance Applications*, DOI 10.1007/978-1-4419-1539-9_4,
© Springer Science+Business Media, LLC 2010

and Mathematicians with some knowledge of functional programming. A basic understanding of hardware is recommended. Some experience with hardware design is recommended, but not required.

1.1 Outline

The remainder of this document is organized into the following sections:

- *Cryptol overview*. This provides an introduction to the Cryptol language and interpreter, describes the modes relevant to the design and verification of hardware circuits, and explains how to use some of the basic features of these modes.
- *Cryptol for hardware design*. This describes how to use the Cryptol language for hardware design; discusses limitations of the language, including features that are not supported by the hardware compiler; and provides examples of using features in the language to make space–time tradeoffs, including how to use Block RAMs, how to reuse circuits over time, and how to pipeline a circuit.
- *AES specification*. This describes the AES algorithm and provides a high-level specification written in Cryptol.
- *AES implementations*. This specializes the reference specification to two different and efficient implementations of 128- and 256-bit AES encryption; rewrites the second implementation into two different implementations, one that is optimized for area and one that is pipelined for high throughput; and refines the second implementation to use a T-Box and pipelines it to obtain a very high throughput.
- *Conclusion*. This concludes with a discussion of summary of the tutorial and future work on Cryptol.

2 Cryptol Overview

2.1 Language Features

This section briefly mentions some of Cryptol's useful features, including available primitives, features of the type system, and other constructs.

2.1.1 Function Values and Anonymous Functions

In a functional language, functions have values just like any other expressions. For example, `f where f x = x + 1` is a function that increments its argument. It can be bound to variable, which can be applied:

```
g = f where f x = x + 1;
y = g 10;
```

Or, it can be applied in place:

```
y = (f where f x = x + 1) 10;
```

Cryptol supports *anonymous functions*, also known as *lambda abstractions*, functions that are defined without any name. The f function above can be defined as a lambda abstraction: \x -> x + 1. We can bind g to this function just as we did above when it was called f:

```
g = \x -> x + 1;
y = g 10;
```

Or, we can apply the anonymous function in place:

```
y = (\x -> x + 1) 10;
```

In Cryptol, \arg -> body is simply syntactic sugar for f where f arg = body.

2.1.2 Types and Polymorphism

Cryptol supports two interesting polymorphic terms: zero and undefined. Both of these have the following type:

```
{a} a
```

This means they can be of *any* type. When zero is a sequence of bits, then each element in the sequence is False. Otherwise, each element in zero is itself zero. For example, zero : [4] [3] is the same as:

```
[ [False False False]
  [False False False]
  [False False False]
  [False False False]
]
```

Or simply this:

```
[0b000 0b000 0b000 0b000]
```

Cryptol supports polymorphism and type-level constraints. The type of a function can be parameterized over other types, and these types can be constrained. For example, consider the following function that gets the fourth element of a sequence:

```
f x = x @ 3;
```

Note that Cryptol utilizes 0-based array indices. The type of this function can be as generic as:

```
f : {a b} [a]b -> b;
```

This can be read as "for any types a and b, the function f takes a sequence of width a, where each element of the sequence is of type b, and returns a single element of type b." However, to ensure that we do not try to index outside the sequence, we should constrain a to be at least 4:

```
f : {a b} (a >= 4) => [a]b -> b;
```

This says that, for any types a and b where a is at least 4, the function takes in a sequence of width a, where each element of the sequence is of type b, and returns a single element of type b. The type variable b could itself be Bit or some sequence of arbitrary length.

Type ascriptions can appear almost anywhere within Cryptol code, not just on a line of their own. This is especially useful to prevent Cryptol from inferring a width that is too small. For example, at the Cryptol interpreter prompt, we can observe the following behavior:

```
Cryptol> 1+1
0x0
```
———————————— *interpreter* ————————————

This is because 1 defaults to a width of 1, the smallest width necessary to represent the value. The type of + is:

```
+ : {a b} ([a]b, [a]b) -> [a]b
```

Therefore, the result is the same as the widths of the inputs, so 0x2 overflows to 0x0. To prevent this, we can ascribe a type to either argument to +, which causes Cryptol to infer the type of the other argument and the result:

```
Cryptol> (1:[4]) + 1
0x2
```
———————————— *interpreter* ————————————

All polymorphic types must be specialized to monomorphic (i.e., specific) types at compile time. Cryptol will infer types as much as possible, including defaulting to the minimum width possibly needed to represent a sequence. If it cannot reduce a function to a monomorphic type, it will refuse to apply or compile it.

The way to force an expression to a monomorphic type is to ascribe a type, either directly to the expression or somewhere else that causes the compiler to infer the type of that expression. For example, consider a polymorphic function that increments its argument:

```
inc : {a} (a >= 1) => [a] -> [a];
inc x = x + 1;
```

In the expression `inc x`, the compiler must be able to infer either the type of
x or the type of `inc x` in order to know the type of this particular instantiation
of `inc`. We can explicitly ascribe either of these types using `inc x :: [10]` or
`inc (x :: [10])`, or we can place an ascription elsewhere in our code that will
cause the compiler to infer one of these types.

Cryptol allows constant terms to be used as types. So, we could define a and the
type of f as follows:

```
a = 16;
f : {b} [a]b -> b;
```

Note that, in this case, the type of f is no longer parameterized over the type
variable a.

Cryptol supports a primitive called `width` that can be used at both the expression
level and the type level. At the expression level, `width x` is the number of elements
in x, and it must be known at compile time; if x is a sequence of bits, then `width x`
is the number of bits needed to represent x. At the type level, `width` can be used
in a constraint.

So, we can write polymorphic functions whose behavior depends on the width
of the inputs, as in the following example that outputs the least significant bit of xs
when xs has more than 10 bits and otherwise outputs the most significant bit. (See
Sect. 2.1.5 for more about the indexing operators @ and !.)

```
f : {a} [a] -> Bit;
f xs = (w > 10 & xs @ 0) | xs ! 0
    where w = width xs;
```

2.1.3 Type Aliases and Records

Cryptol supports type aliases and records. One can define a new type as an alias
to existing types, and it can be parameterized over other types. For example, the
following defines a type `Complex x` as an alias to `(x, x)`.

```
type Complex x = (x, x);
```

We can use the type alias to define a function that multiplies two complex
numbers.

```
multC : {n} (Complex [n], Complex [n]) -> Complex [n];
multC ((a, b), (x, y)) = (a*x - b*y, a*y + b*x);
```

A record contains named fields. We can define a record for complex numbers that
names the real and imaginary fields:

```
type Complex x = { real : x;
                   imag : x;
                 };
```

Then, the `multC` function can be defined as follows:

```
multC : {n} (Complex [n], Complex [n]) -> Complex [n];
multC (x, y) = { real = x.real * y.real - x.imag * y.imag;
                 imag = x.real * y.imag + x.imag * y.real;
               };
```

2.1.4 Enumerations

Cryptol supports finite and infinite enumerations. Following are some examples:

```
[1..10]     == [1 2 3 4 5 6 7 8 9 10]
[1 3..10]   == [1 3 5 7 9]
[1..]       == [1 2 3 4 5 6 7 ..]
[10--1]     == [10 9 8 7 6 5 4 3 2 1]
[10 6--0]   == [10 6 2]
[10--]      == [10 9 8 7 6 ...]
```

2.1.5 Index Operators

Cryptol supports the following index operators: @ @@ ! !!. The @ operator indexes from the least significant element and ! operator indexes from the most significant element. The least significant element is the rightmost for sequences of bits and the leftmost for other sequences.[2] The operators @@ and !! lookup a range of indices. So, the following equalities hold:

```
([0 1 2 3 4 5 6 7] : [8][3]) @@ [1..3] == [1 2 3]
([0 1 2 3 4 5 6 7] : [8][3]) @@ [3--1] == [3 2 1]
([0 1 2 3 4 5 6 7] : [8][3]) !! [1..3] == [6 5 4]
([0 1 2 3 4 5 6 7] : [8][3]) !! [3--1] == [4 5 6]

(0b00011100 : [8]) @@ [1..3] == 0b110
(0b00011100 : [8]) @@ [3--1] == 0b011
(0b00011100 : [8]) !! [1..3] == 0b100
(0b00011100 : [8]) !! [3--1] == 0b001
```

---

[2] This is different in Cryptol 2.0.

All sequence widths are fixed and therefore must be known at compile time. The width of the result of @@ and !! depends on the value of the second argument; therefore, the value of the second argument must be known at compile time.

2.1.6 Sequence Operations and Transformations

Cryptol provides the take and drop primitives:

```
take : {a b c} (fin a,b >= 0) => (a,[a+b]c) -> [a]c
drop : {a b c} (fin a,a >= 0) => (a,[a+b]c) -> [b]c
```

As their names imply, they take or drop a certain number of elements from the beginning of a sequence. Because the width of the result depends on the first argument (number of elements that are taken or dropped), this argument must be constant. For example, we cannot define the following function, because n is not known:

```
f n = take(n, [1 2 3 4 5 6 7 8 9 10]);
```

Cryptol supports several primitives for splitting and combining sequences:

```
split : {a b c} [a*b]c -> [a][b]c
splitBy : {a b c} (a,[a*b]c) -> [a][b]c
groupBy : {a b c} (b,[a*b]c) -> [a][b]c
join : {a b c} [a][b]c -> [a*b]c
transpose : {a b c} [a][b]c -> [b][a]c
```

The split, splitBy, and groupBy functions each convert a sequence into a two-dimensional sequence. Their inverse is the join function. split and splitBy behave the same; the splitBy function is provided as an alternative to split because it allows the user to explicitly choose the size of the first dimension, rather than forcing the compiler to infer it.

The following equalities show how splitBy, groupBy, join, and transpose behave.

```
splitBy (3, [1 2 3 4 5 6 7 8 9 10 11 12]) ==
[[1 2 3 4] [5 6 7 8] [9 10 11 12]]
groupBy (3, [1 2 3 4 5 6 7 8 9 10 11 12]) ==
[[1 2 3] [4 5 6] [7 8 9] [10 11 12]]
join [[1 2 3 4] [5 6 7 8] [9 10 11 12]] ==
[1 2 3 4 5 6 7 8 9 10 11 12]
join [[1 2 3] [4 5 6] [7 8 9] [10 11 12]] ==
[1 2 3 4 5 6 7 8 9 10 11 12]
transpose [[1 2 3 4] [5 6 7 8]] ==
[[1 5] [2 6] [3 7] [4 8]]
transpose [[1 2 3] [4 5 6] [7 8 9]] ==
[[1 4 7] [2 5 8] [3 6 9]]
```

Furthermore, for all n and x, the following equalities hold:

```
join (split x) == x
join (splitBy (n, x)) == x
join (groupBy (n, x)) == x
transpose (transpose x) == x
```

2.2 Cryptol Interpreter

One enters the Cryptol interpreter by typing "cryptol" at the shell command prompt. The interpreter provides a typical command-line interface (a read–eval–print loop).

One can execute a shell command from within the interpreter by placing a ! at the beginning of the command:

```
Cryptol> !ls
```
interpreter

The Cryptol interpreter supports a number of commands, each of which begins with a colon (:). Following are the most common commands used in the Cryptol interpreter. For a more detailed discussion of these and other commands, see the reference manual.

`:load <path>`

Load all definitions from a `.cry` file, bringing them into scope.

`:set <mode>`

Switch to a given mode. Each mode supports a different set of options and performs evaluation on a different intermediate form that is translated from the Abstract Syntax Tree (AST). See Sect. 2.3 for a discussion of the modes that are useful for hardware design.

`:translate <function> [<path>]`

Compile a function to the intermediate form associated with the current mode (see Sect. 2.3). This function is known as the *top-level function*. The optional `<path>` argument is relative to the current working directory, or can be written as an absolute path, and should include the desired extension. If `<path>` is not provided, then Cryptol uses the `outfile` setting, saves the file in `outdir`, and adds the extension automatically.

`:fm <function> [<path>]`

Generate a formal model from a function. The path is optional and will be chosen automatically unless provided by the user. All modes that support the generation of formal models produce the same format, so functions can be checked for equivalence across modes.

`:eq <function or path> <function or path>`

Determines whether two functions are equivalent. Requires two arguments, each of which is either a Cryptol expression in parentheses or the path to a formal model in quotes.

The `<path>` argument to any command above must be surrounded in quotes and may include variables from the interpreter's environment, each proceeded by a dollar sign. For example, to translate a function `f` to a file `foo.vhdl` in the current output directory, issue the following command: `:translate f "$outdir/foo.vhdl"`.

2.3 Cryptol Interpreter Modes for Hardware Design

This section discusses the Cryptol modes that are relevant to hardware design and verification. Enter a mode by typing `:set <mode>`, where `<mode>` is one of the following modes, at the Cryptol interpreter prompt. When an expression is entered at the interpreter prompt, it is translated to the intermediate form associated with the current mode and then evaluated. The `translate` command produces the intermediate form for an expression in the current mode, but does not evaluate it. Regardless of the mode, all concrete syntax is first translated into an abstract syntax tree called the IR (intermediate representation). For some modes, the IR is their intermediate form, while other modes translate from the IR to their own intermediate form.

The relevant Cryptol modes are as follows:

symbolic: This performs symbolic interpretation on the IR. This is useful for prototyping circuits and supports equivalence checking.

LLSPIR: This compiles the IR to Low Level Signal Processing Intermediate Representation (LLSPIR), inlining all function calls into the top-level function and performing timing transformations that optimize the circuit. This provides rough profiling information of the final circuit, including longest path, estimated clockrate, output rate, latency, and size of circuit. This supports equivalence checking. Rather than output LLSPIR, the `translate` command produces a `.dot` file, a graph of the LLSPIR circuit that can be viewed graphically.

VHDL: This compiles to LLSPIR and then translates to VHDL. Evaluation is performed by using external tools to compile the VHDL to a simulation executable and then running the executable. This is useful for generating VHDL that is manually integrated into another design, rather than directly synthesizing the result.

FPGA: This compiles to LLSPIR, translates to VHDL, and uses external tools to synthesize the VHDL to an architecture-dependent netlist. There are several options in this mode that control what the external tools should do next, and they are most easily accessed via the following aliases:

- **FSIM:** This compiles the netlist to a low-level structural VHDL netlist suitable for simulation only. Evaluation is performed by compiling the VHDL to a simulation executable and running the executable. This produces profiling information that does not take into account routing delays. This reports the maximum theoretical clockrate.

- **TSIM:** This is similar to FSIM, but performs *map* and *place-and-route* when generating the VHDL netlist. This process can increase compilation time significantly, but produces very accurate profiling, including a true obtainable clockrate.
- **FPGA_Board:** This compiles the architecture-dependent netlist to a bitstream suitable for loading onto a particular FPGA board.

The :eq command is supported in FSIM and TSIM modes. It changes the synthesis target to a Verilog netlist suitable for equivalence checking and compiles this netlist to a formal model.

The top-level function is compiled to a single VHDL entity whose interface is determined by the type of the function. The top-level function always has some variation of the following type:

```
[inf]a -> [inf]b;
```

For each of a and b, if the type is a tuple, then each element of that tuple becomes a port in VHDL; otherwise the type becomes a bit or bit-vector in VHDL. Tuples nested inside a top-level tuple are appended into single bit-vectors. For example, if the type of the top-level function is

```
[inf](Bit, [4],([8],[12])) -> [inf](Bit, [10]);
```

then the VHDL entity will have the following interface:

```
port (input1 : in std_logic;
      input2 : in std_logic_vector(3 downto 0);
      input3 : in std_logic_vector(19 downto 0);
      output1 : out std_logic;
      output2 : out std_logic_vector(9 downto 0);
      restart : in std_logic;
      clk : in std_logic);
```

2.4 Equivalence Checking

The following modes support equivalence checking: symbolic, LLSPIR, and FPGA (FSIM and TSIM). The user can generate a formal model of a function in any of these modes using the :fm command and check a function for equivalence to a formal model using :eq. Two formal models generated in two different modes may also be compared using :eq.

There are two main uses of equivalence checking: (1) to verify that an implementation is correct with respect to a specification and (2) to verify that Cryptol compiles a particular function correctly, by comparing a function in LLSPIR or FPGA mode to the same function symbolic mode.

Cryptol supports three specific equivalence checkers: jaig, eaig, and abc. The equivalence checker outputs `True` if two functions are equivalent and otherwise outputs `False` along with a counterexample.

3 Cryptol for Hardware Design

This section discusses the Cryptol language with respect to hardware design. First, it discusses features in the language that are not supported by the compiler and other issues that may make it difficult to generate efficient circuits. It also describes techniques for making space–time tradeoffs and applies these techniques to several concrete and simple examples. These techniques are the same that are used to manipulate the AES implementations in Sect. 5.

3.1 *Issues and Limitations*

This section discusses some of the limitations of the Cryptol hardware compiler, including some techniques for avoiding them.

3.1.1 Supported Subset

The hardware compiler only supports division by powers of 2. This is a limitation of the backend tools.

The hardware compiler does not support primitive recursion. Most recursive functions can be easily rewritten using value recursion (i.e., stream recursion), which the compiler supports. For example, consider a recursive implementation of factorial:

```
fact n = if n == 0 then 1 else n * fact(n-1);
```

We can reimplement this using a value that is defined recursively:

```
fact2 n = facts @ n
  where facts = [1] # [| i * prev
                      || i <- [1..]
                      || prev <- facts |];
```

The hardware compiler only partially supports higher-order functions (functions as first-class values). Specifically, it does not allow functions to return functions, but it does allow functions to take another function as input as long as that function can be inlined away at compile time. For example, consider the following function that takes in a function and applies it to each element of a tuple:

```
app_tup : {a b} (a -> b, (a, a)) -> (b, b);
app_tup (f, (x, y)) = (f x, f y);
```

This function itself cannot be compiled to hardware. However, we can apply app_tup to a known function as follows:

```
inc_tup : {a} (a >= 1) => ([a],[a]) -> ([a],[a])
inc_tup t = app_tup (inc, t) where inc x = x + 1;
```

Then, the compiler inlines inc into app_tup to obtain code that no longer contains higher-order functions:

```
inc_tup t = (inc x, inc y) where inc x = x + 1;
```

However, the hardware compiler does not support functions that return nonclosed functions, such as the following:

```
f : [8] -> ([8] -> [8]);
f x = g where g y = x + y;
```

In this definition, g is not closed because it relies on the variable x.

3.1.2 Inefficient Sequence Comprehensions

Hardware performance can vary drastically based on subtle changes in how sequence comprehensions are written. This section explains how to generate efficient circuits from sequence comprehensions.

Although the following two expressions are semantically equivalent, they compile to significantly different circuits:

```
take(N, [1..])
```

```
[1..N]
```

The first one generates code to calculate a sequence of numbers at *run time*. Furthermore, because the sequence is infinite, it is mapped across *time*, so each element is calculated in a different clock cycle. The second expression generates the enumeration at *compile time*. When used in larger circuits, this subtle difference can cause drastic changes in performance.

Consider a function that takes in some fixed number of bytes, N, and pair-wise multiplies each byte by 1 through N, respectively. A naive implementation might look like this:

```
prods : {N} [N][8] -> [N][8];
prods xs = [| x * i || i <- [1..]
                   || x <- xs |];
```

However, the infinite sequence [1..] will force the sequence comprehension to take N cycles to complete. It is possible that the user wants to reuse a single multiplier over N cycles, but that is not what happens. To correctly lay this sequence out over multiple clock cycles and reuse a single multiplier, the user should use the *seq* pragma (see Sect. 3.4).

The correct way of implementing this function is to force the sequence that i draws from to be finite:

```
prods' xs = [| x * i || i <- [1..(width xs)]
                    || x <- xs |];
```

This will instantiate `width xs` multipliers (actually one less, since the multiplication by 1 is optimized away) in parallel. A single multiplier can be reused over multiple clock cycles by using the `seq` pragma (see Sect. 3.4).

3.2 Combinatorial and Sequential Circuits in Cryptol

A *combinatorial* circuit is one whose output is a pure function of its input, so it is no surprise that Cryptol functions model combinatorial circuits very well. Associated with any combinatorial circuit is a propagation delay, the amount of time it takes for the circuit to complete its function and output a value. Cryptol can report the estimated propagation delay of any combinatorial circuit. Combinatorial circuits are basic building blocks for designing complex circuits.

Combinatorial circuits themselves are not clocked,[3] but they may be used as building blocks in a clocked circuit. The various combinatorial components of a circuit can be synchronized by placing clocked registers between them, so that each one operates as one stage in a pipeline (see Sect. 3.5).

A *sequential* circuit depends on past input and/or internal state. The state itself is a function of the initial state and past inputs; thus, the difference between a combinatorial circuit and a sequential circuit in Cryptol is that the output of a combinatorial circuit may only depend on the most recent input, whereas the output of a sequential circuit may depend on multiple inputs across time.

There are two fundamental ways to model a sequential circuit in Cryptol (1) the clocked *stream model* and (2) the unclocked *step model*. A circuit may be defined using either model, or some combination of them, though all circuits modeled in the step model must eventually be lifted to the stream model.

---

[3] Some pure Cryptol functions, which look like combinatorial circuits, may actually map to clocked, sequential FPGA primitives, such as a Block RAM. In this case, they may have a latency of several clock cycles.

In the stream model, sequential circuits are modeled using infinite sequences over *time*, so a function in the stream model has some variation of the following type:

```
[inf]input -> [inf]output;
```

Each element in the input or output corresponds to some number of clock cycles, which is the latency of the circuit. To manage state, the user may define a stream within the circuit that holds the state, as in the definition of `lift_step` below.

One can lift a combinational circuit into the stream model using the following function. It takes in the function for a combinational circuit and an infinite stream and maps the function across all elements in the stream.

```
lift : {a b} (a -> b, [inf]a) -> [inf]b;
lift (f, ins) = [| f x || x <- ins |];
```

Note that this may not cause the circuit to become clocked, especially if there is no stateful information passed from one cycle to the next. It is important that circuits generated by Cryptol always be clocked, otherwise the synthesis tools cannot make use of clock constraints to produce useful timing analyses. In general, it is good to latch onto the inputs and outputs of a circuit by inserting registers after the inputs, before the outputs, or both. The following function lifts a combinatorial function into the stream model *and* places those registers:

```
lift_and_latch : {a b} (a -> b, [inf]a) -> [inf]b;
lift_and_latch (f, ins) = [undefined] #
                          [| f x || x <- [undefined]
                             # ins |];
```

See Sects. 3.3 and 3.5 to learn how to use registers in Cryptol.

In the step model, sequential circuits are modeled as combinatorial circuits that are later lifted into the stream model. A function in the step model is defined as a pure mapping from input and current state to output and next state, so it has some variation of the following type:

```
(input, state) -> (output, state)
```

The top-level function (the argument to the `:translate` command) must be defined in the stream model. One can easily lift a function from the step model to the stream model. The following function lifts any step function `f_step` into the stream model:

```
lift_step : {in out state}
            ((in,state) -> (out,state), state,
             [inf]in) -> [inf]out;
lift_step (f_step, init_state, inputs) =
  [| out || (out, state) <- xs |]
    where xs = [(undefined, init_state)] #
```

```
[| f_step (in, state)
|| in <- inputs
|| (out, state) <- xs |];
```

Like the `lift` function, this function applies a combinational function `f_step` to each element in an infinite stream, `inputs`. However, it also starts with an initial state, `init_state`, and propagates the state as it folds `f_step` over the inputs.

In this example, a local stream has been defined that carries both the states and the outputs. The final output is simply the contents of the stream after discarding the states. `xs` must be prepended with an initial output-state pair because it is defined recursively; it needs an initial value on which to base all further computations. `init_state` should be chosen according to the step function being used.

3.3 Delays and Undelays

Consider a sequence, `s`, in the stream model, so `s` is infinite and mapped over time. Assuming an output rate of one element per cycle, we can delay the stream by n cycles by appending n elements to the beginning of the stream. For example, the following function outputs its inputs unmodified, but each output is delayed by 1 cycle:

```
delay : {a} [inf]a -> [inf]a;
delay ins = [undefined] # ins;
```

And this function delays the output by 2 cycles:

```
delay2 : {a} [inf]a -> [inf]a;
delay2 ins = [undefined undefined] # ins;
```

Alternatively, we could define `delay2` by applying `delay` to the input twice: `delay (delay ins)`.

Note that using `zero` instead of `undefined` adds latency to the circuit because it takes some time to initialize it.

While a "delay" causes its output to occur some number of cycles after the input, an "undelay" causes its output to occur *before* the input. One can cause an undelay to occur using the `drop` construct:

```
undelay : {a} [inf]a -> [inf]a;
undelay ins = drop(1, ins);
```

Undelays are not synthesizable. During an optimization pass, the compiler pushes delays and undelays around in the circuit, introducing new ones and canceling delays with adjacent undelays.

Delays and undelays can be used to synchronize data across time. For example, the following Fibonacci implementation uses `drop` to look back in time in the stream so that we can add the previous two values together:

```
fib : [inf][32];
fib = [ 1 1 ] # [| x + y || x <- fib
                 || y <- drop(1, fib) |];
```

One can also use delays to produce pipelines. A delay synthesizes to a register/latch. Section 3.5 shows how to use registers to pipeline circuits.

3.4 Space–Time Tradeoffs via par and seq Pragmas

Cryptol supports two simple but very powerful pragmas that control space-time tradeoffs in the compiler. The *par* pragma causes circuitry to be replicated, whereas the *seq* pragma causes circuitry to be reused over multiple clock cycles. By default the compiler replicates circuitry as much as possible in exchange for performance, and the user overrides this behavior using `seq`; the `par` pragma is only useful for switching back to the default behavior within an instance of `seq`.

Semantically, both `seq` and `par` are the identity, because the types and semantics of Cryptol have no notion of time:

```
seq : {n t} [n]t -> [n]t;
seq x = x;

par : {n t} [n]t -> [n]t;
par x = x;
```

To understand the basic behavior of the `seq` pragma, consider some combinatorial circuit, `f`, which is implemented as a sequence comprehension and has the following type:

```
f : [n]a -> [n]a;
```

By default, the compiler will unroll and parallelize the sequence comprehension as much as possible. However, `seq (f xs)` requests that the circuitry within `f` be reused over n cycles. This requires extra flip-flops to synchronize each reuse of the circuitry within the comprehension, but can reduce overall logic utilization, resulting in a smaller circuit.

For example, the `prods'` function from Sect. 3.1.2 can be defined to reuse one multiplier over multiple clock cycles:

```
prods' xs = seq [| x * i || i <- [1..(width xs)]
                         || x <- xs |];
```

Sequence comprehensions can be performed either in sequence or in parallel. If the sequence is infinite or if each element depends on the previous element, then it must be implemented sequentially. Otherwise, it may be implemented in parallel. This section provides two examples, one involving a sequence comprehension that can be performed in parallel and one involving a sequence comprehension that must be performed sequentially. In both examples, the `seq` pragma is used to map the sequence over multiple clock cycles, and the performance advantages and disadvantages of doing so are discussed.

For both examples, we assume that there is some function `f` of type `a->b` or `b->b`. The advantages of using `seq` are greater when `f` consumes more area, because reusing `f` will then have a greater impact on logic utilization.

3.4.1 Example 1: Parallel Sequence Comprehension

Consider the `map` function that applies some function to every element in a finite sequence:

```
map : {n a b} (fin n) => (a -> b, [n]a) -> [n]b;
map (f, xs) = [| f x || x <- xs |];
```

By default, `f` will be instantiated n times and applied in parallel to each of the n inputs. A definition and graph of this function are provided in Fig. 1, in which n is set to 4. Once lifted into the stream model, this circuit will accept n elements of type a every cycle and output n elements of type b every cycle. To translate this in LLSPIR mode and view profiling information, we lift the function into the stream model:

```
:translate (\x -> lift(map_f1, x))
                  interpreter
```

This is ideal if `f` is a relatively small circuit. However, if duplicating `f` violates a size constraint, then we should trade time for space by instantiating `f` only once and applying it in sequence using the `seq` pragma. In this case, the output rate is every 4 cycles. The definition and graph of this function are provided in Fig. 2. There is extra logic needed to facilitate the sequencing of `f` over multiple cycles, since the input comes in all at once and the output must be produced all at once. Clearly, defining the circuit this way can be beneficial only if `f` is large.

We can also define this function using a combination of parallel and sequential computations. For example, we may want to instantiate `f` exactly two times. Since

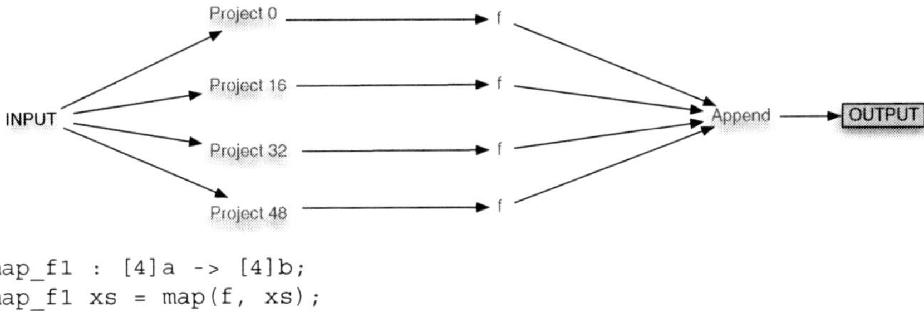


```
map_f1 : [4]a -> [4]b;
map_f1 xs = map(f, xs);
```

Fig. 1 Map f in parallel

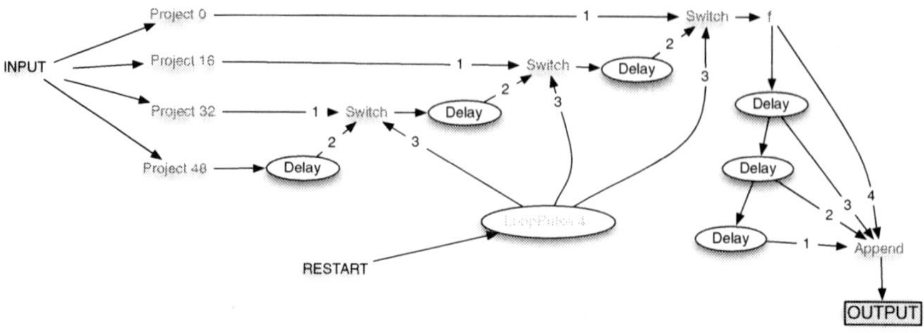

```
map_f2 : [4]a -> [4]b;
map_f2 xs = seq(map(f, xs));
```

Fig. 2 Map f in sequence

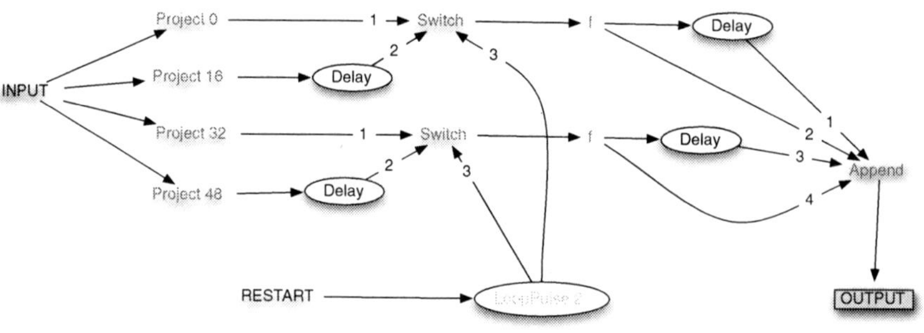

```
map_f3 : [4]a -> [4]b;
map_f3 xs = join [xs0' xs1']
            where {
                xs0, xs1 : [2]a;
                [xs0 xs1] = split xs;
                xs0' = seq(map(f, xs0));
                xs1' = seq(map(f, xs1));
            };
```

Fig. 3 Map f, two in sequence and two in parallel

n is 4, this would cause each instantiation of f to operate on two sequential inputs to obtain an output rate of every 2 cycles. Such a circuit would provide a balance between space utilization and speed. Figure 3 provides the definition and graph of a function that implements this functionality.

Note that the number of elements in each input (currently 4) can be increased without changing the algorithm for map_f3. This function will always generate two instantiations of f.

3.4.2 Example 2: Sequential Sequence Comprehension

Suppose we want to apply the function f : b -> b an arbitrary number of times
to an input, where this number of times is k. We can construct the following sequence, where the element at index k-1 is the one that results from k applications
of f:

```
[ f(x)  f(f(x))  f(f(f(x)))  ... ]
```

The following function constructs this sequence:

```
iterate : b -> [k]b;
iterate x = outs
  where outs = [(f x)] # [| f prev
                         || i <- [1..(k-1)]
                         || prev <- outs |];
```

However, f appears twice in this function. We want f to only appear once (inside
the comprehension), so that when we use the seq pragma f is only instantiated
once. So, we define iterate as follows, where the element at index k is the one
that results from k applications of f:

```
iterate : b -> [(k+1)]b;
iterate x = outs
  where outs = [x] # [| f prev
                     || i <- [1..k]
                     || prev <- outs |];
```

Note that each element depends on the previous, so the sequence will be evaluated
sequentially. Rather than instantiating f in parallel, as in the previous section, the
sequence comprehension will be unrolled.

We define two hardware implementations; in the graphs of these implementations
below, k is fixed at 4. The first implementation simply returns the last element in the
sequence produced by iterate. Its definition and graph are provided in Fig. 4. It
uses k copies of f and chains them together sequentially in one clock cycle.

iterate2 uses the seq pragma to request that f be reused each of k clock
cycles. Its definition and graph are provided in Fig. 5.

The result is that iterate1 will have a lower clockrate and will take up more
area, but will have an input/output rate of one element per clockcycle. iterate2


```
iterate1 : b -> b;
iterate1 x = iterate x ! 0;
```

Fig. 4 iterate (unrolled)

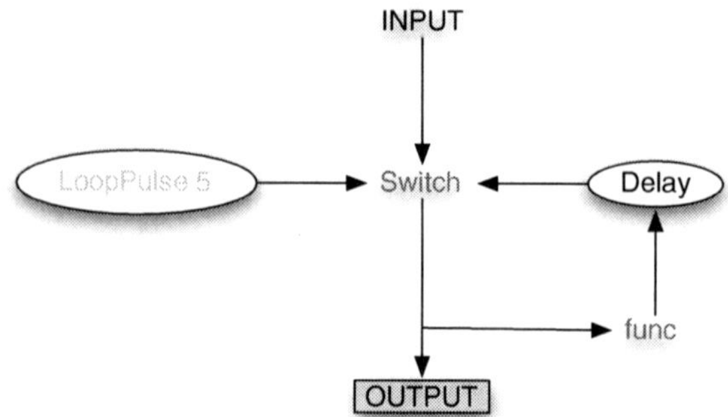


```
iterate2 : b -> b;
iterate2 x = seq (iterate x) ! 0;
```

Fig. 5 `iterate` (reused over time)

will have a higher clockrate and will take up less area, but it will have an input/output rate of one element every k cycles and will require extra flip-flops to latch onto the output of f each cycle.

We can verify that the two implementations are equivalent, using `iterate1` as the reference spec:

```
:set symbolic
:fm iterate1 "iterate1.fm"

:set LLSPIR
:eq iterate1 "iterate1.fm"
:eq iterate2 "iterate1.fm"
                         interpreter
```

3.5 Pipelining

Sequential circuits in the stream model can be pipelined to increase clockrate and throughput. One separates a function into several smaller computational units, each of which is a stage in the pipeline that consumes output from the previous stage and produces output for the next stage. The stages are synchronized by placing registers between them.

Pipelining an implementation typically increases the overall latency and area of a circuit, but can increase the clockrate and total throughput dramatically. Each stage is a relatively small circuit with some propagation delay. The clockrate is limited by the stage in the pipeline with the highest propagation delay, whereas the un-pipelined implementation would be limited by the sum of the propagation delays of all stages. So, rather than performing one large computation on one input during a very long clock cycle, an n-stage pipeline performs n parallel computations on n partial results, each corresponding to a different input to the pipeline. Sequences provide an appropriate abstraction for representing pipeline stages. Each stage can be represented as one or more parallel infinite streams across time.

Recall from Sect. 3.3 that one can insert *delays* and *undelays* by adding and removing elements from an infinite stream. We can also use delays to explicitly insert registers in a pipeline and undelays to drop the initially undefined output from a pipeline. A delay in Cryptol maps directly to a register (also known as a flip-flop or latch) in hardware.

This section introduces pipelining in Cryptol by defining and comparing many different pipelined implementations of two generic reference specifications. The first specification is a simple combinatorial circuit that takes two inputs, applies two arbitrary arithmetic functions to the respective inputs, and sums the result. The second specification uses stream comprehension to apply a function to the input an arbitrary number of times. Finally, a new `reg` pragma is introduced which can be used to produce pipelined designs more easily and that more closely resemble their high-level specifications. We show how each of the examples in the following sections can be rewritten much more simply by using the `reg` pragma.

3.5.1 Example 1: Combinatorial Circuit

Consider the following combinatorial circuit, where g and h are arbitrary combinatorial circuits of type a -> a. This will be our reference specification.

```
add_g_h_spec : ([c],[c]) -> [c];
add_g_h_spec (a, b) = g(a) + h(b);
```

The definitions of g and h are not revealed here, because they are irrelevant to implementing the pipeline unless we want to split g and h themselves into multiple stages (see `add_g_h_4` below). The functions g and h can be polymorphic, but for simplicity we fix the width:

```
c = 8;
```

Below are five separate implementations of `add_g_h` in the stream model. The first simply lifts the original spec into the stream model. The remaining are pipelined implementations. These functions all share the same type:

```
add_g_h_1, add_g_h_2, add_g_h_3, add_g_h_4 :
[inf]([c],[c]) -> [inf][c];
```

The definitions and graphs of each implementation are provided in the figures below.

First, we implement the spec in the stream model as a circuit that consumes input and produces output on every clock cycle. In this case, g(a) and h(b) are performed in parallel, but the addition operation cannot be performed until g(a) and h(b) both finish. Thus, the total propagation delay is the maximum delay of g and h plus the delay of the addition. See Fig. 6 for the definition and graph of this circuit.

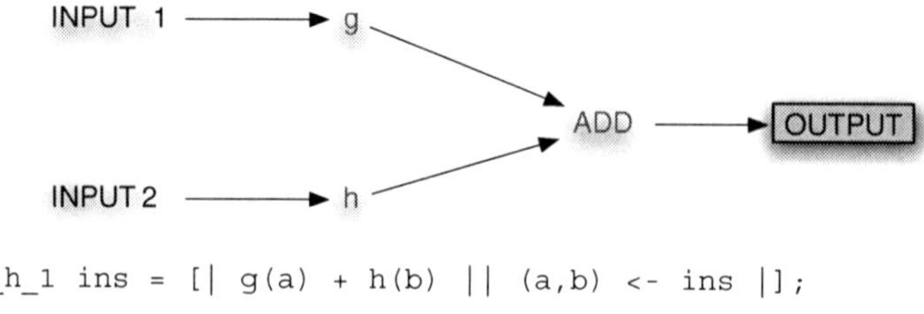

```
add_g_h_1 ins = [| g(a) + h(b) || (a,b) <- ins |];
```

Fig. 6 add_g_h_1 (unpipelined)

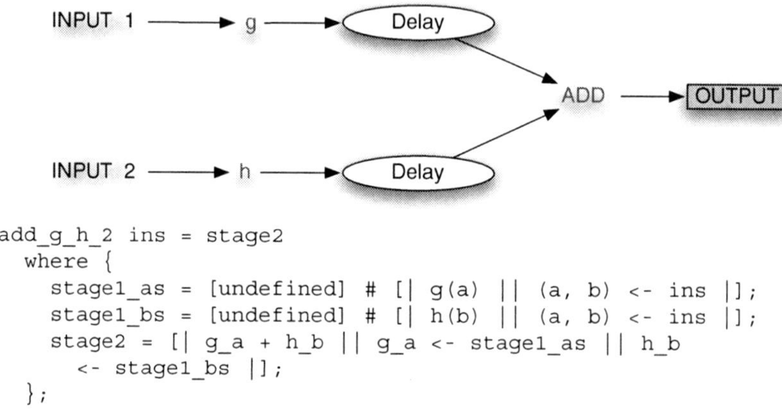

```
add_g_h_2 ins = stage2
  where {
    stage1_as = [undefined] # [| g(a) || (a, b) <- ins |];
    stage1_bs = [undefined] # [| h(b) || (a, b) <- ins |];
    stage2 = [| g_a + h_b || g_a <- stage1_as || h_b
      <- stage1_bs |];
  };
```

Fig. 7 add_g_h_2 (pipelining at the stream level)

We can pipeline this circuit by identifying two distinct stages that can execute in parallel (1) the applications of g and h and (2) the addition operation. To implement the pipeline, we evaluate g(a) and h(b) in parallel and store the results in a state in 1 cycle. On the next cycle, we add the two elements of the state together and make that the output of the circuit. This adds an extra clock cycle of latency so that it now takes 2 cycles to perform the entire computation. However, the clockrate is only limited by maximum delay of g, h, and the addition, whereas in the previous implementation it was the maximum delay of g and h *plus* the delay of the addition. Therefore, each stage of the computation can execute faster, and the throughput increases.

Note that during any given clock cycle, a pipelined implementation operates on data associated with two consecutive and unrelated inputs; it applies h and g to the most recent inputs, and it applies addition to the state which stores h(a) and h(b) associated with the inputs of the previous cycle.

Our first pipelined implementation, add_g_h_2, is provided in Fig. 7.

In each pipelined implementation above, the second stage only performs a single addition operation. Therefore, if either g or h has a propagation delay of more than one addition operation, then the first stage is the bottleneck of the pipeline and

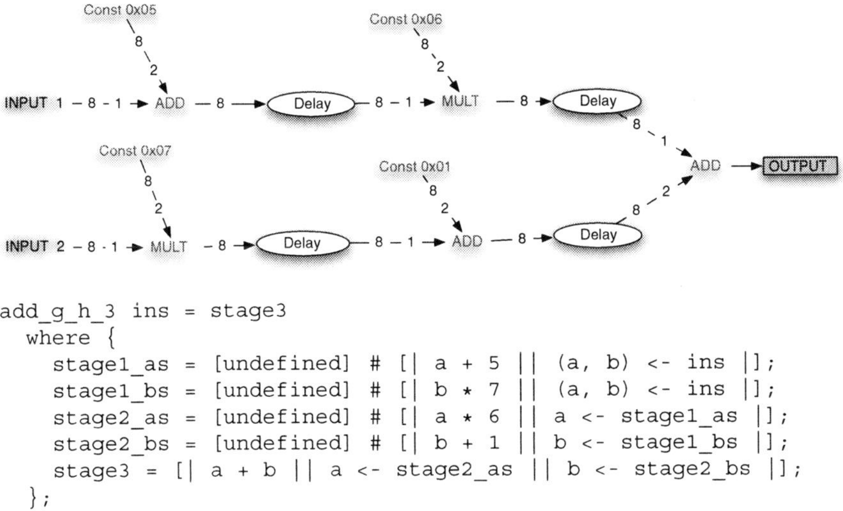


```
add_g_h_3 ins = stage3
  where {
    stage1_as = [undefined] # [| a + 5 || (a, b) <- ins |];
    stage1_bs = [undefined] # [| b * 7 || (a, b) <- ins |];
    stage2_as = [undefined] # [| a * 6 || a <- stage1_as |];
    stage2_bs = [undefined] # [| b + 1 || b <- stage1_bs |];
    stage3 = [| a + b || a <- stage2_as || b <- stage2_bs |];
  };
```

Fig. 8 add_g_h_3 (three-stage pipeline)

should be split into multiple stages if possible. Suppose g and h are defined as follows:

```
g a = (a+5) * 6;
h b = b*7 + 1;
```

We can implement g and h each as a two-stage pipeline, so our entire circuit has three stages (1) perform a+5 and b*7; (2) apply multiplication by 6 and addition by 1 to the results from stage 1; and (3) add the results from stage 2. This three-stage pipeline, add_g_h_3, is defined in Fig. 8.

We can define a utility function that lifts any combinatorial circuit into a stage of a pipeline in the stream model:

```
stage : {a b} (a -> b, [inf]a) -> [inf]b;
stage (f, ins) = [undefined] # [| f x || x <- ins |];
```

Equivalence checking is not possible on infinite streams, so we cannot verify that a sequential circuit is equivalent to its spec for all time. However, we can still provide some level of assurance that the circuit is correct. First, we verify that the function always produces the correct output for the first input. For example, to test add_g_h_1 above, use a function like the following:

```
test_add_g_h : ([c], [c]) -> [c];
test_add_g_h input = add_g_h_1 ([input] # zero) @ 0;
```

This function should be equivalent to the spec, which we can verify with the following command:

```
:eq add_g_h_spec test_add_g_h
```
——————————— interpreter ———————————

One may substitute `add_g_h_1` with any pipelined implementation above and change the index operand to account for circuit latency.

Next, we can verify that for some fixed number of inputs, the circuit generates the correct outputs. The following function asks this question, returning `True` if any n inputs can be found for which the implementation is not equivalent to the spec. The value of n can be adjusted as desired.

```
check_add_g_h inputs = reference != result
  where {
    reference = [| add_g_h_spec x || x <- inputs |];
    result = take (n, add_g_h_1 (inputs # zero));
    n = 1000;
  };
```

The sat solver determines whether any inputs satisfy the query. The circuit is correct if this function is *not* satisfiable.

```
:sat check_add_g_h
```
——————————— interpreter ———————————

Again, one may substitute `add_g_h_1` with any pipelined implementation above, but account for latency by dropping the initial undefined outputs from `result` using the `drop` construct.

3.5.2 Pipelining via the *reg* Pragma

In order to pipeline the above examples, we had to lift each circuit into the stream model. This is because we need to have access to a stream that is mapped over time in order to delay it.

In this section, we introduce a new pragma that allows the user to pipeline combinational code without lifting code into the stream model, and show how it can be applied to the examples above. This allows the user to pipeline code without changing the structure, yielding a pipelined implementation that more closely resembles the original spec.

When used as intended, this `reg` pragma causes a combinational circuit to be divided into smaller combinational circuits with registers between. Each application of `reg` generates a Delay–Undelay pair in the SPIR AST, so the net delay through the circuit is exactly 0. This allows us and the compiler to treat the circuit as combinational and without any notion of time. During the translation from SPIR to LLSPIR, the circuit becomes sequential as the compiler uses specific rewrite rules to move the Delays and Undelays around while keeping the circuit synchronized with respect to time.

Unlike when the user pipelines by appending [undefined], the compiler is aware of the latency that the `reg` pragma introduces. The compiler will report the correct latency of the circuit, and when we lift the circuit into the stream model, we do not have to drop the undefined outputs from the beginning of the stream; the first element will be the first valid output.

Using the `reg` pragma, we can pipeline g and h as combinational circuits without changing the definition of add_g_h_spec:

```
g x = reg(reg(x+5) * 6);
h y = reg(reg(y*7) + 1);
```

Using these definitions of g and h, when add_g_h_spec is lifted into the stream model, it will be identical to the add_g_h_4 circuit that was manually pipelined above.

We can also use the `reg` pragma to pipeline the iterate function introduced in Sect. 3.4.2:

```
iterate : b -> [(k+1)]b;
iterate x = outs
  where outs = [x] # [| reg(f prev)
                     || i <- [1..k]
                     || prev <- outs |];

iterate_pipe' : b -> b;
iterate_pipe' x = iterate x ! 0;
```

4 AES Specification

This section develops AES from the description in FIPS-197 [4] while using Cryptol's advanced features to make the specification clear and amenable to verification techniques. While in many respects the correspondence between the specification and the Cryptol code is easily seen, we have made one major departure. Namely, while the specification is defined in terms of successively applying permutations to the plaintext to yield the ciphertext, this implementation uses higher-order functions to compute a single key-dependent pair of permutations that encrypt and decrypt. The advantage of this is that the decryption function is, by construction, manifestly the inverse of the encryption function.

Most of the sections and subsections herein are numbered as they are in FIPS-197. The intent is that this document be read in tandem with that one.

4.1 API

Within FIPS-197 there are three key sizes (128, 192, and 256) and a common block size (128). These three key sizes correspond to encryption/decryption primitives

aptly named as AES-128, AES-192, and AES-256. It is intended that programs
needing these primitives would access them via the following API:

```
AES128.encrypt(key128,plaintext)
AES128.decrypt(key128,ciphertext)
AES192.encrypt(key192,plaintext)
AES192.decrypt(key192,ciphertext)
AES256.encrypt(key256,plaintext)
AES256.decrypt(key256,ciphertext)
```

where `plaintext` and `ciphertext` are 128-bit blocks and `key128`, `key192`,
and `key256` are 128-, 192-, and 256-bit keys. The return value for all six primitives
has type `[128]`. Naturally the encrypt and decrypt functions for a given size and
key are inverses of each other.

   The Cryptol code for the API follows:

```
AES128 : {encrypt : (Key(4),Block) -> Block;
          decrypt : (Key(4),Block) -> Block};
AES128 = {encrypt = Cipher; decrypt = InvCipher};

AES192 : {encrypt : (Key(6),Block) -> Block;
          decrypt : (Key(6),Block) -> Block};
AES192 = {encrypt = Cipher; decrypt = InvCipher};

AES256 : {encrypt : (Key(8),Block) -> Block;
          decrypt : (Key(8),Block) -> Block};
AES256 = {encrypt = Cipher; decrypt = InvCipher};
```

4.2 Types

This section loosely corresponds to FIPS-197 §3 for concrete types like byte and
word. Cryptol's more advanced types facilitate more appropriate signatures for
many of the functions defined in FIPS-197.

   AES has the familiar data structures byte (`[8]`) and word (`[32]`). It also has
block[4] (`[128]`) and state, a four-by-four[5] matrix of bytes (`[4][4][8]`).

   In code these are:

```
type BitsPerByte = 8;
type BitsPerWord = 32;
```

---

[4] While AES insists that the block size is 128, Rijndael [1], on which AES was based, allows block
sizes of 128, 160, 192, 224, and 256.

[5] If AES had block sizes other than 128, the number of columns would differ from 4.

```
type BitsPerBlock = 128;
type BytesPerColumn = 4;
type ColumnsPerState = 4;

type Byte = [BitsPerByte];
type Word = [BitsPerWord];
type Block = [BitsPerBlock];

type Column = [BytesPerColumn]Byte;
type State = [ColumnsPerState]Column;
type RoundKey = State;
```

The type `Column` is alluded to in FIPS-197 §3.5, but it proves useful enough that we had made it explicit. Also while `RoundKey` is the same type as `State`, for documentation purposes we use it as appropriate.

The key is measured in the number of words, Nk, so a key has type:

```
type Key(nk)  =  [nk * BitsPerWord];
```

AES mandates[6] that nk must be 4, 6, or 8.

Some uses of keys regard them as a collection of words, so we also have:

```
type KeyAsWords(nk)  =  [nk]Word;
```

In addition, depending on Nk, there is a number of per-round keys:

```
type RoundKeys(nr)  =  [nr + 1]RoundKey;
```

Oddly, while the number of rounds, Nr, is 10, 12, or 14 (depending on the key size), there are, respectively, 11, 13, or 15 round keys. (This accounts for the [nr+1] on the right-hand side above.)

Many of the steps in the cryptoalgorithm are permutations from `State` to `State`, so we define:

```
type Permutation = State -> State;
```

Some permutations are also their own inverses so we define:

```
type Involution = State -> State;
```

to document those situations.

Finally, so that the decryption function can be easily seen to be the inverse of the encryption function, we define a type that holds a permutation and its inverse

```
type Duo = {function : Permutation; inverse :
    Permutation};
```

and build the encryption and decryption functions by composing these.

4.3 Conversions Between Types

Following are some useful definitions for converting between types.

```
blockToState : Block -> State;
blockToState(x) = transpose(split(reverse(split(x))));

stateToBlock : State -> Block;
stateToBlock(x) = join(reverse(join(transpose(x))));

keyToWords :{nk}(fin nk) => Key(nk) ->KeyAsWords(nk);
keyToWords(key) = reverse(split(key));
```

4.4 Constructors for the Duo Type

Functions to construct duos from permutations and involutions are given by:

```
makeDuo : (Permutation,Permutation) -> Duo;
makeDuo(t,u) = {function = t; inverse = u};

makeDuoFromInvolution : {a} Involution -> Duo;
makeDuoFromInvolution(i) = makeDuo(i,i);
```

Since code in the sequel is predicated on the components of a Duo being inverses of each other, ideally we would enforce that somehow in code. The brute force approach of ensuring that both compositions of the duo's function and inverse behave as the identity on the $2^{4\times4\times8} = 340\,282\,366\,920\,938\,463\,463\,374\,607\,431\,768\,211\,456$ inputs is impractical. While we lack a general approach, there are some efforts to be made in the following as many of the permutations have a structure that allows for decomposition of the inverse test onto substructures having smaller domains.

4.5 Mathematical Preliminaries

This section implements the necessary functions described in FIPS-197 §4. An attempt is made to follow the section numbering for a strong correspondence between that section and this one. Since Cryptol has Galois field primitives and much of that section deals with arithmetic in $GF(2^8)$, much of this is brief.

4.5.1 Addition

As is well known, addition in $GF(2^8)$ may be implemented via exclusive or which is the ^ operator in Cryptol, so the few instances where addition in $GF(2^8)$ is needed are simply coded via ^. It is handy to have a n-ary addition, so that is defined via:

```
gSum : {a} (fin a) => [a]Byte -> Byte;
gSum(xs) = sums ! 0
 where {
  sums = [zero] # [| x ^ y || x <- xs || y <- sums |];
 };
```

4.6 Multiplication

Multiplication in $GF(2^8)$ is easily defined in Cryptol via:

```
gTimes : (Byte,Byte) -> Byte;
gTimes(x,y) = pmod(pmult(x,y),<| x^8 + x^4 + x^3
    + x + 1 |>);
```

4.6.1 Multiplication by x

Cryptol's ease of defining gTimes above means we do not need the xtime()
function defined in the corresponding section of FIPS-197.

4.7 Polynomials with Coefficients in $GF(2^8)$

Multiplying by $a(x)$ from FIPS-197 §4.3 corresponds to the function:

```
mixColumn : Column -> Column;
mixColumn(column) = gMatrixVectorProduct
                     (aMatrix,column);

aMatrix : State;
aMatrix = transpose(columns)
 where {
  columns = [| [ 0x02 0x01 0x01 0x03 ] >>> i ||
           i <- [0 .. 3] |];
 };
```

And multiplying by $a^{-1}(x)$ from FIPS-197 §4.3 corresponds to the function:

```
invMixColumn : Column -> Column;
invMixColumn(column) = gMatrixVectorProduct
                     (aInverseMatrix,column);

aInverseMatrix : State;
aInverseMatrix = transpose(columns)
```

```
where {
  columns = [| [ 0x0e 0x09 0x0d 0x0b ] >>> i ||
            i <- [0 .. 3] |];
};
```

The routines for performing the matrix-vector product follow:

```
gMatrixVectorProduct : (State,Column) -> Column;
gMatrixVectorProduct(matrix,column) =
  join(gMatrixProduct(matrix,split(column)));

gMatrixProduct : {a b c} (fin a,fin b,fin c) =>
  ([a][b]Byte,[b][c]Byte) -> [a][c]Byte;
gMatrixProduct(xss,yss) = [| [| gDotProduct(row,col)
                            || col <- transpose(yss) |]
                         || row <- xss |];

gDotProduct : {a} (fin a) => ([a]Byte,[a]Byte) -> Byte;
gDotProduct(xs,ys) =
  gSum([| gTimes(x,y) || x <- xs || y <- ys |]);
```

4.8 Algorithm Specification

In this section, we follow the development of the AES algorithm in FIPS-197 §5. Although the more abstract types `State`, `Permutation`, and `Duo` are used throughout, all the functions with capitalized names have essentially the same functionality as those in FIPS-197.

The implementation is quite different in that it focuses on composing permutations to compute a permutation that corresponds to encryption and then applying that composed permutation to the plaintext to get the ciphertext rather than successively applying permutations to get intermediate results whose culmination is the ciphertext. The advantage to composing the permutations is that the `Duo` type carries the inverse through those permutations, so that it is readily apparent that the decryption permutation is inverse of the encryption permutation.

4.9 Cipher

The main function for encryption follows:

```
Cipher : {nk} (fin nk,8 >= width(nk),nk >= 1) =>
  (Key(nk),Block) -> Block;
Cipher(key,plaintext) = stateToBlock(out)
  where {
  in = blockToState(plaintext);
```

```
  roundKeys = KeyExpansion(keyToWords(key));
  duos = duosByRound(roundKeys);
  out = applyPermutation(composeDuos(duos),in);
};
```

In it the duos (permutations paired with their inverses) by round are computed, composed, and applied to the suitably processed plaintext.

```
duosByRound : {nr} (fin nr,nr >= 1) => RoundKeys(nr) ->
  [nr + 1]Duo;
duosByRound(roundKeys)
  = [ (initialRoundDuo(roundKeys @ 0)) ]
  # [| (medialRoundDuo(roundKey))
    || roundKey <- roundKeys @@ [1 .. width(roundKeys) - 2] |]
  # [ (finalRoundDuo(roundKeys ! 0)) ];
```

The structure of the `Cipher` function is most visible in the following:

```
initialRoundDuo : RoundKey -> Duo;
initialRoundDuo(roundKey) = makeAddRoundKeyDuo(roundKey);

medialRoundDuo : RoundKey -> Duo;
medialRoundDuo(roundKey)
  = composeDuos([ subBytesDuo shiftRowsDuo
                  mixColumnsDuo (makeAddRoundKeyDuo
                  (roundKey))]);
finalRoundDuo : RoundKey -> Duo;
finalRoundDuo(roundKey)
  = composeDuos([subBytesDuo
                  shiftRowsDuo(makeAddRoundKeyDuo
                  (roundKey))]);
```

4.9.1 `SubBytes()` Transformation

Following the style of FIPS-197, we define `sBoxSlow` from mathematical principles and use the equivalent `SBox` for the sake of speed.

```
sBoxSlow : [256]Byte;
sBoxSlow = [| f(b) || b <- [ 0 .. 255 ] |]
  where {
   f(x) = b ^ (b >>> 4) ^ (b >>> 5) ^ (b >>> 6)
       ^ (b >>> 7) ^ c
    where {
     b = gQuasiInverse(x);
     c = 0x63;
    };
  };
```

```
SubBytes : Permutation;
SubBytes(state) = mapBytes(f,state)
 where {
  f(b) = SBox @ b;
 };

SBox : [256]Byte;
SBox = [
  0x63 0x7c 0x77 0x7b 0xf2 0x6b 0x6f 0xc5 0x30 0x01 0x67 0x2b
  0xfe 0xd7 0xab 0x76 0xca 0x82 0xc9 0x7d 0xfa 0x59 0x47 0xf0
  0xad 0xd4 0xa2 0xaf 0x9c 0xa4 0x72 0xc0 0xb7 0xfd 0x93 0x26
  0x36 0x3f 0xf7 0xcc 0x34 0xa5 0xe5 0xf1 0x71 0xd8 0x31 0x15
  0x04 0xc7 0x23 0xc3 0x18 0x96 0x05 0x9a 0x07 0x12 0x80 0xe2
  0xeb 0x27 0xb2 0x75 0x09 0x83 0x2c 0x1a 0x1b 0x6e 0x5a 0xa0
  0x52 0x3b 0xd6 0xb3 0x29 0xe3 0x2f 0x84 0x53 0xd1 0x00 0xed
  0x20 0xfc 0xb1 0x5b 0x6a 0xcb 0xbe 0x39 0x4a 0x4c 0x58 0xcf
  0xd0 0xef 0xaa 0xfb 0x43 0x4d 0x33 0x85 0x45 0xf9 0x02 0x7f
  0x50 0x3c 0x9f 0xa8 0x51 0xa3 0x40 0x8f 0x92 0x9d 0x38 0xf5
  0xbc 0xb6 0xda 0x21 0x10 0xff 0xf3 0xd2 0xcd 0x0c 0x13 0xec
  0x5f 0x97 0x44 0x17 0xc4 0xa7 0x7e 0x3d 0x64 0x5d 0x19 0x73
  0x60 0x81 0x4f 0xdc 0x22 0x2a 0x90 0x88 0x46 0xee 0xb8 0x14
  0xde 0x5e 0x0b 0xdb 0xe0 0x32 0x3a 0x0a 0x49 0x06 0x24 0x5c
  0xc2 0xd3 0xac 0x62 0x91 0x95 0xe4 0x79 0xe7 0xc8 0x37 0x6d
  0x8d 0xd5 0x4e 0xa9 0x6c 0x56 0xf4 0xea 0x65 0x7a 0xae 0x08
  0xba 0x78 0x25 0x2e 0x1c 0xa6 0xb4 0xc6 0xe8 0xdd 0x74 0x1f
  0x4b 0xbd 0x8b 0x8a 0x70 0x3e 0xb5 0x66 0x48 0x03 0xf6 0x0e
  0x61 0x35 0x57 0xb9 0x86 0xc1 0x1d 0x9e 0xe1 0xf8 0x98 0x11
  0x69 0xd9 0x8e 0x94 0x9b 0x1e 0x87 0xe9 0xce 0x55 0x28 0xdf
  0x8c 0xa1 0x89 0x0d 0xbf 0xe6 0x42 0x68 0x41 0x99 0x2d 0x0f
  0xb0 0x54 0xbb 0x16 ];
```

Naturally, the expression `sBoxSlow == SBox` evaluates to `True`.

4.9.2 `ShiftRows()` Transformation

This definition is straightforward from the corresponding section of FIPS-197.

```
ShiftRows : Permutation;
ShiftRows state = [| row <<< i || row <- state ||
    i <- [ 0 .. 3 ] |];
```

4.9.3 `MixColumns()` Transformation

This definition is straightforward from the corresponding section of FIPS-197.

```
MixColumns : Permutation;
MixColumns(state) = mapColumns(mixColumn,state);
```

4.9.4 `AddRoundKey()` Transformation

Pedantically speaking, `AddRoundKey` is not a permutation as are `SubBytes`, `ShiftRows`, and `MixColumns`, as it takes an additional parameter of type `RoundKey`. (This is reflected in the type.) In the compositions of permutations elsewhere, we partially apply `AddRoundKey` to a round key to get a function of type `Involution`. (`Involutions` are self-inverse `Permutations`.)

```
AddRoundKey : (RoundKey,State) -> State;
AddRoundKey(roundKey,state) = roundKey ^ state;
```

4.10 Key Expansion

In FIPS-197, `KeyExpansion` returns an array of `Words`, but it is clearer in Cryptol to give the result as an array of `RoundKeys`.

```
KeyExpansion : {nk} (fin nk,8 >= width(nk),nk >= 1) =>
 KeyAsWords(nk) -> RoundKeys((nk + 6));
KeyExpansion(keyAsWords) = [|(transpose s)||s <- ss|]
 where {
 Nk = width(keyAsWords);
 ss = groupBy(4,[| (reverse(splitBy(4,w)))
                 || w <- take(4 * (Nk + 7),ws) |]);
 ws = keyAsWords # [| nextWord(Nk,i,w,w')
                    || i <- [Nk .. ]
                    || w <- ws
                    || w' <- drop(Nk - 1,ws) |];
 nextWord : ([8],[8],Word,Word) -> Word;
 nextWord(Nk,i,w,w') = w ^ temp
  where {
   temp = if 0 == i % Nk
     then SubWord(RotWord(w')) ^ Rcon(i / Nk)
       else if (Nk > 6) & (4 == i % Nk)
         then SubWord(w')
         else w';
  };
 };
```

Nk is one of 4, 6, or 8, which implies that i is drawn from 4–43, 6–51, or 8–59, respectively. This in turn implies that the call to Rcon has a parameter that ranges from 1–10, 1–8, or 1–7. So we defined Rcon via:

```
Rcon : [8] -> Word;
Rcon(i) = rcon @ (i - 1);
```

```
rcon : [10]Word;
rcon = [| zero # p || p <- take(10,ps) |]
 where {
  ps = [ 0x01 ] # [| gTimes(p,0x02) || p <- ps |];
 };
```

The other functions needed by `KeyExpansion` are straightforward:

```
SubWord : Word -> Word;
SubWord(w) = join([| (SBox @ b) || b <- split(w) |]);

RotWord : Word -> Word;
RotWord(w) = w <<< 8;
```

4.11 *Inverse Cipher*

Due to the `Duo` data structure, the `InvCipher` function below reads almost exactly like the `Cipher` function of §4.9.

```
InvCipher : {nk} (fin nk,8 >= width(nk),nk >= 1) =>
  (Key(nk),Block) -> Block;
InvCipher(key,ciphertext) = stateToBlock(out)
 where {
  in = blockToState(ciphertext);
  roundKeys = KeyExpansion(keyToWords(key));
  duos = duosByRound(roundKeys);
  out = applyInversePermutation(composeDuos(duos),in);
 };
```

4.11.1 `InvShiftRows()` Transformation

This definition is straightforward from the corresponding section of FIPS-197.

```
InvShiftRows : Permutation;
InvShiftRows(state) =
  [| row >>> i || row <- state || i <- [ 0 .. 3 ] |];
```

4.11.2 `InvSubBytes()` Transformation

This definition is straightforward from the corresponding section of FIPS-197.

```
InvSubBytes : Permutation;
InvSubBytes(state) = mapBytes(f,state)
```

```
where {
 f(b) = InverseSBox @ b;
};

InverseSBox : [256]Byte;
InverseSBox = [
  0x52 0x09 0x6a 0xd5 0x30 0x36 0xa5 0x38 0xbf 0x40 0xa3 0x9e
  0x81 0xf3 0xd7 0xfb 0x7c 0xe3 0x39 0x82 0x9b 0x2f 0xff 0x87
  0x34 0x8e 0x43 0x44 0xc4 0xde 0xe9 0xcb 0x54 0x7b 0x94 0x32
  0xa6 0xc2 0x23 0x3d 0xee 0x4c 0x95 0x0b 0x42 0xfa 0xc3 0x4e
  0x08 0x2e 0xa1 0x66 0x28 0xd9 0x24 0xb2 0x76 0x5b 0xa2 0x49
  0x6d 0x8b 0xd1 0x25 0x72 0xf8 0xf6 0x64 0x86 0x68 0x98 0x16
  0xd4 0xa4 0x5c 0xcc 0x5d 0x65 0xb6 0x92 0x6c 0x70 0x48 0x50
  0xfd 0xed 0xb9 0xda 0x5e 0x15 0x46 0x57 0xa7 0x8d 0x9d 0x84
  0x90 0xd8 0xab 0x00 0x8c 0xbc 0xd3 0x0a 0xf7 0xe4 0x58 0x05
  0xb8 0xb3 0x45 0x06 0xd0 0x2c 0x1e 0x8f 0xca 0x3f 0x0f 0x02
  0xc1 0xaf 0xbd 0x03 0x01 0x13 0x8a 0x6b 0x3a 0x91 0x11 0x41
  0x4f 0x67 0xdc 0xea 0x97 0xf2 0xcf 0xce 0xf0 0xb4 0xe6 0x73
  0x96 0xac 0x74 0x22 0xe7 0xad 0x35 0x85 0xe2 0xf9 0x37 0xe8
  0x1c 0x75 0xdf 0x6e 0x47 0xf1 0x1a 0x71 0x1d 0x29 0xc5 0x89
  0x6f 0xb7 0x62 0x0e 0xaa 0x18 0xbe 0x1b 0xfc 0x56 0x3e 0x4b
  0xc6 0xd2 0x79 0x20 0x9a 0xdb 0xc0 0xfe 0x78 0xcd 0x5a 0xf4
  0x1f 0xdd 0xa8 0x33 0x88 0x07 0xc7 0x31 0xb1 0x12 0x10 0x59
  0x27 0x80 0xec 0x5f 0x60 0x51 0x7f 0xa9 0x19 0xb5 0x4a 0x0d
  0x2d 0xe5 0x7a 0x9f 0x93 0xc9 0x9c 0xef 0xa0 0xe0 0x3b 0x4d
  0xae 0x2a 0xf5 0xb0 0xc8 0xeb 0xbb 0x3c 0x83 0x53 0x99 0x61
  0x17 0x2b 0x04 0x7e 0xba 0x77 0xd6 0x26 0xe1 0x69 0x14 0x63
  0x55 0x21 0x0c 0x7d ];
```

4.11.3 `InvMixColumns()` Transformation

This definition is straightforward from the corresponding section of FIPS-197.

```
InvMixColumns : Permutation;
InvMixColumns(state) = mapColumns(invMixColumn,state);
```

4.11.4 Equivalent Inverse Cipher

Expressing `EqInvCipher` in terms of the `Duo` structures requires only a little manipulation.

```
EqInvCipher : {nk} (fin nk,8 >= width(nk),nk >= 1) =>
  (Key(nk),Block) -> Block;
EqInvCipher(key,ciphertext) = stateToBlock(out)
 where {
  in = blockToState(ciphertext);
  roundKeys = KeyExpansion(keyToWords(key));
  duos = eqDuosByRound(roundKeys);
  out = applyInversePermutation(composeDuos(duos),in);
 };
```

The expression `InvMixColumns(roundKey)` in the following definition corresponds to the modified key schedule used by `EqInvCipher` in FIPS-197.

```
eqDuosByRound : {nr} (fin nr,nr >= 1) =>
  RoundKeys(nr) -> [nr + 1]Duo;
eqDuosByRound(roundKeys)
  = [ (eqInitialRoundDuo(roundKeys @ 0)) ]
  # [| (eqMedialRoundDuo(InvMixColumns(roundKey)))
    || roundKey <- roundKeys @@ [1 .. width(roundKeys) - 2]|]
  # [ (eqFinalRoundDuo(roundKeys ! 0)) ];
```

The definition style below is encrypt rather than decrypt biased, but the symmetry of the Duos, the shrewd use of `reverse`, and the ordering of the statements make the algorithm of `EqInvCipher` presented in FIPS-197 readily apparent.

```
eqFinalRoundDuo : RoundKey -> Duo;
eqFinalRoundDuo(roundKey) =makeAddRoundKeyDuo(roundKey);


eqMedialRoundDuo : RoundKey -> Duo;
eqMedialRoundDuo(roundKey)
  = composeDuos(reverse([ subBytesDuo shiftRowsDuo
      mixColumnsDuo (makeAddRoundKeyDuo(roundKey)) ]));


eqInitialRoundDuo : RoundKey -> Duo;
eqInitialRoundDuo(roundKey)
  = composeDuos(reverse([ subBytesDuo
    shiftRowsDuo (makeAddRoundKeyDuo(roundKey)) ]));
```

FIPS-197 does not define `EqCipher`, but since the Duo style makes it so easy, we define it for completeness.

```
eqCipher : {nk} (fin nk,8 >= width(nk),nk >= 1) =>
  (Key(nk),Block) -> Block;
eqCipher(key,ciphertext) = stateToBlock(out)
 where {
  in = blockToState(ciphertext);
  roundKeys = KeyExpansion(keyToWords(key));
  duos = eqDuosByRound(roundKeys);
  out = applyPermutation(composeDuos(duos),in);
 };
```

4.12 Auxiliary Definitions

Following are some convenience functions for operating on data of type `State` in a `Byte-by-Byte` or `Column-by-Column` fashion.

```
mapBytes : (Byte -> Byte,State) -> State;
mapBytes(f,state) =
  [| [| f(byte) || byte <- row |] || row <- state |];

mapColumns : (Column -> Column,State) -> State;
mapColumns(f,state) = transpose([| f(column)
                                || column <- transpose
                                      (state) |]);
```

The various Duos are constructed from transformations via:

```
mixColumnsDuo : Duo;
mixColumnsDuo = makeDuo(MixColumns,InvMixColumns);

shiftRowsDuo : Duo;
shiftRowsDuo = makeDuo(ShiftRows,InvShiftRows);

subBytesDuo : Duo;
subBytesDuo = makeDuo(SubBytes,InvSubBytes);

makeAddRoundKeyDuo : RoundKey -> Duo;
makeAddRoundKeyDuo(roundKey) =
  makeDuoFromInvolution(makeAddRoundKeyInvolution
      (roundKey));

makeAddRoundKeyInvolution : RoundKey -> Involution;
makeAddRoundKeyInvolution(roundKey) = f
 where {
  f(state) = AddRoundKey(roundKey,state);
 };
```

Composition of Duos is defined via:

```
composeDuos : {a} (fin a) => [a]Duo -> Duo;
composeDuos(duos) = compositions ! 0
 where {
  compositions = [ identityDuo ] # [| compose(c,d)
                              || c <- compositions
                              || d <- duos |];
  identityInvolution : Involution;
  identityInvolution(state) = state;
  identityDuo : Duo;
  identityDuo=makeDuoFromInvolution(identityInvolution);
  compose : (Duo,Duo) -> Duo;
  compose(p,q)={function(x) =
                q.function(p.function(x));
                inverse(x) = p.inverse(q.inverse(x))};
 };
```

Applying a permutation or its inverse is defined via:

```
applyPermutation : (Duo,State) -> State;
applyPermutation(duo,state) = duo.function(state);

applyInversePermutation : (Duo,State) -> State;
applyInversePermutation(duo,state)=duo.inverse(state);
```

The following two functions are used for verifying the `SBox` table. The cyclicity of the multiplicative group of $GF(2^8)$ is employed by `gQuasiInverse` in that for nonzero x in $GF(2^8)$ $x^{255} = 1$, so $x^{-1} = x^{254}$. (Happily, that definition gives the desired result for zero as well.)

```
gQuasiInverse : Byte -> Byte;
gQuasiInverse(x) = gPower(x,254);

gPower : {a} (fin a) => (Byte,[a]) -> Byte;
gPower(x,n) = rs ! 0
 where {
  ss = [x] # [| gTimes(s,s) || s <- ss |];
  rs = [1] # [| if b then gTimes(r,s) else r
              || b <- n
              || r <- rs
              || s <- ss |];
 };
```

4.13 Key Expansion Example: 128-bit Cipher Key

```
a1_test : Bit;
a1_test = ws == ws'
 where {
  roundKeys : RoundKeys(10);
  roundKeys =
    KeyExpansion(keyToWords
                (0x2b7e151628aed2a6abf7158809cf4f3c));
  ws' = join([| [| join(reverse(row))
              || row <- transpose(roundKey) |]
            || roundKey <- roundKeys |]);
  ws : [44]Word;
  ws = [ 0x2b7e1516 0x28aed2a6 0xabf71588 0x09cf4f3c
         0xa0fafe17 0x88542cb1 0x23a33939 0x2a6c7605
         0xf2c295f2 0x7a96b943 0x5935807a 0x7359f67f
         0x3d80477d 0x4716fe3e 0x1e237e44 0x6d7a883b
         0xef44a541 0xa8525b7f 0xb671253b 0xdb0bad00
         0xd4d1c6f8 0x7c839d87 0xcaf2b8bc 0x11f915bc
```

```
          0x6d88a37a 0x110b3efd 0xdbf98641 0xca0093fd
          0x4e54f70e 0x5f5fc9f3 0x84a64fb2 0x4ea6dc4f
          0xead27321 0xb58dbad2 0x312bf560 0x7f8d292f
          0xac7766f3 0x19fadc21 0x28d12941 0x575c006e
          0xd014f9a8 0xc9ee2589 0xe13f0cc8 0xb6630ca6 ];
  };
```

As expected, evaluating `a1_test` returns `True`.

4.14   Test and Verification

The permutation composition style used herein does not lend itself to verifying the
intermediate steps as in FIPS-197 Appendix B. We content ourselves with the end-
to-end test:

```
b_test : Bit;
b_test = output == output'
 where {
   key : Key(4);
   key = 0x2b7e151628aed2a6abf7158809cf4f3c;
   plaintext : Block;
   plaintext = 0x3243f6a8885a308d313198a2e0370734;
   ciphertext : Block;
   ciphertext = AES128.encrypt(key,plaintext);
   output' : State;
   output' = blockToState(ciphertext);
   output : State;
   output = [ [ 0x39 0x02 0xdc 0x19 ]
              [ 0x25 0xdc 0x11 0x6a ]
              [ 0x84 0x09 0x85 0x0b ]
              [ 0x1d 0xfb 0x97 0x32 ] ];
 };
```

Evaluating `b_test` returns `True` as one would expect.

4.14.1   Example Vectors

Again the permutation composition style used herein does not lend itself to verifying
the intermediate steps as in FIPS-197 Appendix C. We content ourselves with the
following end-to-end tests.

AES-128 (*Nk=4, Nr=10*)

```
cl_test_1: Bit;
cl_test_1= AES128.encrypt(0x000102030405060708090a0b0c0d0e0f,
                          0x00112233445566778899aabbccddeeff)
        == 0x69c4e0d86a7b0430d8cdb78070b4c55a;

cl_test_2: Bit;
cl_test_2= AES128.decrypt(0x000102030405060708090a0b0c0d0e0f,
                          0x69c4e0d86a7b0430d8cdb78070b4c55a)
        == 0x00112233445566778899aabbccddeeff;

cl_test_3: Bit;
cl_test_3= EqInvCipher(0x000102030405060708090a0b0c0d0e0f : [128],
                  0x69c4e0d86a7b0430d8cdb78070b4c55a)
          == 0x00112233445566778899aabbccddeeff;
```

Evaluating (`c1_test_1,c1_test_2,c1_test_3`) returns the expression
(`True, True, True`).

4.14.2   Formal Verification

Formal models of this specification can be generated as follows:

```
:set symbolic
:fm (KeyExpansion :  [4][32]  ->  [11][4][4][8])
  "keyexp128-spec.fm"
:fm (KeyExpansion :  [6][32]  ->  [13][4][4][8])
  "keyexp192-spec.fm"
:fm (KeyExpansion :  [8][32]  ->  [15][4][4][8])
"keyexp256-spec.fm"


:fm (Cipher :  ([128],[128])  ->  [128])  "aes128-spec.fm"
:fm (Cipher :  ([192],[128])  ->  [128])  "aes192-spec.fm"
:fm (Cipher :  ([256],[128])  ->  [128])  "aes256-spec.fm"
                          interpreter
```

Each implementation in Sect. 5 is checked for equivalence against this formal
model.

5   AES Implementations

This section documents the process of refining the AES specification described in
Sect. 4 into synthesizable implementations of AES-128 and AES-256. First, we remove constructs that are unsupported in the hardware compiler, replace many of
the functions with more efficient versions, and specialize the algorithm to two implementations, one for AES-128 and one for AES-256. Then, we define a second

pair of implementations by defining a function that performs one round of key expansion and encryption for each algorithm and using this function to combine the `KeyExpansion` and `Cipher` functions into one top-level function.

We adjust the second pair of implementations to meet space and time requirements. We provide two specific implementations of each algorithm (1) a smaller implementation that uses the `seq` pragma to reuse the same key expansion and encryption circuitry over multiple clock cycles and (2) a high-throughput pipelined implementation.

We then improve the AES-128 implementation by replacing the round of encryption with an equivalent and more efficient T-Box implementation and use the `seq` pragma again to produce a small circuit from the T-Box implementation.

In Sect. 5.2.2, we show how to use the new `reg` pragma to pipeline the AES-128 implementations in this section.

The resulting implementations are checked for equivalence with the original spec in symbolic, LLSPIR, and FPGA modes. We also present performance results with and without Block RAMs in both LLSPIR mode and TSIM mode.

5.1 Implementation #1

In this section, we replace many of the functions from the reference specification with much more efficient versions and specialize the encryption algorithm. Specifically, we perform the following changes:

- Replace `gTimes` with a more efficient implementation from the Rijndael spec, then rewrite using stream recursion
- Eliminate `gPower`, replacing with specialized `gPower2`
- Eliminate `gTimes`, replacing with specialized `gTimes2` and `gTimes3`
- Replace `mixColumn` with much more efficient implementation, through inlining and static evaluation
- Specialize `KeyExpansion` to AES-128 and AES-256 (only AES-128 is shown)

These changes result in an AES implementation with very reasonable performance (see Sect. 5.5).

The following are not supported by the Cryptol hardware compiler and therefore should be removed when producing an implementation from the spec:

- Recursive functions, which can usually be replaced by stream recursion
- Functions that return nonclosed functions

There are no recursive functions in the spec. However, the `Cipher Duo` record contains higher-order functions. We can remove the use of higher-order functions and inline into `Cipher` the functions that use `Duo`, yielding the following definition:

```
Cipher : {nk} (fin nk,8 >= width nk,nk >= 1) =>
                (Key(nk),Block) -> Block;
Cipher(key,plaintext) = stateToBlock(finalRound) where {
  in = blockToState(plaintext);
  roundKeys = KeyExpansion(keyToWords(key));
  initialRound = AddRoundKey (in, roundKeys @ 0);
  medialRounds = [initialRound] #
                  [| AddRoundKey
                       (MixColumns
                         (ShiftRows(SubBytes(state))),roundKey)
                      || state <- medialRounds
                      || roundKey <- roundKeys @@
                                      [1 .. width(roundKeys)-2]  |];
  finalRound = AddRoundKey
                 (ShiftRows
                   (SubBytes(medialRounds ! 0)), roundKeys ! 0);
};
```

We will specialize this function to `Cipher128` below.

One can inline the use of `gMatrixVectorProduct`, `gMatrixProduct`, `gDotProduct`, and `gSum` into the definition of `mixColumn`, yielding a simpler and much more efficient implementation. Note that the value of `aMatrix` is known statically as:

```
[[0x2 0x3 0x1 0x1] [0x1 0x2 0x3 0x1] [0x1 0x1 0x2 0x3]
 [0x3 0x1 0x1 0x2]]
```

Given that the argument to `mixColumn` is `[y0 y1 y2 y3]`, then the matrix product of the above `aMatrix` and this `column` argument is simply:

```
[ (gTimes(2,y0) ^ gTimes(3,y1) ^ gTimes(1,y2) ^ gTimes(1,y3))
  (gTimes(1,y0) ^ gTimes(2,y1) ^ gTimes(3,y2) ^ gTimes(1,y3))
  (gTimes(1,y0) ^ gTimes(1,y1) ^ gTimes(2,y2) ^ gTimes(3,y3))
  (gTimes(3,y0) ^ gTimes(1,y1) ^ gTimes(1,y2) ^ gTimes(2,y3))
]
```

In every call to `gTimes`, the first argument is either 1, 2, or 3. We can replace `gTimes(1,x)` with `x`, `gTimes(2,x)` with `gTimes2(x)`, and `gTimes(3,x)` with `gTimes3(x)`, where `gTimes2` and `gTimes3` are defined as follows:

```
gTimes2 : [8] -> [8];
gTimes2 x = (x << 1) ^ (if x ! 0 then <| x^4 + x^3 + x + 1 |>
   else zero);

gTimes3 : [8] -> [8];
gTimes3(b) = gTimes2(b) ^ b;
```

Thus, `mixColumn` can be rewritten as follows:

```
mixColumn' : [4][8] -> [4][8];
mixColumn'([y0 y1 y2 y3]) = [ (gTimes2(y0)  ^ gTimes3(y1) ^
                               y2           ^ y3)
                              (y0           ^ gTimes2(y1) ^
                               gTimes3(y2)  ^ y3)
                              (y0           ^ y1          ^
                               gTimes2(y2)  ^ gTimes3(y3))
                              (gTimes3(y0)  ^ y1          ^
                               y2           ^ gTimes2(y3))
                            ];
```

The original `gTimes` is now used only by `gPower`. However, `gPower` is called only by `Rcon` and always with a constant first argument, 2. Therefore, we could rewrite `gPower` as `gPower2`:

```
gPower2 : {a} ([a]) -> [8];
gPower2(i) = ps @ i
  where ps = [1] # [| gTimes2(p) || p <- ps |];
```

However, the latency of this implementation is 2^a, where a is the width of the input. This is because Cryptol implements synchronous circuits whose latency must be known statically; therefore, the latency of this circuit is equal to the worst-case latency. We can be more efficient by implementing it as a static lookup table. The following expression generates the sequence for the table:

```
[| gPower(2,x) || x <- [0..15] |]
——————————————— interpreter ———————————————
```

`gPower` takes 8-bit inputs and therefore we would need a 256-element lookup table to represent its entire range. However, the argument to `Rcon` is a 4-bit round number, so only the first 16 elements of the table will ever be accessed. (In fact, because the round number never goes above 11, we could shorten the table even more.) Thus, we write `gPower2_table` as follows:

```
gPower2_table : [16][8];
gPower2_table = [0x01 0x02 0x04 0x08 0x10 0x20 0x40 0x80
                 0x1b 0x36 0x6c 0xd8 0xab 0x4d 0x9a 0x2f];
```

Some utility functions must be rewritten to use `mixColumn'` and `gPower2_table`:

```
Rcon' : [4] -> [32];
Rcon'(i) = zero # (gPower2_table @ (i-1));

MixColumns' : [4][4][8] -> [4][4][8];
MixColumns'(state) = transpose([| mixColumn' column
                                  || column <- transpose(state) |]);
```

For each replaced function above, we can use the equivalence checker to verify that the new version is equivalent to the original. For example, the following commands can be used to verify that `gTimes2` and `gTimes3` are correct with respect to the spec for `gTimes`:

```
:set symbolic
:eq (\x -> gTimes(2,x)) gTimes2
:eq (\x -> gTimes(3,x)) gTimes3
```
interpreter

To produce a more efficient implementation of `KeyExpansion`, we specialize it to a 128-bit key size by making two minor changes. First, we replace `nextWord` with `nextWord_128`, which does not have to check if the key size is more than 6. Second, we replace the infinite intermediate sequences, `ws` and `[Nk ..]` with finite sequences.

In the specification, `KeyExpansion` is implemented using an infinite sequence of words, `ws`, that is defined by drawing `i` from an infinite sequence, `[Nk ..]`. `KeyExpansion` then draws a finite number of elements from that sequence using `take`. The hardware compiler can produce a more efficient implementation if we draw `i` from `[4..43]`, so that its contents can be evaluated at compile time, and implement `ws` as a finite sequence. The new implementation is defined in Fig. 9. The same technique is used to define key expansion for AES-256.

Finally, `Cipher128` is written to use the new definitions (see Fig. 10), and we lift this definition into the stream model as follows:

```
Cipher128_stream : [inf]([128],[128]) -> [inf][128];
Cipher128_stream ins = [| Cipher128 in || in <- ins |];
```

```
KeyExpansion128 : [4][32] -> [11][4][4][8];
KeyExpansion128 keyAsWords = [| transpose s || s <- ss |]
  where {
    ss = groupBy(4,[| reverse(splitBy(4,w)) || w <- ws |]);
    ws = keyAsWords # ([| nextWord_128(i,w,w')
                      || i <- [4 .. 43]
                      || w <- ws
                      || w' <- drop(3,ws)
                      |]);
  };

nextWord_128 : ([8],[32],[32]) -> [32];
nextWord_128(i,w,w') = w ^ temp
  where {
    temp = if i % 4 == 0
    then SubWord(RotWord(w')) ^ Rcon' (take (4, i / 4))
    else w';
  };
```

Fig. 9 `KeyExpansion128`

```
Cipher128 : ([128],[128]) -> [128];
Cipher128(key,plaintext) = stateToBlock(finalRound) where {
  in = blockToState(plaintext);
  roundKeys = KeyExpansion128(keyToWords(key));
  initialRound = AddRoundKey (in, roundKeys @ 0);
  medialRounds = [initialRound] #
                  [| AddRoundKey(MixColumns'(ShiftRows(SubBytes
                     (state))),roundKey)
                  || state <- medialRounds
                  || roundKey <- roundKeys @@ [1 .. width
                     (roundKeys) - 2] |];
  finalRound = AddRoundKey(ShiftRows(SubBytes(medialRounds ! 0)),
    roundKeys ! 0);
};
```

Fig. 10 `Cipher128`

```
nextKey128 : ([4],[4][32]) -> [4][32];
nextKey128(rnd, [w0 w1 w2 w3]) = [w0' w1' w2' w3']
    where {
        w0' = SubWord(RotWord w3) ^ (Rcon' rnd) ^ w0;
        w1' = w0' ^ w1;
        w2' = w1' ^ w2;
        w3' = w2' ^ w3;
    };
```

Fig. 11 Next key for AES-128

5.2 Implementation #2

In this section, we design functions that perform a single round of AES-128
and AES-256 (only the AES-128 implementation is shown), and we combine the
`KeyExpansion` and `Cipher` functions into one. This implementation of AES
makes it easier to apply the `seq` pragma, and is simpler to pipeline than the previous implementation.

Because the `KeyExpansion` function will be eliminated, some of the operations that it performed are moved into a new `AddRoundKey'` function:

```
AddRoundKey' : ([4][4][8],[4][32]) -> [4][4][8];
AddRoundKey'(state, keyAsWords) = state ^ roundKey
  where roundKey = transpose [| reverse (splitBy (4, word))
                             || word <- keyAsWords
                             |];
```

Rather than use `nextWord` to produce one word at a time, we can write a function that produces four words at a time. This function is specialized to AES-128 in
Fig. 11 and is equivalent to four sequential applications of `nextWord`.

It is important to note that the `SubWord` and `RotWord` operations can be ex-
changed. This can be checked as follows:

```
:eq (\x -> SubWord (RotWord x)) (\x -> RotWord (SubWord x))
                          interpreter
```

```
oneRound128 : ([4], ([4][4][8],[4][32])) -> ([4][4][8],[4][32]);
oneRound128 (round, (state, key)) = (next_state, next_key)
  where {
    state' = if round == 1 then state
               else if round == 11 then ShiftRows(SubBytes(state))
               else MixColumns'(ShiftRows(SubBytes(state)));
    next_state = AddRoundKey'(state', key);
    next_key = nextKey128(round, key);
  };
```

Fig. 12 One round of AES-128

```
Cipher128' : ([128],[128]) -> [128];
Cipher128' (key, pt) = stateToBlock final_state
  where {
    init = (blockToState pt, keyToWords key);

    rounds = [init] #
                [| oneRound128 (round, state_and_key)
                || state_and_key <- rounds
                || round <- [1..11]
                |];
    (final_state, dont_care) = rounds ! 0
  };
```

Fig. 13 `Cipher128'`

When pipelining AES in the following sections, exchanging these operations may produce a faster implementation.

We can now implement the `oneRound128` function that performs a single round of encryption and key expansion. It takes in the previous state and the key for this round and produces the key for the next round and a new state that uses that key from this round. This function is provided in Fig. 12.

Our new `Cipher` function simply applies the appropriate "oneRound" function for each of the 11 rounds. This is defined in Fig. 13.

To prevent the `Cipher` function from being laid out over time, we lift it into the stream world:

```
Cipher128'_stream: [inf]([128],[128]) -> [inf][128];
Cipher128'_stream ins= [| Cipher128' in || in <- ins|];
```

This sort of function should be used whenever translating to a hardware implementation.

5.2.1 Implementation #2 Optimized for Area

In this section, we optimize the implementation from Sect. 5.2 for area by reusing one round over multiple clock cycles. `Cipher128'` was written with this goal in mind, so we can reuse the `oneRound128` function and all we have to do is insert

```
Cipher128_seq : ([128],[128]) -> [128];
Cipher128_seq (key, pt) = stateToBlock final_state
  where {
    init_state = blockToState pt;
    init_key = keyToWords key;

    rounds = [(init_state, init_key)] #
             seq [| oneRound128 (round, state_and_key)
                  || state_and_key <- rounds
                  || round <- [1..11] |];
    (final_state, bla) = rounds ! 0
  };
```

Fig. 14 Cipher128_seq

```
Cipher128_reg : ([128],[128]) -> [128];
Cipher128_reg (key, pt) = stateToBlock final_state
  where {
    init_state = blockToState pt;
    init_key = keyToWords key;

    rounds = [(init_state, init_key)] #
                [| add_reg(round, oneRound128 (round,
                   state_and_key))
                 || state_and_key <- rounds
                 || round <- [1..11] |];
    (final_state, bla) = rounds ! 0
    };
```

Fig. 15 Cipher128_reg

the `seq` pragma as shown in Fig. 14. The `seq` pragma has the same effect here as
it did in Sect. 3.4.

To implement in hardware, this function should be lifted into the stream model:

```
Cipher128_seq_stream : [inf]([128],[128]) -> [inf][128];
Cipher128_seq_stream ins = [| Cipher128_seq in || in <- ins |];
```

5.2.2 Implementation #2 Pipelined Using the *reg* Pragma

In this section, we use the `reg` pragma to pipeline the implementation from
Sect. 5.2. All we do is apply the `reg` pragma to each round. The function is defined in Fig. 15.

We use the same method as before to conditionally insert registers after each
stage:

```
add_reg (i, x) = if check_stage i then (reg x) else x;

check_stage : {a} (fin a, a >= 4) => [a] -> Bit;
check_stage i = True; // (i%3 == 1);
```

To implement this in hardware, it should be lifted into the stream model:

```
Cipher128_reg_stream : [inf]([128],[128]) -> [inf][128];
Cipher128_reg_stream ins = [| Cipher128_reg in || in <- ins |];
```

5.3 T-Box Implementation

In this section, we replace `oneRound128` with an equivalent T-Box implementation that is more efficient. In the next section, we will use the `reg` pragma to produce a very high throughput pipeline from this implementation.

One round is defined in Fig. 16; it is a drop-in replacement for `oneRound128`. The implementation uses the following utility functions:

```
T0, T1, T2, T3 : [8] -> [4][8];
T0(a) = T0_table @ a;
T1(a) = T0(a) >>> 1;
T2(a) = T0(a) >>> 2;
T3(a) = T0(a) >>> 3;


T0_table = const [| T0_func(a) || a <- [0..255] |];


T0_func : [8] -> [4][8];
T0_func(a) = [(gTimes2 s) s s (gTimes3 s)] where
    s = SBox @ a;


 oneRound128_Tbox : ([4], ([4][4][8],[4][32])) -> ([4][4][8],
    [4][32]);
 oneRound128_Tbox (round, (state, key)) = (next_state, next_key)
   where {
      state' = if round == 1 then state
               else transpose d;
      next_state = AddRoundKey'(state', key);
      next_key = nextKey128(round, key);

      d = [| if (round == 11)
               then [(t0@1) (t1@2) (t2@3) (t3@0)]
               else t0 ^ t1 ^ t2 ^ t3
          where {
              t0 = T0 (state @ 0 @ (j+0));
              t1 = T1 (state @ 1 @ (j+1));
              t2 = T2 (state @ 2 @ (j+2));
              t3 = T3 (state @ 3 @ (j+3));
          }
      || j <- [0 .. 3] |];
   };
```

Fig. 16 One round of AES-128, using T-Box

The top-level "Cipher" function is the same as before, but we use
`oneRound128_Tbox` instead of `oneRound128`. It is defined in Fig. 17.

This T-Box implementation does not realize a very impressive increase in performance, because key expansion is the bottleneck in the algorithm. To obtain a high clockrate from this T-Box implementation, we can pipeline both the key expansion and encryption rounds using the `reg` pragma (see Sect. 5.3.2).

These functions should be used whenever translating to a hardware implementation:

```
Cipher128_Tbox_stream : [inf]([128],[128]) -> [inf][128];
Cipher128_Tbox_stream ins = [| Cipher128_Tbox in || in <- ins |];
```

5.3.1 T-Box Implementation Optimized for Area

In this section, we use the `seq` pragma to optimize the T-Box implementation for area, reusing the round each cycle. We use the same technique as in Sects. 3.4 and 5.2.1.

The implementation reuses the `oneRound128_Tbox` function from the previous section and is defined in Fig. 18.

```
Cipher128_Tbox : ([128],[128]) -> [128];
Cipher128_Tbox (key, pt) = stateToBlock final_state
  where {
    init = (blockToState pt, keyToWords key);

    rounds = [init] #
                [| oneRound128_Tbox (round, state_and_key)
                || state_and_key <- rounds
                || round <- [1..11] |];
    (final_state, dont_care) = rounds ! 0
  };
```

Fig. 17 AES-128, using T-Box

```
Cipher128_Tbox_seq : ([128],[128]) -> [128];
Cipher128_Tbox_seq (key, pt) = stateToBlock final_state
  where {
    init = (blockToState pt, keyToWords key);

    rounds = [init] #
              seq [| oneRound128_Tbox (round, state_and_key)
                  || state_and_key <- rounds
                  || round <- [1..11] |];
    (final_state, dont_care) = rounds ! 0
  };
```

Fig. 18 AES-128, using T-Box, reused over time

These functions should be used whenever translating to a hardware implementation:

```
Cipher128_Tbox_seq_stream : [inf]([128],[128]) -> [inf][128];
Cipher128_Tbox_seq_stream ins = [| Cipher128_Tbox_seq in
                                 || in <- ins |];
```

5.3.2 T-Box Implementation Pipelined Using the *reg* Pragma

In this section, we use the `reg` pragma to pipeline the AES-128 T-Box implementation. We have decided to pipeline each round to five stages and target a clockrate of about 400 MHz (2.5 ns).

We should attempt to implement the same number of stages for both "nextKey" and "oneRound," because they execute in parallel. If we do not use the same number of stages in each function, the compiler will simply insert registers to keep the circuits synchronized, but the registers will not be optimally placed. Also, we should use Block RAMs with 2-cycle latency; they have a higher clockrate, and the extra latency simply allows us to do more in parallel with the Block RAM. Therefore, to obtain five stages per round we should insert three more registers into both "nextKey" and "oneRound."

Each Block RAM behaves as a register, so the *SBox* and *Tbox* operations already define some of the pipeline stages. We are defining a pipeline with many small stages, so the latency of Block RAMs will dominate. Therefore, the input to a Block RAM should never be calculated in the same stage as the Block RAM because this would increase the delay of that stage. For example, we should place a register between the `RotWord` and `SubWord` operations so that they are performed in separate stages. Alternatively, we can exchange these operations as this may result in a more balanced pipeline.

Because compiling to LLSPIR is relatively quick and provides us with an estimated clockrate, one can experiment with many different combinations and placements of `reg` in search of the fastest possible clockrate.

The "nextKey" circuit performs the following, sequentially:

1. The following, in parallel:

   (a) `SubWord` and `RotWord`
   (b) `Rcon` and xor with w0

2. Xor the previous two results
3. Xor with w1
4. Xor with w2
5. Xor with w3

Because it is implemented using a Block RAM, the `SubWord` operation will take 2 cycles. We can check the propagation delay of the other operations by executing the following commands in LLSPIR mode:

```
nextKey128_reg : ([4],[4][32]) -> [4][32];
nextKey128_reg(rnd, [w0 w1 w2 w3]) = [w0' w1' w2' w3']
    where {
        w0' = RotWord(SubWord w3) ^ (Rcon' rnd) ^ w0;
        w1' = reg(w0' ^ w1);
        w2' = (w1' ^ w2);
        w3' = reg(w2' ^ w3);
    };
```

Fig. 19 Pipelined key expansion for AES-128

```
:translate (\(x, rnd) -> Rcon' rnd ^ x)
:translate (\(x,y,z) -> (x:[32]) ^ y ^ z)
:translate (\(x,y,z,w) -> (x:[32]) ^ y ^ z ^ w)
:translate (\(x,y,z) -> (RotWord x) ^ y ^ z)
```
interpreter

From this we find that 1(b) can fit within 2 cycles and therefore can execute in parallel with `SubWord`. We also discover that two xor operations can fit within 1 cycle, `RotWord` and two xor operations can, but three xor operations cannot. Therefore, we should implement the "nextKey" stages as follows:

1. `SubWord` in parallel with `Rcon' rnd ^ w0` (this takes 2 cycles)
2. `RotWord` and two xor operations
3. Two more xor operations

This pipeline is implemented in Fig. 19.

Pipelining one round of encryption is fairly straightforward. We define the following stages:

1. Four parallel T-Box lookups
2. `t0 ^ t1` in parallel with `t2 ^ t3`, then xor the results
3. `AddRoundKey'`

The function in Fig. 20 implements this pipeline.

Our top-level function is defined in Fig. 21; it is identical to the one in the previous section, except that it uses `oneRound128_Tbox_reg` for the "oneRound" function.

In TSIM mode (after place-and-route), this implementation yields a circuit with a rate of one element per cycle that can be clocked between 350 MHz (44.8 Gbps) and 450 MHz (57.6 Gbps), depending on the implementation technology.

To implement this in hardware, it should be lifted into the stream model:

```
Cipher128_Tbox_reg_stream : [inf]([128],[128]) -> [inf][128];
Cipher128_Tbox_reg_stream ins = [| Cipher128_Tbox_reg in
                                 || in <- ins |];
```

```
oneRound128_Tbox_reg : ([4], ([4][4][8],[4][32])) -> ([4][4][8],
    [4][32]);
oneRound128_Tbox_reg(round, (state, key))= (next_state, next_key)
  where {
    state' = if round== 1 then state
             else transpose d;
    next_state = reg(AddRoundKey'(state', key));
    next_key = nextKey128_reg(round, key);

    d = [| reg(if (round == 11)
              then [(t0@1) (t1@2) (t2@3) (t3@0)]
              else reg(t0 ^ t1) ^ reg(t2 ^ t3))
          where {
          t0 = T0 (state @ 0 @ (j+0));
          t1 = T1 (state @ 1 @ (j+1));
          t2 = T2 (state @ 2 @ (j+2));
          t3 = T3 (state @ 3 @ (j+3));
        }
    || j <- [0 .. 3]  |];
  };
```

Fig. 20 Heavily pipelined round of AES-128, using T-box


```
Cipher128_Tbox_reg : ([128],[128]) -> [128];
Cipher128_Tbox_reg (key, pt) = stateToBlock final_state
  where {
    init = (blockToState pt, keyToWords key);

    rounds = [init] #
                [| oneRound128_Tbox_reg (round, state_and_key)
                || state_and_key <- rounds
                || round <- [1..11] |];
    (final_state, dont_care) = rounds ! 0
  };
```

Fig. 21 Heavily pipelined AES-128, using T-box

5.4 Testing and Verification

In this section, we verify that the implementations in Sects. 5.1 and 5.2 are equivalent to the reference specifications for AES-128 and AES-256.

First, we generate formal models of the reference specification in symbolic mode:

```
:l AES_Revisited.tex


:set symbolic
:fm (Cipher : ([128],[128]) -> [128]) "aes128-spec.fm"
:fm (Cipher : ([256],[128]) -> [128]) "aes256-spec.fm"
```
——————— *interpreter* ———————

Then, we generate formal models for the first implementation in `symbolic` mode and check them against the reference specification. Because the implementation is much more efficient than the spec, we will check all further implementations against the first implementation, rather than against the reference spec.

```
:l aes-impl.tex

:set symbolic
:fm Cipher128 "aes128-impl.fm"
:fm Cipher256 "aes256-impl.fm"


:eq "aes128-impl.fm" "aes128-spec.fm"
:eq "aes256-impl.fm" "aes256-spec.fm"
                        interpreter
```

We also check that the implementation is correct in LLSPIR, FSIM, and `TSIM` modes by issuing the following commands in each of these modes:

```
:eq Cipher128 "aes128-impl.fm"
:eq Cipher256 "aes256-impl.fm"
                     interpreter
```

And we check all other implementations against the first one in symbolic, LL-SPIR, FSIM, and TSIM modes:

```
:eq Cipher128' "aes128-impl.fm"
:eq Cipher256' "aes256-impl.fm"
                    interpreter
```

```
:eq Cipher128_seq "aes128-impl.fm"
:eq Cipher256_seq "aes256-impl.fm"
                     interpreter
```

5.5 *Performance*

This section summarizes the performance obtained by the AES implementations presented previously in this section. We present numerical results for logic utilization, latency, clockrate, and throughput.

The following implementations have been provided in this paper:

1. Section 5.1: `aes-impl.tex` – Unrolled; when using Block RAMs, automatically pipelined to one pipeline stage per round
2. Section 5.2: `aes-imp2.tex` – Using a "oneRound" function, combined key expansion and cipher functions; unrolled; and pipelined when using Block RAMs

    (a) Section 5.2.1: `aes-seq.tex` – Using the `seq` pragma to reuse one round
    (b) Section 5.2.2: `aes-reg.tex` – Pipelined using the `reg` pragma

Version	Logic utilization			Performance						
	LUTs	Flip-flops	BRAMs	Latency (cycles)	Rate (cycles)	Clockrate (MHz) FSIM	TSIM	Throughput (Gbps) FSIM	TSIM
(1) 128	52158	0	0	1	1	26.0	X	3.3	X
	4530	1280	100	11	1	203.0	144.4	26.0	18.5
(1) 256	50343	0	0	1	1	20.0	X	2.6	X
	6132	3456	138	15	1	208.1	150.0	26.6	19.2
(2a) 128	10565	498	0	11	11	109.2	100.3	1.3	1.2
	1405	457	10	12	11	168.5	110.2	2.0	1.3
(2a) 256	10919	670	0	15	15	110.6	100.2	0.9	0.9
	1565	457	10	16	15	168.5	150.9	1.4	1.3
(3) 128	4096	1280	100	11	1	213.2	156.0	27.3	20.0
(3a) 128	1134	461	10	12	11	179.2	143.5	2.09	1.67
(3b) 128	6336	12576	100	53	1	385.5	370.0	49.3	47.4

Fig. 22 Performance results for AES implementations

3. Section 5.3: `aes-tbox.tex` – Replacing each round of encryption from `aes-imp2.tex` with a T-Box implementation (AES-128 only); unrolled; and pipelined when using Block RAMs

   (a) Section 5.3.1: `aes-tbox-seq.tex` – Using the `seq` pragma to reuse one round

   (b) Section 5.3.2: `aes-tbox-reg.tex` – Pipelining the T-Box implementation using the `reg` pragma

The performance results for some of these implementations are provided in Fig. 22.

When not using Block RAMs in an unrolled AES, the backend tools sometimes run out of memory during synthesis and always crash before finishing place-and-route. This is indicated by "X" in the results in the table.

Both FSIM (synthesis only) and TSIM (synthesis with place-and-route) results are presented. All performance results are obtained by synthesizing for a specific FPGA part. This part has 63,168 slices, each with two LUTs and two flip-flops, for a total of 126,336 LUTs and 126,336 flip-flops, and 552 Block RAMs. This choice greatly influences the results. Some other FPGA parts yield smaller and/or faster implementations. The effort level (`fpga_optlevel`) is at its default, minimal setting. A higher effort level could yield better performance in TSIM mode.

6 Conclusion

We have introduced the types and constructs of the Cryptol language, as well as the Cryptol interpreter, and in Sect. 3 we have provided several examples of how to use the language and toolchain to specify, implement, refine, and verify hardware circuits. We then used these techniques to produce and refine several implementations of AES, including implementations with very little resource utilization and pipelined implementations with very high throughput.

The LLSPIR compiler allows us to view the results of optimizations, including the latency, rate, estimated area utilization, and estimated clockrate, without performing lengthy synthesis. This greatly reduces the time to produce new refinements.

Once the user has formulated a circuit the right way – as an appropriate sequence comprehension – the `seq`, `par`, and `reg` pragmas provide simple, effective ways of making space–time tradeoffs and implementing pipelines without sacrificing the high-level, readable format of the source code.

In the future, the Cryptol team plans to increase the accuracy of profiling in LLSPIR, support more hardware architectures, and implement automated and guided pipelining.

References

1. Daemen J, Rijmen V (2000) The block cipher rijndael. In: Quisquater J-J, Schneier B (eds) Smart card research and applications, LNCS 1820. Springer, Berlin, pp 288–296
2. Galois, Inc. (2008a) Cryptol programming guide. http://www.galois.com/files/Cryptol/Cryptol_Programming_Guide.pdf
3. Galois, Inc. (2008b) From cryptol to FPGA: a tutorial. http://www.galois.com/files/Cryptol/Cryptol_Tutorial.pdf
4. National Institute of Standards and Technology (2001) Federal information processing standards publication 197: advanced encryption standard (AES). http://csrc.nist.gov/publications/fips/fips197/fips-197.pdf

Verifying Pipelines with BAT

Panagiotis Manolios and Sudarshan K. Srinivasan

1 Introduction

In this chapter, we show how to use the Bit-level Analysis Tool (BAT) [4, 20–22] for hardware verification. The BAT system has been used in the analysis of systems ranging from cryptographic hash functions to machine code to biological systems to large component-based software systems [18, 23, 24], but here we focus on one application: verification of pipelined hardware systems. This chapter brings together results from previous work in a self-contained way, and is intended as a starting point for someone who is interested in using automatic formal verification tools to prove the correctness of hardware or low-level software. The structure and examples in this chapter are based on previous work by the authors that showed how to use the ACL2 theorem proving system [8] to model and verify pipelined machines [12].

Hardware systems are ubiquitous and are an integral part of safety-critical and security-critical systems. Ensuring the correct functioning of hardware is therefore of paramount importance as failure of deployed systems can lead to loss of life and treasure. A well known example is the bug that was found in the floating point division (FDIV) unit of the Intel Pentium processor and that led to a 475 million dollar write-off by Intel. Estimates show that a similar bug in the current generation of Intel processors would cost the processor company about 12 billion dollars [1].

One of the key optimizations used in hardware systems is pipelining. Pipelining is used extensively in hardware designs, including both mainstream and embedded microprocessor designs, multi-core systems, cache coherence protocols, memory interfaces, etc. Therefore, the verification of pipelines is an important, ubiquitous problem in hardware verification and has received a lot of attention from the research community.

P. Manolios (✉)
Northeastern University, Boston, MA, USA
e-mail: pete@ccs.neu.edu

D.S. Hardin (ed.), *Design and Verification of Microprocessor Systems for High-Assurance Applications*, DOI 10.1007/978-1-4419-1539-9_5,
© Springer Science+Business Media, LLC 2010

Pipelines are essentially assembly lines. Just like it is much more efficient to build cars using an assembly line, it is also much more efficient to break up the execution of processor instructions into well-defined stages, e.g., fetch, decode, and execute. In this way, at any point in time there can be multiple instructions being executed simultaneously, in parallel and in various stages of completion. Furthermore, in order to extract maximum performance from pipelines, synchronization between the various instructions being executed in parallel is required. This synchronization between instructions, memories, and register files is provided by complex pipeline controllers. This added complexity makes the design and verification of pipelines a challenging problem.

We use the BAT system [22] for pipelined machine verification for several reasons. The BAT specification language [21] is designed as a synthesizable HDL with formal semantics and can therefore be used to construct bit-level pipelined machine models amenable to formal analysis. The decision procedure incorporated in BAT includes a memory abstraction algorithm and memory rewriting techniques and can therefore deal with verification problems that involve large memories [20]. Also, the BAT decision procedure uses an efficient circuit to CNF compiler, which drastically improves efficiency [4, 19].

The notion of correctness that we use for pipelined machines is based on Well-founded Equivalence Bisimulation (WEB) refinement [10, 11]. There are several attractive properties of refinement. The instruction set architecture (ISA) is used as the specification. Both safety and liveness are accounted for. The refinement map (a function used to relate states of the pipelined machine with states of its ISA) is a parameter of the framework and can therefore be studied and optimized to improve efficiency [7, 13, 15, 16]. Refinement is a compositional notion, a property that can be exploited to deal with scalability issues [14].

The rest of the chapter is organized as follows. Section 2 describes the BAT system, including the BAT specification language and the BAT decision procedure. Section 3 describes a three-stage pipelined machine example and its ISA, and also shows how to model these machines using BAT. In Sect. 4, we provide an overview of the notion of correctness we use, which is based on refinement. Section 5 shows how to verify pipelines with the BAT system, using the example of the three-stage pipeline. Section 6 provides an overview of techniques to cope with the efficiency and scalability issues that arise when reasoning about more complex pipelined systems. Conclusions are given in Sect. 7.

2 Bit-Level Analysis Tool

The BAT is a system for solving verification problems arising from hardware, software, and security. BAT is designed to be used as a bounded model checker and k-induction engine for register transfer level (RTL) models. At the core of the system is a decision procedure for quantifier-free formulas over the extensional theory of fixed-size bit-vectors and fixed-size bit-vector arrays (memories).

BAT also incorporates a specification language that can be used to model hardware designs at the word-level and to express linear temporal logic (LTL) properties. In this section, we describe the BAT specification language and provide a brief overview of the BAT decision procedure.

2.1 BAT Specification Language

The BAT specification language is strongly typed and includes a type inference algorithm. BAT takes as input a machine description and LTL specification, and tries to either find a counterexample requiring no more steps than a user provided upper bound, or tries to prove no such counterexample exists. While BAT accepts various file formats, a commonly used format for the machine specification requires the following four sections: `:vars`, `:init`, `:trans`, and `:spec`. These correspond to the declaration of the variables making up the machine state, a Boolean formula describing valid initial states, a Boolean formula describing the transition relation, and an LTL formula giving the desired formula, respectively. In this section, we describe the main features of the language. For a complete description, see the BAT Web page [21].

2.1.1   Data Types

The BAT language is strongly typed. Variables are either bit-vectors or memories. The `:vars` section is a list of variable declarations that specify the types of each variable. Each variable declaration is either (1) A symbol corresponding to the variable name, in which case the variable is a bit-vector of one bit (e.g., x). (2) A list with two elements, a variable name and a positive integer, in which case the variable is a bit-vector of the given number of bits (e.g., `(x 4)` is a bit-vector of 4 bits). (3) A list with three elements, a variable name and two positive integers, specifying that the variable is a memory with the given word size and number of words (e.g., `(x 8 4)` is a memory with eight 4-bit words).

A `:vars` section then looks like this: `(:vars (x 2) y (z 8 16))`. In addition to variables, there are bit-vectors and integer constants. Bit-vectors can be given in binary, hex, or octal. For example, numbers in binary start with `0b` and are followed by an arbitrary sequence of 0s and 1s.

Integers are represented by signed bit-vectors. The size of the bit-vector is determined by BAT's type-inferencing mechanism. The appropriate size is determined by the context in which the integer is used. For example, if x is a 4-bit bit-vector, then if we bitwise-and it with 3, it is written as `(and x 3)`. Then in this context, 3 is represented by the bit-vector `0b0011`, since bit-vectors that are bitwise-anded together must be of the same type. The only restriction in this case is that the integer must be representable in signed binary notation (2's complement) in the number of bits dictated by the context.

2.1.2 Primitives

BAT supports primitives for Boolean, arithmetic, and memory operations. All the basic bitwise Boolean functions are provided. The functions `and`, `or`, and `xor` all take an arbitrary number of arguments and perform the appropriate operations. In addition, `->` (implication), and `<->` (iff) take exactly two arguments. The `not` function takes exactly one argument. All of these functions take bit-vectors of the same size, and return a bit-vector of that size.

Arithmetic operations include `=`, `<`, `>`, `<=` (less than or equal to), `>=` (greater than or equal to), `add`, `sub`, `inc`, and `dec`. BAT contains bit-vector related functions as well. These include different kind of shift and rotate operations, concatenation, and (signed) extension. For example, the `cat` function concatenates bit-vectors, returning a bit-vector with size equal to the sum of the inputs to the `cat` function. The most significant bits are to the left so the earlier arguments to the `cat` formula are more significant than the later arguments.

Memories have to be treated with care because the obvious translation that converts formulas involving memories to propositional logic leads to an exponential blow-up. The BAT system introduced a decision procedure for memories that leads to greatly reduced SAT problems [20]. The memory-specific BAT functions are `get` and `set`. The `get` function takes a memory and a bit-vector and returns the word of the memory addressed by the bit-vector. The `set` function takes a memory and two bit-vectors. It returns a memory equivalent to the original memory except that the word addressed by the first bit-vector is set to the value of the second bit-vector. In both cases the size of the addresses must be equal to the ceiling of the log of the number of words in the memory, and in the case of the `set` the size of the last argument must be equal to the words size of the memory. Memories can be directly compared for equality using `=` (type checking makes sure that they have the same type, i.e., that they have the same word size and the same number of elements). In a similar way, they type of an `if` can be a memory (type checking again checks that the then and else cases have the same type).

2.1.3 Expressions

BAT supports several constructs to build bit-vector and bit-vector memory expressions. Conditional statements include `if` and `cond`. The `if` statement takes three arguments: the first is the test and must be a 1-bit bit-vector. The second and third arguments are the then and else clauses respectively and must be the same type. Cond statements are convenient for expressing a series of `if` statements. For example, a `cond` statement that returns `-1` if $x < y$, 1 if $x > y$ and 0 otherwise is shown below:

```
(cond ((< x y) -1)
      ((> x y) 1)
      (0b1     0)))
```

BAT provides a way to return multiple values from an expression (this becomes helpful in conjunction with user-defined functions). This is done simply by wrapping a sequence of values in an mv form:

```
(mv (+ a b) (set m x y))
```

This returns both the sum and the result of the set form.

The most complex construct of the BAT language is local. In its simplest form, it operates like a let* in Lisp. The following implementation of an ALU slice demonstrates one of the more complex abilities of the local.

```
(local ((nb (xor bnegate b))
        (res0 (and a nb))
        (res1 (or a nb))
        (((cout 1) (res2 1)) (fa a nb cin)))
   (cat cout (mux-4 res0 res1 res2 lu op))))
```

Here, the last binding binds variables cout and res2 simultaneously. It declares each to be 1 bit, and binds them to the 2-bit output of the fa function (a user-defined function). This splits up the output of the fa function between cout and res2 according to their sizes. Another feature of the local is illustrated by the following.

```
(local ((c 2))
       (((t0 (c 0))
         (alu-slice (a 0) (b 0) bnegate bnegate op))
        ((t1 (c 1))
         (alu-slice (a 1) (b 1) t0 bnegate op))
        (zero (= c 0)))
    (cat t1 c zero))
```

Here an extra argument appears at the beginning of the local. This is a list of bit-vector variable declarations. The idea is that these variables can be bound by bits and pieces through the bindings. The first binding binds several values, as in the last example. However, in this example the second value being bound is not a variable, but a bit of the variable, c, declared in the first argument to the local. Likewise, the other bit of c is set in the second binding. It is also possible to set a sequence of bits in a similar way by giving two integers: ((c 0 1) (and a b)).

Finally, it is possible to set multiple values to the result of an mv form:

```
(local ((aa mm) (mv (inc a) (set m a b)))
   (set mm c aa))
```

Here the types of the variables being bound are inferred from the type of the mv form.

2.1.4 User-Defined Functions

In addition to the `:vars`, `:init`, `:trans`, and `:spec` sections of a specification, the user can define his or her own functions in the `:functions` section. Consider the following example.

```
(:functions
 (alu-output
  (32)
  ((op 4) (val1 32) (val2 32))
  (cond ((= op 0) (bits (+ val1 val2) 0 31))
        ((= op 1) (bits (- val1 val2) 0 31))
        (1b1 (bits (and val1 val2) 0 31)))))
```

The functions section takes a list of function definitions. In this example, we define one function. A function definition starts with the function name. Our function is called `alu-output`. The second element in the definition is the type. This is a list containing one positive integer for a bit-vector function (`alu-output`, for example returns a 32-bit bit-vector), two positive integers if the return type is a memory, and a list of one integer list and two integer lists if multiple values are returned. For example `((1) (8 4))` would specify that the function returns a 1-bit bit-vector and a memory with eight 4-bit words. The third part of a function definition is a list of its arguments. This is just a list of variable definitions just like the ones in the `:vars` section. In the case of `alu-output`, the inputs are `op` (a 4-bit bit-vector), `val1` (a 32-bit bit-vector), and `val2` (another 32-bit bit-vector). The final element of a function definition is the function body. Its return type must be compatible with that of the function.

2.1.5 Specification Formats

BAT takes specifications in one of the three formats. The first is a machine description for bounded model checking. A file in this format contains three items. The first is the keyword ":machine" (without the quotes). The second is the machine description (described above). The third is a natural number which represents the number of steps you want BAT to check the property for.

The other two formats are very similar. They are used to check if a formula holds for some values of the variables (existential), or if a formula holds for all values of the variables (universal). These files contain four items. The first is either ":exists" or ":forall" (without the quotes). The next is a list of variable declarations for the formula. The third is a list of function definitions for use in the formula (this can be ( ) if there are no functions). The final argument is the formula itself, which is over the variables and functions declared earlier in the file.

For examples of all these formats, see the BAT Web page [21].

2.1.6 Other Language Features

Since the BAT language is an s-expression based language implemented in Lisp, it is easy to develop parametrized models. We routinely use Lisp functions that take in a set of input parameters and generate BAT models.

BAT also has support for defining constants, LTL temporal operators, and a number of other primitive operators not discussed here. We point the reader to the BAT Web page for detailed documentation on the BAT specification language [21].

2.2 BAT Decision Procedure

As we saw in the earlier section, a BAT specification includes a model and a property about the model that BAT attempts to verify. The BAT decision procedure translates the input specification to a Boolean formula in conjunctive normal form (CNF). The CNF formula is then checked using a SAT solver. In the common case, where we are checking validity, if the CNF formula is found to be unsatisfiable, then this corresponds to a formal proof that the user-provided property is valid. If the CNF formula is satisfiable, then the satisfying assignment is used to construct a counterexample for the input property.

The translation from the input specification to CNF is performed using four high-level compilation steps and is based on a novel data structure for representing circuits known as the NICE dag, because it is a dag that contains Negations, Ites (If–Then–Else operators), Conjunctions, and Equivalences [4]. In the first step, functions are inlined, constants are propagated, and a range of other simplifications are performed. The output of the first step is a NICE dag that also includes `next` operators, memory variables, and memory operators. In the second step, the transition relation is unrolled for as many steps as specified by the specification. This eliminates the `next` operators, resulting in a NICE dag with memory variables and memory operators. In the third step, BAT uses its own decision procedure for the extensional theory of arrays to reduce memories [20], which are then eliminated by replacing memory variables and memory operators with Boolean circuits, resulting in a NICE dag. In the fourth step, the NICE dag is translated to a SAT problem in CNF format.

2.2.1 Memory Abstraction

BAT incorporates an automatic, sound, and complete memory abstraction algorithm [20]. The algorithm allows BAT to handle verification problems that involve models with large memories, but with correctness properties that include only a small number of memory references. The verification of pipelined microprocessor models is an example of such verification problems.

The key idea of the abstraction algorithm is to reduce the size of a memory to a size that is comparable to the number of unique accesses (both read and write) to that memory. The insight here is that if in a correctness property, there are only 10 unique accesses to a memory with say 2^{32} words, it is enough to reason about a reduced version of the memory whose resulting size is just larger than 10, to check the property. Therefore, the original memory size can be drastically reduced. Note however that care has to taken when performing the reduction because a memory access could be a symbolic reference, i.e., an access that could reference any one of a large number of words in the memory. Another complication is that we allow memories to be directly compared in any context, i.e., we have to support an extensional theory of arrays.

The efficiency of memory abstraction depends on the size of the reduced memories, which in turn depends on the number of unique memory access. However, because of nested memory operations, it is often hard to determine if two different memory references correspond to the same symbolic reference. To improve the efficiency of the abstraction, BAT incorporates automated term-rewriting techniques employing a number of rewrite rules that are used to simplify expressions with memory operators. The simplifications performed by rewriting help to identify equivalent memory references thereby improving the efficiency of memory abstraction.

2.2.2   Efficient Translation To CNF

CNF generation can significantly affect SAT solving times. BAT introduced a new linear-time CNF generation algorithm, and extensive experiments, have shown that our algorithm leads to faster SAT solving times and smaller CNF than existing approaches. Our CNF generation algorithm is based on NICE dags, which subsume and-inverter graphs (AIGs) and are designed to provide better normal forms at linear complexity. The details are beyond the scope of this chapter, but are described in detail elsewhere [4].

3   ISA and Pipelined Machine Models

In this section, we show how to model a simple instruction set architecture and a three-stage pipelined implementation of this instruction set architecture using the BAT specification language. We start by defining ISA, a sequential machine that directly implements the instruction set architecture. We then define MA, a three-stage pipelined implementation (the microarchitecture machine). As stated previously, the models are based on our previous work on using ACL2 for hardware verification [12]. Those models in turn are based on Sawada's simple machine [27] and our subsequent related machines [9].

The instructions in the ISA have four components, including an opcode, a destination register, and two source registers. The pipelined MA machine is shown

Fig. 1 Our simple
three-stage pipelined machine

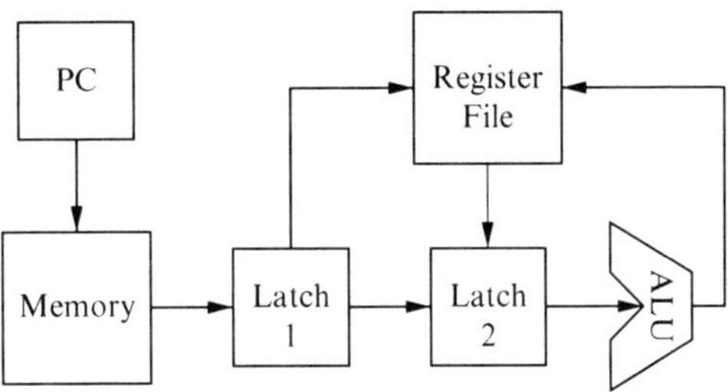

in Fig. 1. The functionality of the ISA is split into three stages so that each of the stages can operate in parallel on different instructions. Registers, known as pipeline latches, are used to separate the stages. The pipeline latches hold the intermediate results generated in a stage. The MA machine has two pipeline latches, latch 1 and latch 2 as shown in the figure. The three stages of our MA machine are fetch, set up, and write. In the fetch stage, an instruction is fetched from memory using the program counter as the address, and is stored in latch 1. In the set-up stage, the source operands are retrieved from the register file and stored in latch 2, along with the rest of the instruction. In the write stage, the appropriate operation is performed by the ALU (arithmetic and logic unit), and the result of the ALU operation is stored in the destination register specified by the destination address of the instruction.

Consider a simple example, where the contents of the memory is as follows.

Inst

0	add	rb	ra	ra
1	add	ra	rb	ra

The following traces are obtained when the two-line code segment is executed on the ISA and MA machines. Note that we only show the values of the program counter and the contents of registers ra and rb.

The rows correspond to steps of the machines, e.g., row Clock 0 corresponds to the initial state, Clock 1 to the next state, and so on. The ISA and MA columns contain the relevant parts of the state of the machines: a pair consisting of the Program Counter (PC) and the register file (itself a pair consisting of registers ra and rb). The final two columns indicate what stage the instructions are in (only applicable to the MA machine).

The PC in the initial state (in row Clock 0) of the ISA machine is 0. The values of the registers ra and rb are 1. The next state of the ISA machine (row Clock 1) is obtained after executing instruction "Inst 0." In this state, the PC is incremented to 1, and the sum of the values stored in registers ra and rb (2) is computed and stored in rb. In the second clock cycle, instruction "Inst 1" is executed. The PC is again incremented to 2. The sum of the values stored in registers ra and rb (3) is computed and stored in ra.

In the initial state of the MA machine, the PC is 0. We assume that the two latches are initially empty. In the first clock cycle, "Inst 0" is fetched and the PC is incremented. In the second clock cycle, "Inst 1" is fetched, the PC is incremented again,

Clock	ISA	MA	Inst 0	Inst 1
0	(0, (1,1))	(0, (1,1))		
1	(1, (1,2))	(1, (1,1))	Fetch	
2	(2, (3,2))	(2, (1,1))	Set–up	Fetch
3		(2, (1,2))	Write	Stall
4		(−, (1,2))		Set–up
5		(−, (3,2))		Write

and "Inst 0" proceeds to the set-up stage. In the third clock cycle, "Inst 0" completes and updates register rb with the correct value (as can be seen from the MA column). However, during this cycle, "Inst 1" cannot proceed, as it requires the rb value computed by "Inst 0," and therefore is stalled and remains in the fetch stage. In the next clock cycle, "Inst 1" moves to set-up, as it can obtain the the rb value it requires from the register file, which has now been updated by "Inst 0." In the fifth clock cycle, "Inst 1" completes and updates register ra.

3.1 ISA Definition

We now consider how to define the ISA and MA machines using BAT. The first machine we define is a 32-bit ISA, i.e., the data path is 32 bits. The main function is isa-step, a function that steps the ISA machine, i.e., it takes an ISA state and returns the next ISA state. The definition of isa-step follows.

```
(isa-step
 ((32) (4294967296 32)
  (4294967296 100) (4294967296 32))
 ((pc 32) (regs 4294967296 32)
  (imem 4294967296 100) (dmem 4294967296 32))
 (local
  ((inst (get imem pc))
   (op (opcode inst))
   (rc (dest-c inst))
   (ra (src-a inst))
   (rb (src-b inst)))
  (cond ((= op 0) (isa-add rc ra rb pc regs imem dmem))
        ;; REGS[rc] := REGS[ra] + REGS[rb]
        ((= op 1) (isa-sub rc ra rb pc regs imem dmem))
        ;; REGS[rc] := REGS[ra] - REGS[rb]
        ((= op 2) (isa-and rc ra rb pc regs imem dmem))
        ;; REGS[rc] := REGS[ra] and REGS[rb]
        ((= op 3) (isa-load rc ra pc regs imem dmem))
        ;; REGS[rc] := MEM[ra]
        ((= op 4) (isa-loadi rc ra pc regs imem dmem))
```

```
         ;; REGS[rc] := MEM[REGS[ra]]
       ((= op 5) (isa-store ra rb pc regs imem dmem))
         ;; MEM[REGS[ra]] := REGS[rb]
       ((= op 6) (isa-bez ra rb pc regs imem dmem))
         ;; REGS[ra]=0 -> pc:=pc+REGS[rb]
       ((= op 7) (isa-jump ra pc regs imem dmem))
         ;; pc:=REGS[ra]
       (1b1 (isa-default pc regs imem dmem)))))
```

In the above function `regs` refers to the register file, `imem` is the instruction memory, and `dmem` is the data memory. The function fetches the instruction from the instruction memory, which is a bit-vector. Then it uses decode functions `opcode`, `dest-c`, `src-a`, and `src-b` to decode the instruction. The opcode is then used to figure out what action to take. For example, in the case of an add instruction, the next ISA state is `(isa-add rc ra rb pc regs imem dmem)`, where `isa-add` provides the semantics of add instructions. The definition of `isa-add` is given below.

```
(isa-add
  ((32)(4294967296 32)
      (4294967296 100) (4294967296 32))
  ((rc 32) (ra 32) (rb 32) (pc 32) (regs 4294967296 32)
    (imem 4294967296 100) (dmem 4294967296 32))
  (mv (bits (+ pc 1) 0 31)
      (add-rc ra rb rc regs)
      imem
      dmem))


(add-rc (4294967296 32)
   ((ra 32) (rb 32) (rc 32) (regs 4294967296 32))
   (set regs
        rc
        (bits (+ (get regs ra) (get regs rb)) 0 31)))
```

Notice that the program counter is incremented and the register file is updated by setting the value of register rc to the sum of the values in registers ra and rb. This happens in function `add-rc`.

The other ALU instructions are similarly defined. We now show how to define the semantics of the rest of the instructions. The semantics of the load instructions are shown next.

```
(isa-loadi
  ((32) (4294967296 32)
   (4294967296 100) (4294967296 32))
  ((rc 32) (ra 32) (pc 32) (regs 4294967296 32)
   (imem 4294967296 100) (dmem 4294967296 32))
```

```
(mv (bits (+ pc 1) 0 31)
    (load-rc (get regs ra) rc regs dmem)
    imem
    dmem))


(load-rc
 (4294967296 32)
 ((ad 32) (rc 32) (regs 4294967296 32)
  (dmem 4294967296 32))
 (set regs rc (get dmem ad)))


(isa-load
 ((32) (4294967296 32)
  (4294967296 100) (4294967296 32))
 ((rc 32) (ad 32) (pc 32) (regs 4294967296 32)
  (imem 4294967296 100) (dmem 4294967296 32))
 (mv (bits (+ pc 1) 0 31)
     (load-rc ad rc regs dmem)
     imem
     dmem))
```

The semantics of the store instruction is given by `isa-store`.

```
(isa-store
 ((32) (4294967296 32)
  (4294967296 100) (4294967296 32))
 ((ra 32) (rb 32) (pc 32) (regs 4294967296 32)
  (imem 4294967296 100) (dmem 4294967296 32))
 (mv (bits (+ pc 1) 0 31)
     regs
     imem
     (store ra rb regs dmem)))


(store
 (4294967296 32)
 ((ra 32) (rb 32) (regs 4294967296 32)
  (dmem 4294967296 32))
 (set dmem (get regs ra) (get regs rb)))
```

Jump and branch instructions follow.

```
(isa-jump
 ((32) (4294967296 32)
```

```
    (4294967296 100) (4294967296 32))
  ((ra 32) (pc 32) (regs 4294967296 32)
   (imem 4294967296 100) (dmem 4294967296 32))
  (mv (bits (get regs ra) 0 31)
      regs
      imem
      dmem))
(isa-bez ((32) (4294967296 32)
 (4294967296 100) (4294967296 32))
  ((ra 32) (rb 32) (pc 32) (regs 4294967296 32)
   (imem 4294967296 100) (dmem 4294967296 32))
  (mv (bez ra rb regs pc)
      regs
      imem
      dmem))


(bez
 (32)
 ((ra 32) (rb 32) (regs 4294967296 32) (pc 32))
 (cond ((= (get regs ra) 0)
        (bits (+ pc (bits (get regs rb) 0 31)) 0 31))
       (1b1 (bits (+ pc 1) 0 31))))
```

No-ops are handled by `isa-default`.

```
(isa-default
 ((32) (4294967296 32)
  (4294967296 100) (4294967296 32))
 ((pc 32) (regs 4294967296 32)
  (imem 4294967296 100) (dmem 4294967296 32))
 (mv (bits (+ pc 1) 0 31)
     regs
     imem
     dmem))
```

3.2 MA Definition

The MA machine is a pipelined machine with three stages that implements the instruction set architecture of the ISA machine. Therefore, the ISA machine can be
thought of as a specification of the MA machine. The MA machine contains a PC, a
register file, a memory, and two pipeline latches. The latches are used to implement

pipelining and stores intermediate results generated in each stage. The first latch
contains a flag which indicates if the latch is valid, an opcode, the target register, and
two source registers. The second latch contains a flag as before, an opcode, the target register, and the values of the two source registers. The definition of ma-step
follows.

```
(ma-step
  ((298) (4294967296 32)
   (4294967296 100) (4294967296 32))
  ((ma 298) (regs 4294967296 32)
   (imem 4294967296 100) (dmem 4294967296 32))
  (mv
   (cat
    (step-latch2 ma regs)
    (step-latch1 ma imem)
    (step-pc ma regs imem))
    (step-regs ma regs dmem)
    imem
    (step-dmem ma dmem)))
```

The ma-step function works by calling functions that given one of the MA
components return the next state value of that component. Note that this is very
different from isa-step, which calls functions, based on the type of the next
instruction, that return the complete next isa state.

Below, we show how the register file is updated. If latch2 is valid, then if we have
an ALU instruction, the output of the ALU is used to update register rc. Otherwise,
if we have a load instruction, then we update register rc with the appropriate word
from memory.

```
(step-regs
  (4294967296 32)
  ((ma 298) (regs 4294967296 32) (dmem 4294967296 32))
    (local
      ((validp (getvalidp2 ma))
       (op (getop2 ma))
       (rc (getrc2 ma))
       (ra-val (getra-val2 ma))
       (rb-val (getrb-val2 ma)))
      (cond ((and validp (alu-opp op))
              (set regs rc (alu-output op ra-val rb-val)))
            ((and validp (load-opp op))
              (set regs rc (get dmem ra-val)))
            (1b1 regs)))))
```

```
(alu-opp
 (1)
 ((op 4))
 (or (= op 0) (= op 1) (= op 2)))

(load-opp
 (1)
 ((op 4))
 (or (= op 3) (= op 4)))

(alu-output
 (32)
 ((op 4) (val1 32) (val2 32))
   (cond ((= op 0) (bits (+ val1 val2) 0 31))
         ((= op 1) (bits (- val1 val2) 0 31))
         (1b1 (bits (and val1 val2) 0 31))))
```

Next, we describe how latch 2 is updated. Latch 2 is invalidated if latch 1 will be stalled or if latch 1 is not valid. Otherwise, we copy the opcode and rc fields from latch1 and read the contents of registers rb and ra, except for load instructions. We use a field pch2 to record the value of the PC value corresponding to the instruction in latch 2. A similar field, pch1, is used in latch 1 to record the PC value corresponding to the instruction in latch 1. Note that the fields pch1 and pch2 do not affect the computation of the machine. They are used as history variables to primarily to aid the proof process.

```
(step-latch2
 (133)
 ((ma 298) (regs 4294967296 32))
   (local ((l1op (getop1 ma)))
    (cond ((= (or (not (getvalidp1 ma))
                  (stall-l1p ma)) 1b1)
              (cat (getpch2 ma)
                   (getrb-val2 ma)
                   (getra-val2 ma)
                   (getrc2 ma)
                   (getop2 ma)
                   1b0))
            (1b1
              (cat (getpch1 ma)
                   (get regs (getrb1 ma))
                   (cond ((= l1op 3) (getra1 ma))
                         (1b1 (get regs (getra1 ma))))
                   (getrc1 ma)
                   l1op
                   1b1)))))
```

Latch 1 is updated as follows. If it is stalled, it retains its previous contents. If it is invalidated, its flag is set to false. Otherwise, the next instruction is fetched from memory and stored in latch 1. The PC of the instruction is stored in pch1. Latch 1 is stalled when the instruction in latch 1 requires a value computed by the instruction in latch 2. Latch 1 is invalidated if it contains any branch instruction (because the jump address cannot be determined yet) or if latch 2 contains a bez instruction (again, the jump address cannot be determined for bez instructions until the instruction has made its way through the pipeline, whereas the jump address for jump instructions can be computed during the second stage of the machine).

```
(step-latch1 (133) ((ma 298)  (imem 4294967296 100))
   (local
     ((latch1 (getlatch1 ma))
      (inst (get imem (getppc ma))))
     (cond ((= (stall-l1p ma) 1b1) latch1)
           ((= (invalidate-l1p ma) 1b1)
            (cat (getpch1 ma)
                 (getrb1 ma)
                 (getra1 ma)
                 (getrc1 ma)
                 (getop1 ma)
                 1b0))
           (1b1
             (cat (getppc ma)
                  (src-b inst)
                  (src-a inst)
                  (dest-c inst)
                  (opcode inst)
                  1b1))))))
```

The function stall-l1p determines when to stall latch 1.

```
(stall-l1p (1) ((ma 298))
  (local
    ((l1validp (getvalidp1 ma))
     (l1op (getop1 ma))
     (l2op (getop2 ma))
     (l2validp (getvalidp2 ma))
     (l2rc (getrc2 ma))
     (l1ra (getra1 ma))
     (l1rb (getrb1 ma)))
    (and l2validp l1validp (rc-activep l2op)
         (or (= l1ra l2rc)
             (and (uses-rbp l1op) (= l1rb l2rc))))))
```

```
(rc-activep (1) ((op 4))
 (or (alu-opp op) (load-opp op)))


(uses-rbp (1) ((op 4))
 (or (alu-opp op) (= op 5) (= op 6)))
```

The function `invalidate-l1p` determines when latch 1 should be invalidated.

```
(invalidate-l1p (1) ((ma 298))
 (local
  ((l1validp (getvalidp1 ma))
   (l1op (getop1 ma))
   (l2op (getop2 ma))
   (l2validp (getvalidp2 ma)))
  (or (and l1validp (or (= l1op 6) (= l1op 7)))
      (and l2validp (= l2op 6)))))
```

Memory is updated only when we have a store instruction, in which case we update the memory appropriately.

```
(step-dmem
 (4294967296 32)
 ((ma 298) (dmem 4294967296 32))
   (local
    ((l2validp (getvalidp2 ma))
     (l2op (getop2 ma))
     (l2ra-val (getra-val2 ma))
     (l2rb-val (getrb-val2 ma)))
    (cond ((= (and l2validp (= l2op 5)) 1b1)
           (set dmem l2ra-val l2rb-val))
          (1b1 dmem)))))
```

Finally, the PC is updated as follows. If latch 1 stalls, then the PC is not modified. Otherwise, if latch 1 is invalidated, then if this is due to a bez instruction in latch2, the jump address can be now be determined, so the program counter is updated as per the semantics of the bez instruction. Otherwise, if the invalidation is due to a jump instruction in latch 1, the jump address can be computed and the program counter is set to this address. The only other possibility is that the invalidation is due to a bez instruction in latch 1; in this case the jump address has not yet been determined, so the pc is not modified. Note, this simple machine does not have a branch predictor. If the invalidate signal does not hold, then we increment the program counter unless we are fetching a branch instruction.

```
(step-pc (32)
 ((ma 298) (regs 4294967296 32) (imem 4294967296 100))
   (local
```

```
((pc (getppc ma))
 (inst (get imem pc))
 (op (opcode inst))
 (l1op (getop1 ma))
 (l2op (getop2 ma))
 (l2validp (getvalidp2 ma))
 (l2ra-val (getra-val2 ma))
 (l2rb-val (getrb-val2 ma)))
 (cond ((stall-l1p ma) pc)
       ((invalidate-l1p ma)
        (cond
          ((and l2validp (= l2op 6))
           (cond
            ((= l2ra-val 0)
             (bits (alu-output 0 pc l2rb-val) 0 31))
            (1b1 (bits (+ pc 1) 0 31))))
          ((= l1op 7)
           (bits (get regs (getra1 ma)) 0 31))
          (1b1 pc)))
       ((or (= op 6) (= op 7)) pc)
       (1b1 (bits (+ pc 1) 0 31))))))
```

4 Refinement

In the previous section, we saw how one can model a pipelined machine and its
instruction set architecture in BAT. We now discuss how to verify such machines.
Consider the partial traces of the ISA and MA machines on the simple two-line code
fragment from the previous section (add rb ra ra followed by add ra rb ra). We
are only showing the value of the program counter and the contents of registers ra
and rb.

ISA	MA		MA		MA
(0, (1,1))	(0, (1,1))		(0, (1,1))		(0, (1,1))
(1, (1,2))	(1, (1,1))	$\longrightarrow$	(0, (1,1))	$\longrightarrow$	(1, (1,2))
(2, (3,2))	(2, (1,1))	Commit	(0, (1,1))	Remove	(2, (3,2))
	(2, (1,2))	PC	(1, (1,2))	Stutter	
	(−, (1,2))		(1, (1,2))		
	(−, (3,2))		(2, (3,2))		

Notice that the PC differs in the two traces and this occurs because the pipeline,
initially empty, is being filled and the PC points to the next instruction to fetch.
If the PC were to point to the next instruction to commit (i.e., the next instruction to
complete), then we would get the trace shown in column 3. Notice that in column 3,

the PC does not change from 0 to 1 until Inst 0 is committed in which case the next instruction to commit is Inst 1. We now have a trace that is the same as the ISA trace except for stuttering; after removing the stuttering we have, in column 4, the ISA trace.

We now formalize the above and start with the notion of a refinement map, a function that maps MA states to ISA states. In the above example we mapped MA states to ISA states by transforming the PC. Proving correctness amounts to relating MA states with the ISA states they map to under the refinement map and proving a WEB. Proving a WEB guarantees that MA states and related ISA states have related computations up to finite stuttering. This is a strong notion of equivalence, e.g., a consequence is that the two machines satisfy the same $\text{CTL}^* \setminus \text{X}$.[1] This includes the class of next-time free safety and liveness (including fairness) properties, e.g., one such property is that the MA machine cannot deadlock (because the ISA machine cannot deadlock).

Why "up to finite stuttering"? Because we are comparing machines at different levels of abstraction: the pipelined machine is a low-level implementation of the high-level ISA specification. When comparing systems at different levels of abstraction, it is often the case that the low-level system requires several steps to match a single step of the high-level system.

Why use a refinement map? Because there may be components in one system that do not appear in the other, e.g., the MA machine has latches but the ISA machine does not. In addition, data can be represented in different ways, e.g., a pipelined machine might use binary numbers whereas its instruction set architecture might use a decimal representation. Yet another reason is that components present in both systems may have different behaviors, as is the case with the PC above. Notice that the refinement map affects how MA and ISA states are related, not the behavior of the MA machine. The theory of refinement we present is based on transition systems (TSs). A TS, $\mathcal{M}$, is a triple $\langle S, \dashrightarrow, L \rangle$, consisting of a set of states, S, a left-total transition relation, $\dashrightarrow\, \subseteq S^2$, and a labeling function L whose domain is S and where $L.s$ (we sometimes use an infix dot to denote function application) corresponds to what is "visible" at state s. Clearly, the ISA and MA machines can be thought of as transition systems (TS).

Our notion of refinement is based on the following definition of stuttering bisimulation [2], where by $fp(\sigma, s)$ we mean that σ is a fullpath (infinite path) starting at s, and by $match(B, \sigma, \delta)$ we mean that the fullpaths σ and δ are equivalent sequences up to finite stuttering (repetition of states).

Definition 1. $B \subseteq S \times S$ is a stuttering bisimulation (STB) on TS $\mathcal{M} = \langle S, \dashrightarrow, L \rangle$ iff B is an equivalence relation and for all s, w such that sBw:

$$\text{(Stb1)} \quad L.s = L.w$$

$$\text{(Stb2)} \quad \langle \forall \sigma : fp(\sigma, s) : \langle \exists \delta : fp(\delta, w) : match(B, \sigma, \delta) \rangle \rangle$$

---

[1] CTL^* is a branching-time temporal logic; $\text{CTL}^* \setminus \text{X}$ is CTL^* without the next-time operator X.

Browne et al. have shown that states that are stuttering bisimilar satisfy the same next-time-free temporal logic formulas [2].

Lemma 1. *Let B be an STB on $\mathcal{M}$ and let sBw. For any* $\mathrm{CTL}^* \setminus \mathrm{X}$ *formula f, $\mathcal{M}, w \models f$ iff $\mathcal{M}, s \models f$.*

We note that stuttering bisimulation differs from weak bisimulation [25] in that weak bisimulation allows infinite stuttering. Stuttering is a common phenomenon when comparing systems at different levels of abstraction, e.g., if the pipeline is empty, MA will require several steps to complete an instruction, whereas ISA completes an instruction during every step. Distinguishing between infinite and finite stuttering is important, because (among other things) we want to distinguish deadlock from stutter.

When we say that MA refines ISA, we mean that in the disjoint union ($\uplus$) of the two systems, there is an STB that relates every pair of states w, s such that w is an MA state and $r(w) = s$.

Definition 2. (STB Refinement) Let $\mathcal{M} = \langle S, \dashrightarrow, L \rangle$, $\mathcal{M}' = \langle S', \dashrightarrow', L' \rangle$, and $r : S \to S'$. We say that $\mathcal{M}$ is a STB refinement of $\mathcal{M}'$ with respect to refinement map r, written $\mathcal{M} \approx_r \mathcal{M}'$, if there exists a relation, B, such that $\langle \forall s \in S :: sBr.s \rangle$ and B is an STB on the TS $\langle S \uplus S', \dashrightarrow \uplus \dashrightarrow', \mathcal{L} \rangle$, where $\mathcal{L}.s = L'.s$ for s an S' state and $\mathcal{L}.s = L'(r.s)$ otherwise.

STB refinement is a generally applicable notion. However, since it is based on bisimulation, it is often too strong a notion and in this case refinement based on stuttering simulation should be used (see [10, 11]). The reader may be surprised that STB refinement theorems can be proved in the context of pipelined machine verification; after all, features such as branch prediction can lead to non-deterministic pipelined machines, whereas the ISA is deterministic. While this is true, the pipelined machine is related to the ISA via a refinement map that hides the pipeline; when viewed in this way, the nondeterminism is masked and we can prove that the two systems are stuttering bisimilar (with respect to the ISA visible components).

A major shortcoming of the above formulation of refinement is that it requires reasoning about infinite paths, something that is difficult to automate [26]. In [10], WEB-refinement, an equivalent formulation is given that requires only local reasoning, involving only MA states, the ISA states they map to under the refinement map, and their successor states.

Definition 3. $B \subseteq S \times S$ is a WEB on TS $\mathcal{M} = \langle S, \dashrightarrow, L \rangle$ iff:

    (1) B is an equivalence relation on S; and

    (2) $\langle \forall s, w \in S :: sBw \Rightarrow L(s) = L(w) \rangle$; and

    (3) There exist functions $erankl : S \times S \to \mathbb{N}$, $erankt : S \to W$,

such that $\langle W, \lessdot \rangle$ is well-founded, and

$$\langle \forall s, u, w \in S :: sBw \ \land \ s \dashrightarrow u \ \Rightarrow$$

    (a) $\langle \exists v :: w \dashrightarrow v \ \land \ uBv \rangle \ \lor$

    (b) $(uBw \ \land \ erankt(u) \lessdot erankt(s)) \ \lor$

    (c) $\langle \exists v :: w \dashrightarrow v \ \land \ sBv \ \land \ erankl(v, u) < erankl(w, u) \rangle \rangle$

We call a pair $\langle rank, \langle W, \lessdot \rangle \rangle$ satisfying condition 3 in the above definition, a well-founded witness. The third WEB condition guarantees that related states have the same computations up to stuttering. If states s and w are in the same class and s can transit to u, then one of the following holds:

1. The transition can be matched with no stutter, in which case, u is matched by a step from w.
2. The transition can be matched but there is stutter on the left (from s), in which case, u and w are in the same class and the rank function decreases (to guarantee that w is forced to take a step eventually).
3. The transition can be matched but there is stutter on the right (from w), in which case, there is some successor v of w in the same class as s and the rank function decreases (to guarantee that u is eventually matched).

To prove a relation is a WEB, note that reasoning about single steps of $\dashrightarrow$ suffices. In addition, we can often get by with a rank function of one argument.

Note that the notion of WEB refinement is independent of the refinement map used. For example, we can use the standard flushing refinement map [3], where MA states are mapped to ISA states by executing all partially completed instructions without fetching any new instructions, and then projecting out the ISA visible components. In previous work, we have explored the use of other refinement maps, e.g., in [7, 15, 16], we present new classes of refinement maps that can provide several orders of magnitude improvements in verification times over the standard flushing-based refinement maps. In this paper, however, we use the commitment refinement map, introduced in [9].

A very important property of WEB refinement is that it is compositional, something that we have exploited in several different contexts [14, 17].

Theorem 1. *(Composition) If* $\mathcal{M} \approx_r \mathcal{M}'$ *and* $\mathcal{M}' \approx_q \mathcal{M}''$ *then* $\mathcal{M} \approx_{r;q} \mathcal{M}''$.

Above, $r; q$ denotes composition, i.e., $(r; q)(s) = q(r.s)$.

From the above theorem we can derive several other composition results; for example:

Theorem 2. *(Composition)*

$$\frac{\text{MA} \approx_r \cdots \approx_q \text{ISA} \qquad \text{ISA} \parallel P \vdash \varphi}{\text{MA} \parallel P \vdash \varphi}$$

In this form, the above rule exactly matches the compositional proof rules in [5]. The above theorem states that to prove MA $\|$ P $\vdash$ φ (that MA, the pipelined machine, executing program P satisfies property φ, a property over the ISA visible state), it suffices to prove MA $\approx$ ISA and ISA $\|$ P $\vdash$ φ: that MA refines ISA (which can be done using a sequence of refinement proofs) and that ISA, executing P, satisfies φ. That is, we can prove that code running on the pipelined machine is correct, by first proving that the pipelined machine refines the instruction set architecture and then proving that the software running on the instruction set – not on the pipelined machine – is correct.

5 Verification

This section describes how BAT is used to verify the three-stage pipelined machine given in Sect. 3. Note that the definition of WEBs given in Sect. 4 cannot be directly expressed in the BAT specification language. Therefore, we first strengthen the WEB refinement proof obligation such that we obtain a statement that is expressible as a quantifier-free formula over the extensional theory of fixed-size bit-vectors and fixed-size bit-vector arrays (memories), the kind of formulas that BAT decides.

We first define the equivalence classes of B to consist of an ISA state and all the MA states whose image under the refinement map r is the ISA state. As a result, condition 2 of the WEB refinement definition clearly holds. Since an ISA machine never stutters with respect to the MA machine, the second disjunct of the third condition in the WEB definition can be ignored. Also, the ISA machine is deterministic, and the MA machine if not deterministic, can be transformed to a deterministic machine using oracle variables [11]. Using these simplifications and after some symbolic manipulation, Condition 3 of the WEB definition can be strengthened to the following core refinement-based correctness formula, where *rank* is a function that maps states of MA into the natural numbers.

$$\langle \forall w \in \text{MA} :: \langle \forall s, u, v :: \quad s = r(w) \ \wedge \ u = \text{ISA-step}(s) \ \wedge$$
$$v = \text{MA-step}(w) \ \wedge \ u \neq r(v)$$
$$\implies \quad s = r(v) \ \wedge \ rank(v) < rank(w) \rangle\rangle$$

The correctness formula shown above is also depicted in Fig. 2. In the formula above, if MA is the set of all reachable MA states, MA-step is a step of the MA machine, and ISA-step is a step of the ISA machine, then proving the above formula guarantees that the MA machine refines the ISA machine. In the formula above, w is an MA state and v (also an MA state) is a successor of w. s is an ISA state obtained by applying the refinement map r to w and u (also an ISA state) is a successor of s. The formula states that if applying the refinement map r to v does not result in the ISA state u, then $r(v)$ must be equal to s and the rank of v should decrease w.r.t. the rank of w. Also, the proof obligation relating s and v can

Fig. 2 Diagram shows
the core theorem

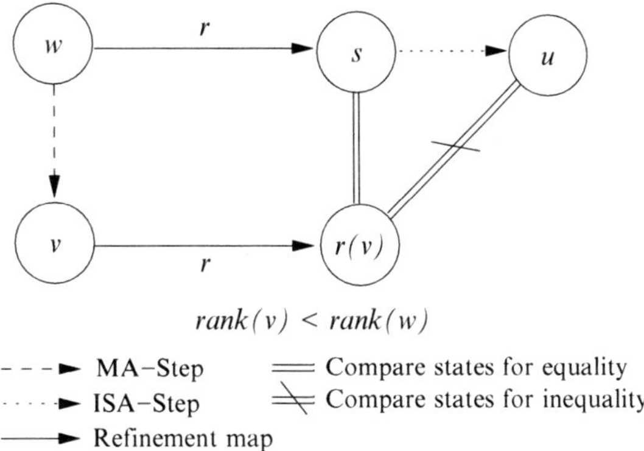

be thought of as the safety component, and the proof obligation $rank(v) < rank(w)$ can be thought of as the liveness component.

If the ISA and MA models are described at the bit-level, then the core refinement-based correctness formula relating these models is in fact expressible in the logic that BAT decides.

5.1 Refinement Map Definitions

To check the core refinement-based correctness formula using BAT, two witness functions are required, a refinement map and a rank function. There are many different ways in which these witness functions can be defined. In this section, we describe one approach.

The following function is a recognizer for "good" MA states.

```
(good-ma (1)
 ((ma 298) (regs 4294967296 32) (imem 4294967296 100)
  (dmem 4294967296 32))
 (local
  (((nma nregs nimem ndmem)
    (committed-ma ma regs imem dmem))
   ((nma1 nregs1 nimem1 ndmem1)
    (ma-step nma nregs nimem ndmem))
   ((nma2 nregs2 nimem2 ndmem2)
    (ma-step nma1 nregs1 nimem1 ndmem1)))
  (cond ((getvalidp2 ma)
         (equiv-ma nma2 nregs2 nimem2 ndmem2
                   ma regs imem dmem))
        ((getvalidp1 ma)
         (equiv-ma nma1 nregs1 nimem1 ndmem1
                   ma regs imem dmem))
        (1b1 1b1)))))
```

The "good" MA states (also known as reachable states) are states that are reachable from the reset states (states in which the pipeline latches are empty). The reason for using a recognizer for reachable states is that unreachable states can be inconsistent and interfere with verification by raising spurious counterexamples. A state in which a pipeline latch has an add instruction, when there are no add instructions in memory is an example of an inconsistent unreachable state. We check for reachable states by stepping the *committed state*, the state obtained by invalidating all partially completed instructions and altering the program counter so that it points to the next instruction to commit.

```
(committed-ma
 ((298) (4294967296 32)
  (4294967296 100) (4294967296 32))
 ((ma 298) (regs 4294967296 32) (imem 4294967296 100)
  (dmem 4294967296 32))
 (local ((inst (get imem (getppc ma)))))
 (mv
  (cat
    (getpch2 ma) (getrb-val2 ma)
    (getra-val2 ma) (getrc2 ma)
    (getop2 ma) 1b0
    (getppc ma) (src-b inst)
    (src-a inst) (dest-c inst)
    (opcode inst) 1b0
    (committed-pc ma))
   regs imem dmem)))
```

The program counter (PC) of the committed state is the PC of the instruction in the first valid latch. Each latch has a history variable that stores the PC value corresponding to the instruction in that latch. Therefore, the PC of the committed state can be obtained from the history variables.

```
(committed-pc (32) ((ma 298))
 (cond ((getvalidp2 ma) (getpch2 ma))
       ((getvalidp1 ma) (getpch1 ma))
       (1b1 (getppc ma))))
```

The equiv-MA function is used to check if two MA states are equal. Note however that if latch 1 in both states are invalid, then the contents of latch 1 in both states are not compared for equality. Latch 2 is also compared similarly.

The committed-MA function invalidates partially executed instructions in the pipeline and essentially rolls back the program counter to correspond with the next instruction to be committed. The consistent states of MA are determined by checking that they are reachable from the committed states within two steps. The refinement map is defined as follows.

```
(ma-to-isa
 ((32) (4294967296 32)
  (4294967296 100) (4294967296 32))
```

```
((ma 298) (regs 4294967296 32)
 (imem 4294967296 100) (dmem 4294967296 32))
(local (((nma nregs nimem ndmem)
          (committed-ma ma regs imem dmem)))
 (mv (getppc nma) nregs nimem ndmem)))
```

We also need a rank function to check for liveness, which is given by the
ma-rank function. Note that this rank function is designed to work with the refinement map we defined. If another refinement map is used, then another rank may
be required. ma-rank defines the rank of an MA state as the number of steps required to reach a state in which MA is ready to commit an instruction. If latch 2 is
valid, an instruction will be committed in the next step. If latch 2 is invalid and latch
1 is valid, MA will commit an in two steps. If both latches are invalid, then then MA
should commit an instruction in three steps.

```
(ma-rank (3) ((ma 298))
 (cond ((getvalidp2 ma) 0)
       ((getvalidp1 ma) 1)
       (1b1 2))))
```

Now, we can state the core theorem for the 3-stage pipelined machine, which is
given by the function commitment-theorem.

```
(commitment-theorem (1)
 ((w-ma 298) (w-regs 4294967296 32)
  (w-imem 4294967296 100) (w-dmem 4294967296 32))
 (local
  (((s-pc s-regs s-imem s-dmem)
    (ma-to-isa w-ma w-regs w-imem w-dmem))
   ((v-ma v-regs v-imem v-dmem)
    (ma-step w-ma w-regs w-imem w-dmem))
   ((u-pc u-regs u-imem u-dmem)
    (isa-step s-pc s-regs s-imem s-dmem))
   ((rv-pc rv-regs rv-imem rv-dmem)
    (ma-to-isa v-ma v-regs v-imem v-dmem)))
  (-> (good-ma w-ma w-regs w-imem w-dmem)
   (and (good-ma v-ma v-regs v-imem v-dmem)
        (or (and (= rv-pc u-pc)
                 (= rv-regs u-regs)
                 (= rv-imem u-imem)
                 (= rv-dmem u-dmem))
            (and (= rv-pc s-pc)
                 (= rv-regs s-regs)
                 (= rv-imem s-imem)
                 (= rv-dmem s-dmem)
                 (< (ma-rank v-ma)
                    (ma-rank w-ma)))))))))))
```

Table 1 Verification times and CNF statistics

Processor model	Verification times (s)		CNF statistics		
	MiniSat	Total (BAT)	Variables	Clauses	Literals
DLX3-2	0.10	0.32	363	1,862	9,914
DLX3-4	0.20	0.49	790	3,972	23,743
DLX3-8	0.47	1.01	1,486	7,536	46,599
DLX3-16	2.04	3.20	2,878	14,664	93,559
DLX3-32	6.03	8.63	5,662	28,920	192,471

The commitment theorem also includes an inductive proof for the "good" MA invariant, i.e., we check that if we step MA from any good state, then the successor of that state will also be good. Next, the property that we ask BAT to check is shown below. We declare a symbolic MA state in the (:vars) section. The symbolic state essentially corresponds to the set of all syntactically possible MA states. In the (:spec) section, we ask BAT to check if the commitment-theorem for all the MA states, which corresponds to the core theorem applied to the "good" MA states and an inductive invariance proof for the "good" MA invariant.

```
(:vars (mastate 298) (regs 4294967296 32)
       (imem 4294967296 100) (dmem 4294967296 32))
(:spec (commitment-theorem mastate regs imem dmem)))
```

Table 1 shows the verification times and CNF statistics for the verification of five three-stage processor models using BAT. The models are obtained by varying the size of the data path and the number of words in the register file and memories. Note that the original three-stage model was parametrized, and the models for the experiments were generated by varying the parameters. The models are given the name "DLX3-n," where "n" indicates the size of the data path and the size of the program counter. The instruction memory, the data memory, and the register file each have 2^n words. The experiments were conducted on a 1.8-GHz Intel (R) Core(TM) Duo CPU, with an L1 cache size of 2,048 KB. The SAT problems generated by BAT were checked using version 1.14 of the MiniSat SAT solver [6].

6 Scaling to More Complex Designs

The formal proof of correctness for the three-stage pipelined machine required stating the refinement correctness formula in the BAT specification language. BAT was then able to automatically prove the refinement theorem relating the three-stage pipelined machine and its ISA. However, a big challenge in verifying pipelined machines using decision procedures is that as the complexity of the machine increases, the verification times are known to increase exponentially [18]. An alternate approach to verifying pipelined machines is based on using general-purpose theorem provers. More complex designs can be handled using theorem provers, but a heroic effort is typically required on the part of the expert user effort to carry out refinement-based correctness proofs for pipelined machines [17]. In this section, we discuss some techniques for handling the scalability issues when using decision procedures for pipelined machine verification.

6.1 Efficient Refinement Maps

One of the advantages of using the WEB refinement framework is that the refinement map is factored out and can be studied independently. In Sect. 5, the commitment refinement map was described. There are other approaches to define the refinement map as well. Another well-known approach to define the refinement map is based on flushing, the idea being that partially executed instructions in the pipeline latches of a pipelined machine state are forced to complete without fetching any new instructions. Projecting out the programmer-visible components in the resulting state gives the ISA state.

There are several more approaches to define the refinement map that have been found to be computationally more efficient. One approach is the Greatest Fixpoint invariant based commitment [15]. The idea here is to define the invariant that characterizes the set of reachable states in a computationally more efficient way. A second approach is collapsed flushing, which is an optimization of the flushing refinement map [7]. A third approach is intermediate refinement maps, that combine both flushing and commitment by choosing a point midway in the pipeline and committing all the latches before that point and flushing all the latches after that point [16]. This approach is also known to improve scalability and efficiency.

6.2 Compositional Reasoning

Refinement is a compositional notion as described in Sect. 4. The idea with compositional reasoning is to decompose the refinement correctness proof into smaller manageable pieces that can be efficiently handled using a decision procedure such as BAT. Another advantage of compositional reasoning is that the counter examples generated are smaller and more localized, making it easier to debug the design. A method for decomposing refinement proofs for pipelined machines has been developed in [14]. Proof rules are also provided to combine the smaller decomposed proofs to construct the refinement proof for the pipelined machine being verified.

6.3 Combining Theorem Proving and Decision Procedures

The BAT decision procedures directly handles the verification problem at the RTL. One approach to handle scalability issues to to abstract and verify the pipelined machine at the term-level. The drawback however is that the final correctness result is only about the abstract model and the formal connection with the RTL model is lost. Hybrid approaches that exploit the refinement framework and use both theorem proving and decision procedures have been developed to address this problem [17]. The idea is to use the theorem prover to formally reduce the verification problem at the RTL to an abstract verification problem, which can then be handled by a decision procedure. The approach scales better for some complex machines, but is much less automatic than using a decision procedure like BAT.

6.4 Parametrization

An advantage of using BAT is that the models can be easily parametrized. This provides an effective debugging mechanism. The idea is based on the fact that models with smaller data path widths lead to computationally more tractable verification problems. For example, the verification of a 32-bit pipelined machine with many pipeline stages may not be tractable, but BAT could probably verify a 2-bit or 4-bit version of the model. While verifying a 4-bit version of the model does not guarantee correctness, a majority of the bugs (for example, control bugs that do not depend on the width of the data path) will be exposed. Generating a 4-bit version of a 32-bit model is easy to accomplish if the model is parametrized.

7 Conclusions

In this chapter, we described how to use the BAT system to verify that pipelined machines refine their instruction set architectures. The notion of correctness that we used is based on WEB refinement. We showed how to strengthen the WEB refinement condition to obtain a statement in the BAT specification language, for which BAT includes a decision procedure. This allows us to automatically check that the pipelined machine satisfies the same safety and liveness properties as its specification, the instruction set architecture. If there is a bug, then BAT will provide a counterexample. We also discussed various techniques to deal with more complex designs.

While much of the focus of pipelined machine verification has been in verifying microprocessor pipelines, these techniques can also be used to reason about other domains in which pipelines occur. Examples include cache coherence protocols and memory interfaces that use load and store buffers.

BAT is not limited to proving properties of pipelines. Any system that can be modeled using BAT's synthesizable hardware description language can be analyzed using BAT. This includes verification problems arising in both hardware and software, embedded systems, cryptographic hash functions, biological systems, and the assembly of large component-based software systems.

Acknowledgments This research was funded in part by NASA Cooperative Agreement NNX08AE37A and NSF grants CCF-0429924, IIS-0417413, and CCF-0438871.

References

1. Bentley B (2005) Validating a modern microprocessor. See URL http://www.-cav2005.inf.ed.ac.uk/bentley_CAV_07_08_2005.ppt
2. Browne M, Clarke EM, Grumberg O (1998) Characterizing finite Kripke structures in propositional temporal logic. Theor Comput Sci 59(1–2):115–131
3. Burch JR, Dill DL (1994) Automatic verification of pipelined microprocessor control. In: Computer-aided verification (CAV '94), vol 818 of LNCS. Springer, Berlin, pp 68–80
4. Chambers B, Manolios P, Vroon D (2009) Faster sat solving with better CNF generation. In: Design, automation and test in Europe. IEEE, New York, pp 1590–1595
5. Clarke EM, Grumberg O, Peled D (1999) Model checking. MIT, Cambridge, MA
6. Een N, Sorensson N (2007) The minisat page. See URL http://minisat.se/-Main.html
7. Kane R, Manolios P, Srinivasan SK (2006) Monolithic verification of deep pipelines with collapsed flushing. In: Gielen GGE (ed) Design, automation and test in Europe, (DATE'06). European Design and Automation Association, Leuven, Belgium, pp 1234–1239
8. Kaufmann M, Manolios P, Moore JS (2000) Computer-aided reasoning: an approach. Kluwer, Dordrecht
9. Manolios P (2000) Correctness of pipelined machines. In: Hunt WA Jr, Johnson SD (eds) Formal methods in computer-aided design – FMCAD 2000, vol 1954 of LNCS. Springer, Berlin, pp 161–178
10. Manolios P (2001) Mechanical verification of reactive systems. PhD thesis, University of Texas at Austin. See URL http://www.ccs.neu.edu/~pete/research.html
11. Manolios P (2003) A compositional theory of refinement for branching time. In: Geist D, Tronci E (eds) 12th IFIP WG 10.5 advanced research working conference, CHARME 2003, vol 2860 of LNCS. Springer, Berlin, pp 304–318
12. Manolios P (2006) Refinement and theorem proving. International school on formal methods for the design of computer, communication, and software systems: hardware verification, LNCS Series. Springer, Berlin
13. Manolios P, Srinivasan SK (2004) Automatic verification of safety and liveness for xscale-like processor models using web refinements. In: Design, automation and test in Europe conference and exposition (DATE'04). IEEE Computer Society, Washington, DC, pp 168–175
14. Manolios P, Srinivasan SK (2005a) A complete compositional reasoning framework for the efficient verification of pipelined machines. In: International conference on computer-aided design (ICCAD'05). IEEE Computer Society, Washington, DC, pp 863–870
15. Manolios P, Srinivasan SK (2005b) A computationally efficient method based on commitment refinement maps for verifying pipelined machines. In: Formal methods and models for co-design (MEMOCODE'05). IEEE, New York, pp 188–197
16. Manolios P, Srinivasan SK (2005c) Refinement maps for efficient verification of processor models. In: Design, automation and test in Europe (DATE'05). IEEE Computer Society, Washington, DC, pp 1304–1309
17. Manolios P, Srinivasan SK (2005d) Verification of executable pipelined machines with bit-level interfaces. In: International conference on computer-aided design (ICCAD'05). IEEE Computer Society, Washington, DC, pp 855–862
18. Manolios P, Srinivasan SK (2008) A refinement-based compositional reasoning framework for pipelined machine verification. IEEE Trans VLSI Syst 16(4):353–364
19. Manolios P, Vroon D (2007) Efficient circuit to cnf conversion. In: Marques-Silva J, Sakallah KA (eds) International conference theory and applications of satisfiability testing (SAT'07), vol 4501 of Lecture notes in computer science. Springer, Berlin, pp 4–9

20. Manolios P, Srinivasan SK, Vroon D (2006a) Automatic memory reductions for RTL model verification. In: Hassoun S (ed) International conference on computer-aided design (ICCAD'06). ACM, New York, NY, pp 786–793
21. Manolios P, Srinivasan SK, Vroon D (2006b) BAT: the bit-level analysis tool. Available from `http://www.ccs.neu.edu/~pete/bat/`
22. Manolios P, Srinivasan SK, Vroon D (2007a) BAT: the bit-level analysis tool. In: International conference computer aided verification (CAV'07)
23. Manolios P, Oms MG, Valls SO (2007b) Checking pedigree consistency with PCS. In: Tools and algorithms for the construction and analysis of systems, TACAS, vol 4424 of Lecture notes in computer science. Springer, Berlin, pp 339–342
24. Manolios P, Vroon D, Subramanian G (2007c) Automating component-based system assembly. In: Proceedings of the ACM/SIGSOFT international symposium on software testing and analysis, ISSTA. ACM, New York, NY, pp 61–72
25. Milner R (1990) Communication and concurrency. Prentice-Hall, Upper Saddle River, NJ
26. Namjoshi KS (1997) A simple characterization of stuttering bisimulation. In: 17th Conference on foundations of software technology and theoretical computer science, vol 1346 of LNCS, pp 284–296
27. Sawada J (2000) Verification of a simple pipelined machine model. In: Kaufmann M, Manolios P, Moore JS (eds) Computer-aided reasoning: ACL2 case studies. Kluwer, Dordrecht, pp 137–150

Formal Verification of Partition Management for the AAMP7G Microprocessor

Matthew M. Wilding, David A. Greve, Raymond J. Richards, and David S. Hardin

1 Introduction

Hardware designed for use in high-assurance applications must be developed according to rigorous standards [3, 19]. For example, security-critical applications at the highest Evaluation Assurance Level (EAL) of the Common Criteria require formal proofs of correctness in order to achieve certification [3]. At the highest EAL, EAL 7, the application must be formally specified, and the application must be proven to implement its specification. This can be a very expensive and time-consuming process. One of the main goals for our research group at Rockwell Collins is to improve secure system evaluation – measured in terms of completeness, human effort required, time, and cost – through the use of highly automated formal methods. In support of this goal, we have developed practical techniques for creating executable formal computing platform models that can both be proved correct and function as high-speed simulators [6, 9]. This allows us to both verify the correctness of the models and also validate that the formalizations accurately model what was actually designed and built.

The AAMP7G microprocessor [17], currently in use in Rockwell Collins high-assurance system products, supports strict time and space partitioning in hardware and has received an NSA certificate based in part of a formal proof of correctness of its separation kernel microcode [18]. The partitioning mechanisms in the AAMP7G's microarchitecture are implemented in a relatively straightforward manner since a design goal for the microprocessor was to support verification and certification activities associated with critical systems developed using the AAMP7G device. In the sections that follow, we present an overview of the AAMP7G verification, built upon a formal model of the AAMP7G microcode written in ACL2 [11]. We begin by describing the formally verified partitioning features of the AAMP7G.

M.M. Wilding (✉)
Rockwell Collins, Inc., Cedar Rapids, IA, USA
e-mail: mmwildin@rockwellcollins.com

D.S. Hardin (ed.), *Design and Verification of Microprocessor Systems for High-Assurance Applications*, DOI 10.1007/978-1-4419-1539-9_6,
© Springer Science+Business Media, LLC 2010

2 The AAMP7G Microprocessor

The AAMP7G is the latest in the line of Collins Adaptive Processing System (CAPS) processors and AAMP microprocessors developed by Rockwell Collins for use in military and civil avionics since the early 1970s [2]. AAMP designs have historically been tailored to embedded avionics product requirements, accruing size, weight, power, cost, and specialized feature advantage over alternate solutions. Each new AAMP makes use of the same multitasking stack-based instruction set while adding state-of-the-art technology in the design of each new CPU and peripheral set. AAMP7G adds built-in partitioning technology among other improvements.

AAMP processors feature a stack-based architecture with 32-bit segmented, as well as linear, addressing. AAMP supports 16/32-bit integer and fractional and 32/48-bit floating point operations. The lack of user-visible registers improves code density (many instructions are a single byte), which is significant in embedded applications where code typically executes directly from slow Read-Only Memory. The AAMP provides a unified call and operand stack, and the architecture defines both user and executive modes, with separate stacks for each user "thread," as well as a separate stack for executive mode operation. The transition from user to executive mode occurs via traps; these traps may be programmed or may occur as the result of erroneous execution (illegal instruction, stack overflow, etc.). The AAMP architecture also provides for exception handlers that are automatically invoked in the context of the current stack for certain computational errors (divide by zero, arithmetic overflow). The AAMP instruction set is of the CISC variety, with over 200 instructions, supporting a rich set of memory data types and addressing modes.

2.1 AAMP7G Intrinsic Partitioning

The AAMP7G provides a feature called "Intrinsic Partitioning" that allows it to host several safety-critical or security-critical applications on the same CPU. The intention is to provide a system developer an architectural approach that will simplify the overall complexity of an integrated system, such as is described in [22]. The transition from multiple CPUs to a single multifunction CPU is shown in Fig. 1. On the left, three federated processors provide three separate functions, A, B, and C. It is straightforward to show that these three functions have no unintended interaction.

On the right of Fig. 1, an integrated processor provides for all three functions. The processor executes code from A, B, and C; its memory contains all data and I/O for A, B, and C. A partition is a container for each function on a multifunction partitioned CPU like the AAMP7G. AAMP7G follows two rules to ensure partition independence:

1. *Time partitioning*. Each partition must be allowed sufficient time to execute the intended function.
2. *Space partitioning*. Each partition must have exclusive-use space for storage.

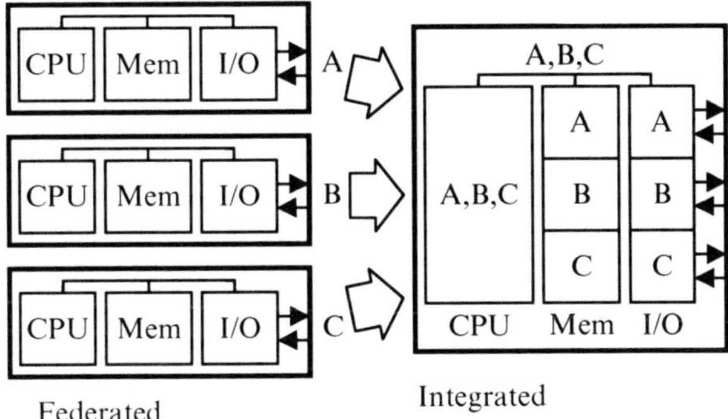

Fig. 1 Transition to multifunction CPU

2.1.1 Time Partitioning

Each partition must be allowed sufficient time to execute the intended function. The AAMP7G uses strict time partitioning to ensure this requirement. Each partition is allotted certain time slices during which time the active function has exclusive use and control of the CPU and related hardware.

For the most secure systems, time slices are allocated at system design time and not allowed to change. For dynamic reconfiguration, a "privileged" partition may be allowed to set time slices. AAMP7G supports both of these approaches as determined by the system designer.

The asynchronous nature of interrupts poses interesting challenges for time-partitioned systems. AAMP7G has partition-aware interrupt capture logic. Each interrupt is assigned to a partition; the interrupt is only recognized during its partition's time slice. Of course, multiple interrupts may be assigned to a partition. In addition, an interrupt may be shared by more than one partition if needed.

System-wide interrupts, such as power loss eminent or tamper detect, also need to be addressed in a partitioned processor. In these cases, AAMP7G will suspend current execution, abandon the current list of partition control, and start up a list of partition interrupt handlers. Each partition's interrupt handler will then run, performing finalization or zeroization as required by the application.

2.1.2 Space Partitioning

Each partition must have exclusive-use space for storage. The AAMP7G uses memory management to enforce space partitioning. Each partition is assigned areas in memory that it may access. Each data and code transfer for that partition is checked to see if the address of the transfer is legal for the current partition. If the transfer is legal, it is allowed to complete. If the transfer is not legal, the AAMP7G Partition Management Unit (PMU) disallows the CPU from accessing read data or code fetch data; the PMU also preempts write control to the addressed memory device.

Memory address ranges may overlap, in order to enable interpartition communication. In this case, interaction between partitions is allowed since it is intended by system design. For maximum partition independence, overlapping access ranges should be kept to a minimum.

As with time slices, memory ranges may be allocated at system design time and not allowed to change. Or, for dynamic reconfiguration, a "privileged" partition may be allowed to set memory ranges. AAMP7G supports both of these approaches as determined by the system designer who employs the AAMP7G.

2.2 *Partition Control*

Only a small amount of memory is needed for the AAMP7G partition control structures (summarized in Fig. 2). This data space is typically not intended to be included in any partition's memory access ranges. The partition control structures include each partition's control includes time allotment, memory space rights, initial state, and test access key stored in ROM. Each partition's saved state is stored in RAM. Partition control blocks are linked together defining a partition activation schedule. AAMP7G partition initialization and partition switching are defined entirely by these structures.

The partition control structures are interpreted entirely in microcode, so no software access is needed to the AAMP7G partitioning structures. This limits the verification of AAMP7G partitioning to proving that the partitioning microcode performs the expected function and that no other microcode accesses the partitioning structures.

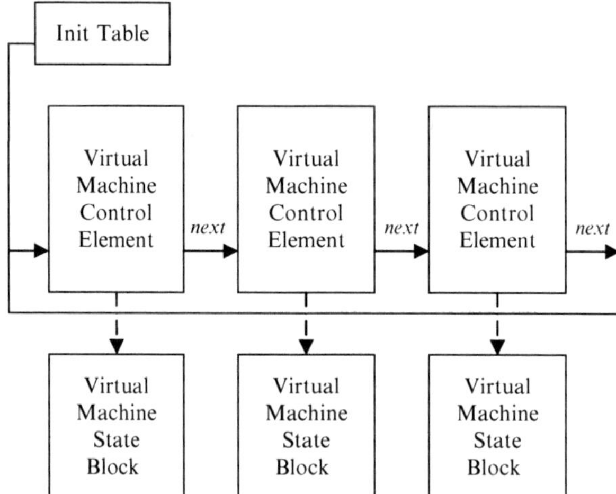

Fig. 2 AAMP7G partition control data structure overview

3 AAMP7G Formal Processing Model

The AAMP7G formal processing model is shown in Fig. 3. Actual AAMP7G processing layers are shown in nonitalic text, while layers introduced for the sake of formal reasoning are shown in italics.

We generally prove correspondence between a concrete model at a given level and a more abstract model. Sequences of microcode implement a given instruction; sequences of abstract instruction steps form basic blocks; a machine code subroutine is made up of a collection of basic blocks. Subroutine invocations are performed in the context of an AAMP thread, and multiple user threads plus the executive mode constitute an AAMP7G partition. Our model supports the entire context switching machinery defined by the AAMP architecture, including traps, outer procedure returns, executive mode error handlers, and so on.

Some aspects of the AAMP7G model are useful for general comprehension of the AAMP7G architecture and for organizing the proof effort. In particular, the correctness theorem we proved about the AAMP7G partitioning mechanism relates the behavior of the microcode of the microprocessor to an abstract notion of AAMP7G partitioning operation, so understanding many of the layers of the model is not strictly necessary to understanding what has been proved about the AAMP7G.

Rockwell Collins has performed a formal verification of the AAMP7G partitioning system using the ACL2 theorem prover [11]. This work was part of an evaluation effort which led the AAMP7G to receive a certification from NSA, enabling a single AAMP7G to concurrently process Unclassified through Top Secret codeword information. We first established a formal security specification, as described in [8], and summarized in Sect. 5. We produced an abstract model of the AAMP7G's partitioning system as well as a low-level model that directly corresponded to the AAMP7G microcode. We then used ACL2 to automatically produce the following:

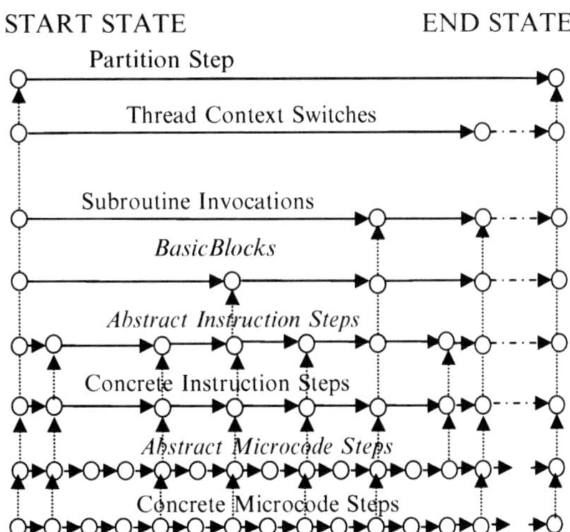

Fig. 3 AAMP7G formal processing model

1. Proofs validating the security model
2. Proof that the abstract model enforces the security specification
3. Proof that the low-level model corresponds to the abstract model.

The use of ACL2 to meet high-assurance Common Criteria requirements at EAL 7 is discussed in [16]. One interpretation of the requirement for a low-level design model is that the low-level design model be sufficiently detailed and concrete so that an implementation can be derived from it with no further design decisions. Because there are no design decisions remaining, one can easily validate the model against the implementation. Note that this low level of abstraction of such a model can make a proof about it challenging. In Sect. 4, we develop the abstract information flow correctness theorem that we proved about the AAMP7G.

4 A Formal Security Specification

High-assurance product evaluation requires precise, unambiguous specifications. For high-assurance products that are relied upon to process information containing military or commercial secrets, it is important to ensure that no unauthorized interference or eavesdropping can occur. A formal security specification prescribes what the system allows and guards against. The construction of a formal security specification that describes the desired behavior of a security-critical system under evaluation is now commonly required for high-level certification.

A computing system that supports multiple independent levels of security (MILS) provides protections to guarantee that information at different security levels is handled appropriately. The design of MILS systems that must perform correctly with respect to a formal security specification is a daunting challenge.

The goal in building a partitioning mechanism is to limit what must be evaluated during the certification process. For example, secure systems can be developed that use partitions to enforce separation between processes at different security levels. A small, trusted "separation kernel" mediates all communication between partitions, thereby assuring that unauthorized communication does not occur. Assuming that the partitioning system is implemented properly and that the communication policy between partitions is loaded correctly, there is no need to evaluate the applications running in different partitions to show that the communication policy is enforced. Safety-critical applications can also exploit intrinsic partitioning: by hosting different applications in separate partitions, it is possible to architect a system so that applications need be evaluated at only the required level of rigor. This system architecting philosophy is described by John Rushby in [20, 21].

The correct implementation of the partitioning mechanism is of course vital to assure the correctness of a larger system that depends upon it. Furthermore, some of the applications of the AAMP7G are security applications that are architected to exploit intrinsic partitioning and require stringent evaluation of all mechanisms being relied upon to separate data at different classification levels.

Architecting a MILS system using a separation kernel breaks the security challenge into two smaller challenges:

1. Building and verifying a dependable separation kernel
2. Building applications that, relying upon protections afforded by the separation kernel, enforce sensible system security policies.

A good specification of a system component has two characteristics. First, it can be mapped to concrete component implementations using convenient and reliable methods. That is, the specification can be proved about a particular system component. Second, a good specification encapsulates needed behavior so that the larger system can benefit from an assurance that the specification holds of the component. In other words, the specification can be used in the larger system that contains the component about which the specification has been proved.

Section 4.1 presents a security property that has both of these desired properties and which has been proved as part of the AAMP7G MILS certification.

4.1 The Formal Security Specification in ACL2

We have chosen the ACL2 logic, an enhancement of the Common Lisp programming language [11], to describe our security specification. ACL2 is a good choice for this work because of its usefulness in modeling and reasoning about computing systems [6, 9] as well as the substantial automation afforded by the ACL2 theorem proving system.

The formal security specification describes abstractly what a separation kernel does. The machine being modeled supports a number of partitions whose names are provided by the constant function `allparts`. We use the notation of ACL2's encapsulate command to indicate a function of no arguments that returns a single value.

```
((allparts)=>*)
```

One of the partitions is designated the "current" partition. The function current calculates the current partition given a machine state.

```
((current *)=>*)
```

We use the notation of ACL2's defthm command, which presents a theorem expressed in Common Lisp notation, to indicate a property about the functions `current` and `allparts`.

```
(defthm current-is-partition
    (member (current st) (allparts)))
```

Associated with partitions are memory segments. Memory segments have names and are intended to model portions of the machine state. The names of the memory segments associated with a particular partition are available from the function `segs`,

which takes as an argument the name of the partition. (Note that since segs is a
function only of partition name and not, for example, a function of machine state,
the assignment of segments to partitions is implicitly invariant.)

```
((segs *)=>*)
```

The values in a machine state that are associated with a memory segment are
extracted by the function select. select takes two arguments: the name of
the memory segment and the machine state.

```
((select * *)=>*)
```

The separation kernel enforces a communication policy on the memory segments. This policy is modeled with the function dia (for Direct Interaction
Allowed), which represents the pairs of memory segments for which direct interaction is allowed. The function takes as an argument a memory segment name
and returns a list of memory segments that are allowed to affect it. (Note that
since dia is a function only of the memory segment name, the formalization here
implicitly requires that the communication policy is invariant.)

```
((dia *)=>*)
```

The last function constrained in the security specification is next, which models one step of computation of the machine state. The function next takes as an
argument a machine state and returns a machine state that represents the effect of
the single step.

```
((next *)=>*)
```

The aforementioned constrained functions are used to construct several
additional functions. selectlist takes a list of segments and returns a list
of segment values; segslist takes a list of partition names and returns the list of
memory segments associated with the partitions; and run takes an initial machine
state and number of steps and returns an initial machine state updated by executing
the number of steps indicated.

```
(defun selectlist (segs st)
     (if (consp segs)
         (cons
               (select (car segs) st)
               (selectlist (cdr segs) st))
          nil))


(defun segslist (partnamelist)
     (if (consp partnamelist)
         (append
               (segs (car partnamelist))
               (segslist (cdr partnamelist)))
          nil))
```

```
    (let ((segs (intersection-equal (dia seg)
                                    (segs (current st1))))))
       (implies
          (and
              (equal (selectlist segs st1) (selectlist segs st2))
              (equal (current st1) (current st2))
              (equal (select seg st1) (select seg st2)))
          (equal
              (select seg (next st1))
              (select seg (next st2))))))
```

Fig. 4 The formal security specification

```
(defun run (st n)
    (if (zp n) st
        (run (next st) (1- n))))
```

The formal security specification theorem, now referred to as "GWV" after its authors, is shown in Fig. 4. The GWV theorem utilizes a "two worlds" formulation, in which two arbitrary states of the system (in this case, `st1` and `st2`) are both hypothesized to satisfy some predicate(s). The two states are both stepped, and some predicate is shown to hold on the two resulting states. In particular, the GWV theorem says that the effect of a single step on the system state on an arbitrary segment of the state, `seg`, is a function only of the segments associated with the current partition that are allowed to interact with `seg`.

We have established the utility of this formal security specification by stating the policy as an axiom and then attempting to prove several well-known security-related theorems using the axiom. In this way, we have proved exfiltration, infiltration, and mediation theorems, as well as proved the functional correctness of a simple firewall, all using the ACL2 system. Subsequently, we have formally shown how GWV relates to classical noninterference [5].

Alves-Foss and Taylor in [1] point out that it is possible to thwart the intent of proving the GWV specification about a particular system by improperly defining the functions of the theorem for that system. Specific instances cited in [1] include omitting relevant system state from the model and modeling the system step with less time resolution than that is appropriate.

Improper instantiation of the specification is indeed important to avoid not only for the functions mentioned explicitly in [1], but also for all the functions in the specification. It is not just those instances of poor modeling that need be avoided. Effective use of GWV as a specification requires care to insure that the system of interest is modeled appropriately. In Sect. 4.2, we present GWV instantiated for the AAMP7G. Fidelity of the functions that instantiate this theorem for the AAMP7G was checked through a process of inspection and review. Work to formalize the notion of appropriate "properties functions" is discussed in [5].

4.2 An AAMP7G Instantiation of GWV

Figure 5 shows the theorem proved about the AAMP7G, which is an instance of the abstract security specification of Fig. 4, made concrete so as to show that the abstract notion of separation holds of a concretely described model of the AAMP7G. The reification of the theorem in Fig. 4 to the AAMP7G theorem in Fig. 5 is not obvious, not the least because the theorem includes an operational model of the AAMP7G, which must be relatable to the actual device in order for the correctness proof to have practical usefulness as part of a certification process.

Recall the different levels of AAMP7G models described in Fig. 3. The correctness theorem involves models of the AAMP7G at two levels, at both the functional level and abstract level. The functional model closely corresponds with the actual AAMP7G microarchitecture implementation. For example, in the functional model, RAM is modeled as an array of values. The abstract model represents the data of the AAMP7G in a manner more convenient for describing properties. For example, the

```
(implies
    (and
        (secure-configuration spex)
        (spex-hyp spex fun::st1)
        (spex-hyp spex fun::st2))
    (implies
        (let ((abs::st1 (lift-raw spex fun::st1))
              (abs::st2 (lift-raw spex fun::st2)))
            (and
                (let ((segs
                        (intersection-equal
                          (dia-fs seg abs::st1)
                          (segs-fs (current abs::st1) abs::st1))))
                  (equal (raw-selectlist segs abs::st1)
                         (raw-selectlist segs abs::st2)))
                (equal (current abs::st1)
                       (current abs::st2))
                (equal (raw-select seg abs::st1)
                       (raw-select seg abs::st2))))
        (equal
            (raw-select seg
              (lift-raw spex (fun::next spex fun::st1)))
            (raw-select seg
              (lift-raw spex (fun::next spex fun::st2))))))))
```

Fig. 5 Theorem that the AAMP7G implements the security specification

list of partitions in the partition execution schedule is represented as a list. The theorem is about the behavior of the functional model, but we express the theorem about an abstract model of the AAMP7G that has been "lifted" from a functional model. In this way, we simplify the expression of the theorem (since the abstract model functions are simpler) but we instantiate the theorem with the behavior of the most concrete model of the AAMP7G to help ensure that the theorem is about the "real" AAMP7G.

Each function of the correctness theorem is defined precisely using the ACL2 defthm command and its function result is summarized below.

`secure-configuration` – Data structures expected during nominal AAMP7G execution are described symbolically by the list `spex`. For example, `spex` includes a description of the data structure summarized in Fig. 2.

`spex-hyp` – An AAMP7G state description includes all the data structures described by `spex`. Also, a number of other elements of the state have the correct type, and relevant microarchitectural registers have expected values.

`lift-raw` – A functional AAMP7G state is lifted to an abstract state with respect to the data structures of `spex`.

`dia-fs` – The elements of the state that can effect `seg` given the configuration of state `st`. This function is defined by recursively searching the partition schedule list in order to detect possible authorized information flow. For example, if `seg` is an interrupt `i`, and the partition schedule includes a partition p that is permitted to change interrupt `i`'s value, and partition p is permitted to read a RAM location r, then RAM location r is returned by this function. Additionally, elements of the control data structures of the AAMP7G, such as the partition schedule itself, are returned by this function for all segments.

`segs-fs` – All the segs are associated with a partition. This includes its RAM memory, interrupt values, and all partitioning system relevant registers.

`current` – The currently executing element of the partition schedule.

`raw-select` – The value of a segment seg in AAMP7G state st.

`raw-selectlist` – The values of a list of segments in `st`.

`next` – The state of the AAMP7G after one execution step starting from processor state `st`.

Note that the notion of "segs" in the AAMP7G includes all the states relevant to the AAMP7G's execution. Note also that `segs-fs` and `dia-fs` are functions not only of a segment as might be expected from a review of the specification in Fig. 4, but also of the state of the machine. We have proved that these functions are invariant with respect to the `next` function in order to justify their use in applying GWV to the AAMP7G.

The `next` function is by far the most complex function used in the correctness theorem in the AAMP7G theorem. It represents in entirety the low-level design of the partitioning-relevant trusted operation of the AAMP7G. "Trusted" microcode is microcode that operates with memory protection turned off, thereby providing access to the data structures maintained by the AAMP7G to support intrinsic partitioning. All the partitioning-relevant microcode runs in this trusted mode, and the low-level design model of the AAMP7G models all the microcode that runs in trusted mode.

Considerable thought was put into defining the "step" of the AAMP7G microprocessor for the purpose of formalizing it in the next function. Broadly speaking, a step is a high-level partition step indicated at the highest level of Fig. 3. For example, in the nominal case (where there is no unusual event such as a power-down warning), a step is the loading of a partition including relevant protections, an execution of a user partition, and the saving of the state of that partition.

While the notion of "step" implemented by the next function is abstract, its definition in ACL2 is most assuredly not. It corresponds to the most concrete, lowest level of Fig. 3. This is because another crucial consideration when developing the model of the AAMP7G contained in the next function is how to validate this hand-written model against the actual AAMP7G. The AAMP7G is a microcoded microprocessor, and much of the functionality of the machine is encoded in its microcode. The low-level design model is written specifically to make a code-to-spec review that relates the model to the actual implementation relatively straightforward. An ACL2 macro allows an imperative-style description that eases comparison with microcode. Also, very importantly, the model is written with the model of memory that the microcode programmer uses. That is, memory has only two operations: read and write. The simplicity of the memory model makes the code-to-spec review easier but adds a great deal of complexity to the proof. Since the proof is machine checked while the model validation process requires evaluation, this is a good trade-off. It provides a high level of assurance with a reasonable level of evaluation. Much of the effort on the project was spent constructing the proofs, but the proofs were reviewed relatively easily by the evaluators because they could be replayed using ACL2.

Figure 6 presents an example fragment of the low-level functional design model. It is typical of the ACL2 microcode model in that each line of microcode is modeled by how it updates the state of the partition-relevant machine. A small program was written that identifies all microcode that can be run in trusted mode, and the results were used to check informally that the ACL2 model in fact models every line of microcode that runs in trusted mode. An additional manual check was performed to insure that the output of this tool correctly identified entry/exit points of the trusted microcode. The entire AAMP7G model is approximately 3,000 lines of ACL2 definitions.

4.3 Proof of the AAMP7G Instantiation of GWV

The AAMP7G GWV theorem was proved using ACL2. The proof architecture decomposes the proof into three main pieces:

1. Proofs validating the correctness theorem (as described in [7])
2. Proof that the abstract model meets the security specification
3. Proof that the low-level model corresponds with the abstract model.

```
;=== ADDR: 052F
(st. ie = nil)
(Tx = (read32 (vce_reg st) (VCE.VM_Number)))


;=== ADDR: 0530
(st. Partition = Tx)


;=== ADDR: 0531
(TimeCount = (read32 (vce_reg st) (VCE.TimeCount)))


;=== ADDR: 0532
(PSL[0]= TimeCount st)
```

Fig. 6 A fragment of the AAMP7G formal microcode-level model

In addition to libraries provided in the standard ACL2 release, several libraries of ACL2 lemmas were developed for the AAMP7G partitioning proofs. As previously discussed, an important challenge of this proof was developing a method for reasoning about read and write operations over a linear address space. The partitioning-relevant data structures of the AAMP7G, summarized in Fig. 2, have nodes containing dozens of elements representing status and protection information, and the list structures are of arbitrary length. Pointer-laden data structure reasoning requires considerable automation, and most of the effort required to prove the correctness theorems about the AAMP7G involved developing an approach for modeling and analyzing operations on such data structures.

The approach used to reason about the AAMP7G data structures, termed GACC for Generalized Accessor, provides a systematic approach for describing data structures and a template for proving a few helpful facts about each operation. The version of GACC used for the AAMP7G proofs is described in [4]. This reasoning infrastructure continues to evolve, and a later version of this capability was used in [15].

5 Use of Formal Analysis in a Certification

The analysis described in Sects. 3 and 4 was part of the evaluation that led to the AAMP7's certification [18]. Machine checking provides a high level of confidence that the theorem was in fact a theorem. But how to ensure that proving the theorem really means that the AAMP7G has the appropriate behavior? An important step in the process was to conduct a code-to-spec review with a National Security Agency evaluation team. This review validated the theorem. Each of the functions in the formal specification was reviewed. The most complex of these functions is the representation of the AAMP7G's design, as the link between the model of the design and the actual implementation must be established. As discussed earlier, the model was designed specifically to facilitate this kind of scrutiny.

The documentation package that was created for this review included:

- Material explaining the semantics of ACL2 and AAMP7G microcode
- Listings of the AAMP7G microcode and the ACL2 low-level model
- The source code listing of a tool that identifies trusted-mode microcode sequences, and a listing of such sequences in the AAMP7G microcode
- Cross-references between microcode line numbers, addresses, and formal model line numbers
- The ACL2-checkable proofs in electronic form.

The exhaustive review accounted for each line of trusted microcode and each model of a line of trusted microcode, ensuring that there was nothing left unmodeled, that there was nothing in the model that was not in the actual device, and that each line of the model represented the actual behavior of the microcode.

This review was made possible because the model of the AAMP7G was designed to correspond to the actual device, in particular its concrete microprocessor model that maintained line-for-line correspondence with the microcode and employed a linear address space model. Although the proof was considerably more challenging to construct because this approach was taken, the proofs were all machine checked so little of that effort was borne by the evaluators. The machine-checked formal analysis allowed the evaluators to focus on validation that the security policy and model described what they were interested in – operation of a separation kernel and the AAMP7G – rather than trying to determine through inspection or testing that the device implementation always did the right thing.

6 Formalization within a Partition: The AAMP7G Instruction Set Model

Having established the correctness of the AAMP7G's partitioning system by proving a theorem that relates an abstract security specification to a concrete model of microcode, we next wished to provide a formal model of the instruction set processing that occurs within a partition's time slice. This model is not relevant to the proof of the partitioning mechanism; as previously explained, software running in a partition cannot effect the operation of the partitioning system. However, having such a model enables us to perform machine code proofs of correctness that can be used in high-assurance evaluations. We therefore describe this model in this section and outline one method for machine code proof based on symbolic simulation.

The instruction set model and all the necessary support books consist of some 100,000 lines of ACL2 code [10]. The AAMP7G instruction set model is shown in the architectural context in Fig. 3. We begin with a concrete instruction model, written in a sequential manner that is reflective of how the machine actually operates. The AAMP memory model is based on the GACC linear address space library previously used in the AAMP7G partitioning proofs [4]. The AAMP7G machine state, including the architecturally defined registers, is represented as an ACL2 single-threaded object (stobj) [14] for simulator performance reasons.

6.1 *Instruction Set Model Validation*

Since we model the AAMP7G instruction set in its entirety, we can analyze AAMP7G machine code from any source, including compilers and assemblers. Additionally, since we directly model memory, we merely translate the binary file for a given AAMP7G machine code program into a list of (address, data) pairs that can be loaded into ACL2. We load the code, reset the model, and the execution of the machine code then proceeds, under the control of an eclipse-based user interface that was originally written to control the actual AAMP7G.

We then validate the AAMP7G instruction set model by executing instruction set diagnostics on the model that are used for AAMP processor acceptance testing. A typical diagnostic exercises each instruction, plus context switching, exception handling, etc.

6.2 *Abstract Instruction Set Modeling and Symbolic Simulation*

For a given AAMP instruction X, the "abstract" function OP-X-PRECONDITIONS collects those conditions that need to hold for "normal" execution of the instruction. VM-X-EXPECTED-RESULT gives the expected output state as a modification of the input state. As this function is an abstraction of the concrete instruction set execution, it does not characterize the *steps* of the computation, but rather the *result* in a very compact and readable form.

Finally, proving the theorem VM-X-REWRITE establishes that stepping the AAMP instruction set model on this instruction will yield exactly the expected result as the abstract formulation, assuming that the preconditions are satisfied. If the instruction semantics involves interesting exceptions, such as overflow or divide-by-zero, these are characterized by additional expected result functions and additional branches in the right-hand side of this rewrite rule.

We have provided similar treatment for each of the AAMP7G instructions. Assuming that we can relieve the preconditions at each step, this allows us efficiently to symbolically step through even very long sequences of AAMP7G instructions. After each step, the rewriter effectively canonicalizes the result into a very compact and readable form. This is practical within the context of a theorem prover only because the ACL2 rewriter can be constrained to be quite fast. A more detailed description of the AAMP7G instruction set modeling process can be found in [10].

6.3 Compositional Code Proof

Our verification of AAMP7G programs is done compositionally. That is, we verify programs in one subroutine at a time. We try to ensure that, after we verify a subroutine, we never have to analyze it again. To prove the correctness theorem for a subroutine we have used a proof methodology called "compositional cutpoints." Our method borrows parts of the method put forth in [12]; both methods are inspired by observations first made by Moore [13].

Cutpoint proofs require annotating the subroutine to be verified by placing assertions at some of its program locations; those locations are called "cutpoints." Every cutpoint has a corresponding assertion which is taken to apply to those states that arise just before the instruction at a given program location is executed. The resulting full set of cutpoints is sufficient if it "cuts every loop," that is, if every cycle in the routine's control flow graph contains a cutpoint. We need not consider cycles in the code of called subroutines; any subroutine call should either be a call to an already-verified routine or be a recursive call (which we handle specially).

The "cutpoint to cutpoint" proof for a routine involves symbolic simulation of the machine model. The simulation starts at a cutpoint and assumes that the assertion for that cutpoint holds. We simulate the machine until it either reaches another cutpoint or exits by executing a return instruction. At the resulting state, we must show that its corresponding assertion holds. Thus, functional proofs of correctness of AAMP7G machine code can proceed largely automatically once the necessary cutpoint assertions have been introduced. Details of this method can be found in [10].

7 Conclusion

We have presented a summary of the formal modeling and verification that led to a MILS certificate for the AAMP7G microprocessor, enabling a single AAMP7G to concurrently process Unclassified through Top Secret codeword information. We discussed the formal model architecture of the AAMP7G at several levels, including the microcode and instruction set levels. We described how the ACL2 theorem prover was used to develop a formal security specification, the GWV theorem, and outlined a mathematical proof (machine-checked using ACL2) which established that the AAMP7G trusted microcode implemented that security specification, in accordance with EAL 7 requirements. We discussed the evaluation process that validated the formal verification evidence through a code-to-spec review. Finally, we detailed a technique for compositional reasoning at the instruction set level, using a symbolic simulation based technique.

Acknowledgments Many thanks to three excellent teams that contributed greatly to this effort: the AAMP7G development team at Rockwell Collins, the ACL2 development team at the University of Texas at Austin, and the AAMP7G security evaluation team from the US DoD.

References

1. Alves-Foss J, Taylor C (2004) An analysis of the GWV security policy. In: Proceedings of the fifth international workshop on ACL2 and its applications, Austin, TX, Nov. 2004
2. Best D, Kress C, Mykris N, Russell J, Smith W (1982) An advanced-architecture CMOS/SOS microprocessor. IEEE Micro 2(3):11–26
3. Common Criteria for Information Technology Security Evaluation (CCITSE) (1999) Available at http://www.radium.ncsc.mil/tpep/library/ccitse/ccitse.html
4. Greve D (2004) Address enumeration and reasoning over linear address spaces. In: Proceedings of ACL2'04, Austin, TX, Nov. 2004
5. Greve D (2010) Information security modeling and analysis. In Hardin D (ed) Design and verification of microprocessor systems for high-assurance applications. Springer, Berlin, pp 249–299
6. Greve D, Wilding M, Hardin D (2000) High-speed, analyzable simulators. In: Kaufmann M, Manolios P, Moore JS (eds) Computer-aided reasoning: ACL2 case studies. Kluwer, Dordrecht, pp 89–106
7. Greve D, Wilding M, Vanfleet M (2003) A separation kernel formal security policy. In: Proceedings of ACL2'03
8. Greve D, Richards R, Wilding M (2004) A summary of intrinsic partitioning verification. In: Proceedings of ACL2'04, Austin, TX, Nov. 2004
9. Hardin D, Wilding M, Greve D (1998), Transforming the theorem prover into a digital design tool: from concept car to off-road vehicle. In: Hu A, Vardi M (eds) CAV'98, vol 1427 of LNCS. Springer, Berlin, pp 39–44
10. Hardin D, Smith E, Young W (2006) A robust machine code proof framework for highly secure applications. In: Proceedings of ACL2'06, Seattle, WA, Aug. 2006
11. Kaufmann M, Manolios P, Moore JS (2000) Computer-aided reasoning: an approach. Kluwer, Dordrecht
12. Matthews J, Moore JS, Ray S, Vroon D (2006) Verification condition generation via theorem proving. In: Proceedings of LPAR'06, vol 4246 of LNCS, pp 362–376
13. Moore JS (2003) Inductive assertions and operational semantics. In Geist D (ed) CHARME 2003, vol 2860 of LNCS. Springer, Berlin, pp 289–303
14. Moore JS, Boyer R (2002) Single-threaded objects in ACL2. In: Proceedings of PADL 2002, vol 2257 of LNCS. Springer, Berlin, pp 9–27
15. Richards R (2010) Modeling and security analysis of a commercial real-time operating system kernel. In Hardin D (ed) Design and verification of microprocessor systems for high-assurance applications. Springer, Berlin, pp 301–322
16. Richards R, Greve D, Wilding M, Vanfleet M (2004) The common criteria, formal methods, and ACL2. In: Proceedings of the fifth international workshop on ACL2 and its applications, Austin, TX, Nov. 2004
17. Rockwell Collins, Inc. (2003) AAMP7r1 reference manual
18. Rockwell Collins, Inc. (2005) Rockwell Collins receives MILS certification from NSA on microprocessor. Rockwell Collins press release, 24 August 2005. http://www.rockwellcollins.com/news/page6237.html
19. RTCA, Inc. (2000) Design assurance guidance for airborne electronic hardware, RTCA/DO-254
20. Rushby J (1981) Design and verification of secure systems. In: Proceedings of the eighth symposium on operating systems principles, vol 15, December 1981
21. Rushby J (1999) Partitioning for safety and security: requirements, mechanisms, and assurance. NASA contractor report CR-1999-209347
22. Wilding M, Hardin D, Greve D (1999) Invariant performance: a statement of task isolation useful for embedded application integration. In: Weinstock C, Rushby J (eds) Proceedings of dependable computing for critical applications – DCCA-7. IEEE Computer Society Dependable Computing Series

Compiling Higher Order Logic by Proof

Konrad Slind, Guodong Li, and Scott Owens

1 Compilation and Logic

There has recently been a surge of research on verified compilers for languages like C and Java, conducted with the aid of proof assistants [22, 24, 25]. In work of this kind, the syntax and semantics of the levels of translation – from the source language to various intermediate representations and finally to the object code – are defined explicitly by datatypes and inductively defined evaluation relations. Verification of a program transformation is then typically performed by proving semantics preservation, e.g. by proving that a simulation relation holds, usually by rule induction over the evaluation relation modelling the operational semantics. This *deep-embedding* approach, in which the compiler under study is a logical function from a source datatype to a target datatype, both represented in the logic, is a by-now classical methodology, which advances in proof environments support increasingly well. A major benefit of such a formalized compiler is that all datatypes and algorithms comprising the compiler are explicitly represented in the logic and are therefore available for a range of formal analyses and manipulations. For example, compilation algorithms are being re-scrutinized from the perspective of formal verification, with the result that some are being precisely specified for the first time [41] and are even being simplified. Finally, the technique applies to a wide range of functional and imperative programming languages.[1]

---

[1] There are a couple of choices when it comes to the deployment of such a verified compiler. Since it is a deep embedding, an actual working compiler exists in the logic and deductive steps can be used to compile and execute programs. This is not apt to scale, so it is much more likely that the formalized compiler is automatically written out into the concrete syntax of an existing programming language, compiled, and subsequently deployed in a standard fashion.

K. Slind (✉)
Rockwell Collins, Inc., Bloomington, MN, USA
e-mail: klslind@rockwellcollins.com

D.S. Hardin (ed.), *Design and Verification of Microprocessor Systems for High-Assurance Applications*, DOI 10.1007/978-1-4419-1539-9_7,
© Springer Science+Business Media, LLC 2010

However, deep embeddings have some drawbacks. Transformations performed by a compiler may be hard to isolate, to verify, and to re-verify: a slight modification of the compilation algorithm can lead to a heavy burden on the revision of previously enacted proofs. As a result, quite often only simple compilers are verified or only parts of a compiler are verified (the recent work on CompCert [26] stands in contrast to this trend). Moreover, to even determine the operational semantics for realistic high-level languages can be an experimental task [36]; in this regard, the semantics of assembly languages is arguably more precise. Finally, a deeply embedded compiler cannot directly access support provided in the implementation of the host logic. For example, operations such as substitution and automatic renaming of bound variables are heavily used components in a compiler – and must be formalized in a deep embedding – but they are usually already available in the host theorem prover. To have to implement and verify this functionality can be seen as an uninteresting and time-consuming chore. Indeed, the use of higher order abstract syntax [39] in reasoning about the meta-theory of programming languages is motivated by just this observation.

1.1 A Gap Between Program Verification and Compiler Verification

A common formal modelling technique is to translate programs to mathematical functions, which can be formally expressed in a logic. Since the majority of algorithms can be directly represented as logical functions,[2] the properties of which can be transparently stated and proved correct using ordinary mathematics and contemporary proof assistants, e.g. ACL2, HOL4, PVS, Coq, and Isabelle/HOL [5, 21, 33, 37, 45], are often used as highly automated tools for reasoning about such programs. This identification of programs with functions is routinely exploited in verifications using these systems.

However, a gap occurs: the issue is how to formally connect a property proved about a particular logical function to the code generated by a verified compiler. For example, suppose one has proved that a logical function f : num list $\rightarrow$ num list sorts lists of numbers, i.e. that f terminates and returns a sorted permutation of its input. One can map f to an abstract syntax tree (AST) and compile it with a verified compiler. We would then know that the compiled code executes in accordance to the operational semantics given to the AST, but any connection with the properties of f is only informal. So even in a setting where we have a trusted compiler, proving properties of the code generated from a program may not be eased, even if properties of the corresponding function are already established! Similarly, analyses of source code do not immediately apply to the compiled machine code; see [2] for discussion.

---

[2] There are some exceptions; for example, to the authors' knowledge, no efficient purely functional Union-Find algorithm has yet been found. However, see [10] for an imperative Union-Find implementation that can be used in a purely functional manner.

There are a variety of solutions to this problem. For example, one could prove an equivalence between f and the operational semantics applied to the AST corresponding to f, or one could attempt to prove the desired functional properties of the AST using the operational semantics directly [42]. We will not investigate these paths. Instead, we will avoid the verified compiler and compile the logical functions.

1.2 Compilation by Proof

We have been pursuing an approach, based on the use of verified rewrite rules, to construct a verifying compiler for the functional programming language inherent in a general-purpose logical framework. In particular, a subset of the term language dwelling within higher order logic (HOL) is taken as the source language; thus, there is no AST type in our approach. Intermediate languages introduced during compilation are not embodied in new types, only as particular kinds of terms. This means that source programs and intermediate forms are simply functions in HOL enjoying exactly the same semantics. Thus, a major novelty of our compiler is that program syntax and operational semantics need not be formalized. In addition, program transformations are isolated clearly and specified declaratively, as term rewrites; a different order of applying rewrites can lead to a different certifying compiler. For a rewriting step, a theorem that establishes the equality for the input and result of the transformation is given as by-product. We call this technique *compilation by proof*, and it can be seen as a fine-grained version of translation validation [40] in which the much of the validation is conducted offline, when proving the rewrite rules.

Each intermediate language is derived from the source language by restricting its syntax to certain formats and introducing new administrative terms to facilitate compilation and validation. Thus, an intermediate language is a restricted instance of the source language. One advantage of this approach is that intermediate forms can be reasoned about using ordinary facilities supplied by the logic implementation, e.g. β-conversion. Our compiler applies translations such as normalization, inline expansion, closure conversion, polymorphism elimination, register allocation and structured assembly generation in order to translate a source program into a form that is suitable for machine code generation. Before examining these more closely, we first provide a self-contained discussion of the HOL logic and its implementation in the HOL4 system [45].

2 Higher Order Logic

The logic implemented by HOL systems [12] (there are several mature implementations besides HOL4, including ProofPower and HOL Light)[3] is a typed higher order predicate calculus derived from Church's Simple Theory of Types [9]. HOL is a

---

[3] Isabelle/HOL is a similar system; it extends the HOL logic with a Haskell-like type-class system.

classical logic and has a set theoretic semantics, in which types denote non-empty sets and the function space denotes total functions. Formulas are built on a lambda calculus with an ML-style system of types with polymorphic type variables.

Formally, the syntax is based on signatures for types (Ω) and terms (Σ_Ω). The type signature assigns arities to type operators, while the term signature assigns constants their types. These signatures are extended by principle of definition for types and terms, as discussed later.

Definition 1 (HOL types). The set of types is the least set closed under the following rules:

Type variable. There is a countable set of type variables, which are represented with Greek letters, e.g. α, β, etc..

Compound type. If *op* in Ω has arity n, and each of $ty_1, \ldots, ty_n$ is a type, then $(ty_1, \ldots, ty_n)op$ is a type.

A type constant is represented by a 0-ary compound type. Types are definitionally constructed in HOL, building on the initial types found in Ω: truth values (bool), function space (written $\alpha \to \beta$), and an infinite set of individuals (ind).

Definition 2 (HOL terms). The set of terms is the least set closed under the following rules:

Variable. if v is a string and ty is a type built from Ω, then $v : ty$ is a term.

Constant. $(c : ty)$ is a term if $c : \tau$ is in Σ_Ω and ty is an instance of τ, i.e. there exists a substitution for type variables θ, such that each element of the range of θ is a type in Ω and $\theta(\tau) = ty$.

Combination. $(M\ N)$ is a term of type β if M is a term of type $\alpha \to \beta$ and N is a term of type α.

Abstraction. $(\lambda v.\ M)$ is a term of type $\alpha \to \beta$ if v is a variable of type α and M is a term of type β.

Initially, Σ_Ω contains constants denoting equality, implication, and an indefinite description operator ε:

Operator	Type	Concrete syntax
$=$	$\alpha \to \alpha \to$ bool $\quad p$	$A = B$
$\Rightarrow$	bool $\to$ bool $\to$ bool	$A \Rightarrow B$
ε	$(\alpha \to$ bool$) \to \alpha$	$\varepsilon x.\ P\ x$

Types and terms form the basis of the *prelogic*, in which basic algorithmic manipulations on types and terms are defined: e.g. the free variables of a type or term, α-convertibility, substitution, and β-conversion. For describing substitution in the following, the notation $[M_1 \mapsto M_2]\,N$ is used to represent the term N where all free occurrences of M_1 have been replaced by M_2. Of course, M_1 and M_2 must have the same type in this operation. During substitution, every binding occurrence of a variable in N that would capture a free variable in M_2 is renamed to avoid the capture taking place.

2.1 Deductive System

In Fig. 1, a mostly conventional set of predicate logic inference rules is outlined, along with the axioms of the HOL logic. The derivable theorems in HOL are just those that can be generated by using the axioms and inference rules of Fig. 1. More parsimonious presentations of this deductive system can be found in [12] or Appendix A of [19], but the rules presented here provide a familiar basis on which to work.

$$\Rightarrow\text{-intro} \quad \frac{\Gamma \vdash Q}{\Gamma - \{P\} \vdash P \Rightarrow Q} \qquad \frac{\Gamma \vdash P \Rightarrow Q \quad \Delta \vdash P}{\Gamma \cup \Delta \vdash Q} \quad \Rightarrow\text{-elim}$$

$$\wedge\text{-intro} \quad \frac{\Gamma \vdash P \quad \Delta \vdash Q}{\Gamma \cup \Delta \vdash P \wedge Q} \qquad \frac{\Gamma \vdash P \wedge Q}{\Gamma \vdash P \quad \Gamma \vdash Q} \quad \wedge\text{-elim}$$

$$\vee\text{-intro} \quad \frac{\Gamma \vdash P}{\Gamma \vdash P \vee Q,\ \Gamma \vdash Q \vee P} \qquad \frac{\Gamma_1 \vdash P \vee Q \quad \Gamma_2, P \vdash M \quad \Gamma_3, Q \vdash M}{\Gamma_1 \cup \Gamma_2 \cup \Gamma_3 \vdash M} \quad \vee\text{-elim}$$

$$\forall\text{-intro}^* \quad \frac{\Gamma \vdash P}{\Gamma \vdash \forall x.\ P} \qquad \frac{\Gamma \vdash \forall x.\ P}{\Gamma \vdash [x \mapsto N]P} \quad \forall\text{-elim}$$

$$\exists\text{-intro}^* \quad \frac{\Gamma \vdash P}{\Gamma \vdash \exists x.\ [N \mapsto x]P} \qquad \frac{\Gamma \vdash \exists x.\ P \quad \Delta, [x \mapsto v]P \vdash Q}{\Gamma \cup \Delta \vdash Q} \quad \exists\text{-elim}^*$$

$$\text{Assume} \quad P \vdash P \qquad \vdash M = M \quad \text{Refl}$$

$$\text{Sym} \quad \frac{\Gamma \vdash M = N}{\Gamma \vdash N = M} \qquad \frac{\Gamma \vdash M = N,\ \Delta \vdash N = P}{\Gamma \cup \Delta \vdash M = P} \quad \text{Trans}$$

$$\text{Comb} \quad \frac{\Gamma \vdash M = N,\ \Delta \vdash P = Q}{\Gamma \cup \Delta \vdash M\ P = N\ Q} \qquad \frac{\Gamma \vdash M = N}{\Gamma \vdash (\lambda v.M) = (\lambda v.N)} \quad \text{Abs}^*$$

$$\text{tyInst}^* \quad \frac{\Gamma \vdash M}{\theta(\Gamma) \vdash \theta(M)} \qquad \vdash (\lambda v.M)N = [v \mapsto N]M \quad \beta\text{-conv}$$

$$
\begin{array}{ll}
\text{Bool} & \vdash P \vee \neg P \\
\text{Eta} & \vdash (\lambda v.\ M v) = M \\
\text{Select} & \vdash P x \Rightarrow P(\varepsilon x.\ P x) \\
\text{Infinity} & \vdash \exists f : \mathsf{ind} \to \mathsf{ind}.\ (\forall x\ y.\ (f x = f y) \Rightarrow (x = y)) \wedge \exists y.\forall x.\ \neg(y = f x)
\end{array}
$$

Fig. 1 HOL deductive system

A theorem with hypotheses $P_1, \ldots, P_k$ and conclusion Q (all of type bool) is written $[P_1, \ldots, P_k] \vdash Q$. In the presentation of some rules, e.g. $\vee$-elim, the following idiom is used: $\Gamma, P \vdash Q$. This denotes a theorem where P occurs as a hypothesis. A later reference to Γ then actually means $\Gamma - \{P\}$, i.e. had P already been among the elements of Γ, it would now be removed.

Some rules, noted by use of the asterisk in Fig. 1, have restrictions on their use or require special comment:

- $\forall$-intro. The rule application fails if x occurs free in Γ.
- $\exists$-intro. The rule application fails if N does not occur free in P. Moreover, only *some* designated occurrences of N need be replaced by x. The details of how occurrences are designated vary from implementation to implementation.
- $\exists$-elim. The rule application fails if the variable v occurs free in $\Gamma \cup \Delta \cup \{P, Q\}$.
- Abs. The rule application fails if v occurs free in Γ.
- tyInst. A substitution θ mapping type variables to types is applied to each hypothesis and also to the conclusion.

An important feature of the HOL logic is $\varepsilon : (\alpha \to \text{bool}) \to \alpha$, Hilbert's *indefinite* description operator. A description term $\varepsilon x : \tau.\, Px$ is interpreted as follows: it represents an element e of type τ such that Pe holds. If there is no object that P holds of, then $\varepsilon x : \tau.\, Px$ denotes an arbitrary element of τ. This is summarized in the Select axiom $\vdash Px \Rightarrow P(\varepsilon x.\, Px)$.

2.2 Definitions

One of the most influential methodological developments in verification has been the adoption of *principles of definition* as logical prophylaxis, and implementations of HOL therefore tend to eschew the assertion of axioms in favour of making soundness-preserving definitions.

Definition 3 (Principle of Constant Definition). Given terms $x : \tau$ and $M : \tau$ in signature Σ_Ω, check that

1. x is a variable and the name of x is not the name of a constant in Σ_Ω.
2. M is a term in Σ_Ω with no free variables.
3. Every type variable occurring in M occurs in τ.

If all these checks are passed, a constant $\mathsf{x} : \tau$ is added to Σ_Ω and an axiom $\vdash \mathsf{x} = M$ is asserted. $\qquad\square$

Invocation of the principle of definition with c and M meeting the above requirements introduces c as an abbreviation for M. There is also a closely related principle of *constant specification* which we shall not discuss. There is also a principle of definition for types. Given a witness theorem showing that a subset P of an existing type is non-empty, the principle of type definition adds a new type constant to the signature and asserts the existence of a bijection between the new type and P.

Definition 4 (Principle of Type Definition). Given string op and theorem $\exists x : \tau.\ P\,x$ in signature Σ_Ω, check that

1. op is not the name of a type in Ω.
2. P is a term in Σ_Ω with no free variables.
3. Every type variable occurring in P occurs in τ.

If these requirements are satisfied, a type operator $(\alpha_1, \ldots, \alpha_n)op$, where $\alpha_1, \ldots, \alpha_n$ are the distinct type variables occurring in $\exists x : \tau.\ P\,x$, is added to Ω and an axiom stating the existence of a bijection between $(\alpha_1, \ldots, \alpha_n)op$ and the subset of τ denoted by P is asserted. We omit the details of the axiom. $\qquad\square$

This concludes the formal description of the basic HOL logic. For more extensive discussion, see [35].

2.3 The HOL4 Proof Environment

HOL4 is an implementation of the logic described above. It is the latest in a sequence of implementations going back to the mid-1980s. It conforms to a characterizing theme in HOL systems: adherence to the derivation judgement of the logic. In order to ensure this, only the few simple axioms and rules of inference enumerated above are encapsulated in the abstract type of theorems implemented in the logic kernel. As a consequence of this design decision, all theorems have to be obtained by actually performing proofs in HOL.[4]

Given such a primitive basis, serious verification efforts would be impossible were it not for the fact that SML is a programmable meta-language for the proof system. Effectively, the theorems provable in HOL are the closure of the base logic under SML programming! Thus, both derived inference rules that take bigger steps and high-level definition principles that introduce powerful abstractions are achieved by programming: the required complex logical derivations are reduced to a sequence of kernel inferences. For example, packages supporting various kind of high-level type definition facilities, e.g. quotient types or ML-style datatypes, have been programmed over the years. Datatypes can be mutually and nested recursive and may also use record notation; thus, they are quite similar to those of SML. For example, a type of polymorphic lists can be defined by invoking the following:

```
Hol_datatype `list = Nil | Cons of 'a => list`
```

which introduces the type `'a list` plus the constructors `Nil` and `Cons`. Many other consequences are automatically derived from this invocation as well: induction, case analysis, and primitive recursion theorems, theorems proving the uniqueness and

---

[4] However, in some cases, it is practical to allow external proof tools to be treated as oracles delivering HOL theorems *sans* proof. Such theorems are tagged in such a way that the provenance of subsequent theorems can be ascertained.

injectivity of the constructors, plus definitions supporting ML-style `case` constructs and pattern matching. The soundness of the results of making such a definition is ensured since the package can only use mechanisms that ultimately use the primitive inferences and definition principles.

At the term level, a number of definitional packages are heavily used. For example, an inductive definition package [18] (by John Harrison) defines inductively specified predicates and relations; mutual recursion and infinitary premises are allowed. Another package supports the definition of total recursive functions specified as recursion equations in ML pattern-matching style [44]. This package manipulates its input into terms that are used to instantiate formally proven well-founded induction and recursion theorems. Mutual and nested recursions are supported. Simple termination proofs have been automated; however, more serious termination proofs have of course to be performed interactively. We discuss this further in Sect. 3.

2.3.1 Proof Techniques

The view of proof in HOL4 is that the user interactively develops the proof at a high level, leaving subsidiary proofs to automated reasoners. Towards this, the system provides an underlying database of theorems (case analysis, induction, etc.) which supports user control of decisive proof steps. In combination with a few 'declarative proof' facilities, this allows many proofs to be conducted at a high level.

HOL4 provides a suite of automated reasoners. All produce HOL proofs. Propositional logic formulas can be sent off to external SAT tools and the resulting resolution-style proofs are backtranslated into HOL proofs. For formulas involving $\mathbb{N}$, $\mathbb{Z}$, or $\mathbb{R}$, decision procedures for linear arithmetic may be used. A decision procedure for n-bit words has recently been released. For formulas falling (roughly) into first-order logic, a robust implementation of ordered resolution, implemented by Joe Hurd, is commonly used.

Finally, probably the most commonly used proof technique in HOL (in common with other proof systems) is simplification. There are several simplification proof tools. For example, there is a call-by-value evaluation mechanism which reduces ground, and some symbolic, terms to normal form [3]. A more general, and more heavily used, tool – the simplifier – provides conditional and contextual ordered rewriting, using matching for higher order patterns. The simplifier may be extended with arbitrary context-aware decision procedures.

2.3.2 Custom Proof Procedures

As mentioned, most simple proofs in HOL can be accomplished via a small amount of interactive guidance (specifying induction or case-analysis, for example) followed by application of the simplifier and first-order proof search. However, it is common for proof tool developers to write their own inference procedures, specialized to the task at hand. Such work is typically based on *tactics* and *conversions*.

Tactics are a goal-directed decomposition technique invented by Milner in the original LCF system [13]. Conversions are due to Paulson [38] and provide a useful set of combinators for making custom simplification tools. Both tactics and conversions are commonly used in order to build specialized proof tools in HOL. Because of the so-called 'LCF-style' design adopted by HOL systems, such specialized extensions to the prover are completely safe in that they cannot introduce inconsistency. Many of the transformations discussed in the following can be seen as applications of the technique of conversions.

2.3.3 Theories and Libraries

The HOL4 system provides a wide collection of theories on which to base further verifications: booleans, pairs, sums, options, numbers ($\mathbb{N}$, $\mathbb{Z}$, $\mathbb{Q}$, $\mathbb{R}$, fixed point, floating point, and n-bit words), lists, lazy lists, character strings, partial orders, monad instances, predicate sets, multi-sets, finite maps, polynomials, probability, abstract algebra, elliptic curves, lambda calculus, program logics (Hoare logic, separation logic), machine models (ARM, PPC, and IA32), temporal logics (ω-automata, CTL, μ-calculus, and PSL), and so on. All theories have been built up definitionally and together represent hundreds of man-years of effort by researchers and students.

HOL4 also has an informal notion of a *library*, which is a collection of theories, APIs, and proof procedures supporting a particular domain. For example, the library for $\mathbb{N}$ provides theories formalizing Peano Arithmetic and extensions (numerals, gcd, and simple number theory), a decision procedure, simplification sets for arithmetic expressions, and an extensive collection of syntactic procedures for manipulating arithmetic terms. Loading a library extends the logical context with the types, constants, definitions, and theorems of the comprised theories; it also automatically extends general proof tools, such as the simplifier and the evaluator, with library-specific contributions.

Both theories and libraries are persistent: this is achieved by representing them as separately compiled ML structures. A 'make'-like dependency maintenance tool is used to automatically rebuild formalizations involving disparate collections of HOL4 libraries and theories, as well as ML or external source code in other programming languages.

2.3.4 External Interfaces

There is a variety of ways for a logic implementation to interface with external tools. On the input side, purported theorems coming from external tools need to be accompanied with enough information to reconstruct a HOL proof of the theorem. An example of this is the interface with SAT solvers, such as `minisat`, which can supply proof objects.

Another approach is illustrated by the integration of a BDD library into HOL. This has been used to support the formalization and application of model-checking algorithms for temporal logic. Since HOL theorems are eventually derived from operations on BDDs representing HOL terms, the oracle mechanism mentioned earlier is used to tag such theorems as having been constructed extra-logically.

On the output side, HOL formalizations confining themselves to the 'functional programming' subset of HOL may be exported to ML. This gives a pathway from formalizations to executables. The generated code is exported as separately compilable ML (SML or OCaml) source with no dependencies on the HOL4 implementation. Thus, the theory hierarchy of HOL4 is paralleled by a hierarchy of ML modules containing exported definitions of datatypes and computable functions formalized in HOL.

Finally, HOL can be used as a meta-logic to formalize another logic; such has been done for ACL2 [15, 16]. In this work, HOL4 was used to show that ACL2 is sound. This justifies a connection between the two systems in which a HOL formalization may be translated to the HOL theory of ACL2; this formalization is then transported to the ACL2 system and processed in some way (e.g. reduced using the powerful ACL2 evaluation engine) and then the result is transported back to HOL4 and backtranslated to the original HOL theory.

2.3.5 Applications

Peter Sewell and colleagues have used HOL4 to give the first detailed formal specifications of commonly used network infrastructure (UDP, TCP) [6]. This work has heavily used the tools available in HOL4 for operational semantics. They also implemented a derived inference rule which tested the conformance of real-world traces with their semantics.

As an application of the HOL4 backend of the Ott tool [43], Scott Owens has formalized the operational semantics of a large subset of OCaml and proved type soundness [36]. The formalization heavily relied upon the definition packages for datatypes, inductive relations, and recursive functions. Most of the proofs proceeded by rule induction, case analysis, simplification, and first-order proof search with user-selected lemmas. In recent work, Norrish has formalized the semantics of C++ [34].

An extremely detailed formalization of the ARM due to Anthony Fox sits at the centre of much current work in HOL4 focusing on the verification of low-level software. The development is based on a proof that a micro-architecture implements the ARM instruction set architecture. In turn, the ISA has been extended with so-called 'Thumb' instructions (which support compact code) and co-processor instructions. On top of the ISA semantics, Myreen has built a separation logic for the ARM and provided proof automation [31].

3 Source Language: TFL

The source language for our compiler is a subset of the HOL term language. This subset, called TFL in [44], amounts to a polymorphic, simply typed, higher order, pure functional programming language supporting pattern matching over algebraic datatypes. A 'program' in this language is simply a (total) mathematical function, and its semantics are obtained by applying the semantics of classical HOL [12]. Thus, notions of program execution, including evaluation order, are absent. This approach has several benefits:

1. Proofs about TFL programs may be conducted in the ordinary mathematics supported by HOL. Reasoning about a TFL program is typically based on the induction theorem arising from the recursion structure of the program, rather than induction along the evaluation relation of an operational semantics.
2. Many front end tasks in a compiler are already provided by HOL4: lexical analysis, parsing, type inference, overloading resolution, function definition, and termination proof.
3. The syntax of the language resembles the pure core subset of widely used functional programming languages such as SML and OCAML. Thus, our results can be easily extended to these practical languages.

The syntax of TFL is shown in Fig. 2, where $[term]_{separator}$ means a sequence of *term*s separated by the *separator*.

For example, Quicksort can be defined by the following invocations of a package implementing the automatic definition of TFL functions:

τ	$::= T \mid t \mid \tau\,D$	primitive type, type variable and algebraic type
	$\mid \tau\,\#\,\tau \mid \tau \rightarrow \tau$	tuple type and arrow(function) type
at_c	$::= id \mid id$ of $[\tau]_{\Rightarrow}$	algebraic datatype clause
at	$::=$ datatype $id = [at_c]_\mid$	algebraic datatype
	$\mid [at]_;$	mutually recursive datatype
pat	$::= v \mid \mathsf{C}\,\overrightarrow{pat}$	pattern
e	$::= i : T \mid v : \tau$	constant and variable
	$\mid \overrightarrow{e}$	tuple, i.e.$[e]_,$
	$\mid p\,\overrightarrow{e}$	primitive application
	$\mid \mathsf{C}\,\overrightarrow{e}$	constructor application
	$\mid f_{id}$	function identifier
	$\mid e\,e$	composite application
	$\mid$ if e then e else e	conditional
	$\mid$ case e of $[(c\ e) \rightarrow e]_\mid$	case analysis
	$\mid$ let $v = e$ in e	let binding
	$\mid \lambda v.e$	anonymous function
f_{decl}	$::= f_{id}\,([pat]_,) = e$	pattern matching clause
	$\mid [f_{decl}]_\wedge$	function declaration
	$\mid v = e$	top level variable declaration

Fig. 2 Syntax of TFL

```
Define
  '(PART P [] l1 l2 = (l1,l2)) /\
   (PART P (h::rst) l1 l2 =
       if P h then PART P rst (h::l1) l2
              else PART P rst  l1  (h::l2))';
tDefine
  "QSORT"
  '(QSORT ord [] = []) /\
   (QSORT ord (h::t) =
       let (l1,l2) = PART (\y. ord y h) t [] []
       in
         QSORT ord l1 ++ [h] ++ QSORT ord l2)'
 (WF_REL_TAC 'measure (LENGTH o SND)' THEN
  < ... rest of termination proof ... >);
```

The definition of the partition function PART is by primitive recursion, using pattern matching over Nil and Cons[5] Similarly, the definition of QSORT is recursive; however, an explicit termination proof (mostly elided) is needed in this case. Thus, Define automatically performs a simple but useful class of termination proofs while tDefine has the termination argument explicitly supplied as a tactic. Reasoning about QSORT is performed by using an induction theorem automatically derived after termination is proved.

In the following, we will sometimes not distinguish between **fun** and **Define** or between **datatype** and **Hol_datatype**. As well, we use some ASCII renderings of mathematical symbols. For example $\lambda v.M$ is replaced by \v. M and $\wedge$ is rendered as /\.

4 Compilation by Rewriting

The compilation process performs transformations that are familiar from existing functional language compilers except that transformations are implemented by deductive steps. TFL's high-level features such as polymorphism, higher order functions, pattern matching, and composite expressions need to be expressed in terms of much lower level structures. Briefly, the translator

- Converts pattern matching first into nested case expressions and eventually into explicit conditional expressions
- Removes polymorphism from TFL programs by making duplications of polymorphic datatype declarations and functions for each distinct combination of instantiating types
- Names intermediate computation results and imposes an evaluation order in the course of performing a continuation-passing-style (CPS) transformation

---

[5] The ML-like notation [] and infix :: is surface syntax for Nil and Cons.

- Applies defunctionalization to remove higher order functions by creating algebraic datatypes to represent function closures and type-based dispatch functions to direct the control to top-level function definitions
- Allocates registers using a naming convention in order to express the status (e.g. spilled or not) of variables representing machine registers

The transformations in our compiler are specified by formally proven rewrite rules the application of which is guided by programs written in the meta-language (Standard ML) used to implement the host logic. Generally, decisions of when and where to apply rewrites is syntax directed. Each rewrite rule has been formally proven so a rewriting step ensures that the transformed code is equivalent to the source. However, we can distinguish two different notions of a transformation:

1. *Prove beforehand.* The correctness of a rewrite rule is proven once and for all: a single object logic theorem establishes that all successful applications of this rule always generates a result that is equivalent to original program. Such a rule is applied in a local manner, and the necessary applicability checks are phrased as antecedents of the theorem.
2. *Prove dynamically.* A per-run correctness check is performed. The result of a rewrite is verified each time it is applied to a program. Rules of this form are necessary when the transformation and its applicability checks are not expressible in a single theorem.

4.1 Pattern Matching

The conversion of a function described by ML-style pattern matching to nested case expressions is based on Augustsson's original work [1], which was adapted by the first author [44] to support pattern-matching function definitions in logic. A pre-processing pass is first performed to deal with incomplete and overlapping patterns: incomplete patterns are made complete by iteratively adding rows for all missing constructors; overlapping patterns are handled by replacing a value with possible constructors. Note that this approach may make the pattern exponentially larger because no heuristics are used to choose the 'best' order in which sub-terms of any term are to be examined. In fact, recent work by Krauss has shown that heuristics may be the best that one can do: pattern minimization is a computationally difficult task, being as hard as QSAT for example [23].

The translation depends on the fact that for the declaration of a logical datatype ty with constructors $c_1, \ldots, c_n$, a $ty_$case function of the form

$$ty\_\text{case}\ f_1 \ldots f_n (c_1\ \overrightarrow{x}) = f_1\ \overrightarrow{x}$$

$$\vdots$$

$$ty\_\text{case}\ f_1 \ldots f_n (c_n\ \overrightarrow{x}) = f_n\ \overrightarrow{x}$$

$$\Delta \begin{pmatrix} \dfrac{z :: stack}{\begin{array}{l} v_1 :: pats_1 = rhs_1, \\ \dots \\ v_n :: pats_n = rhs_n \end{array}} \end{pmatrix} = \Delta \begin{pmatrix} \dfrac{stack}{\begin{array}{l} pats_1 = rhs_1[z \leftarrow v_1], \\ \dots \\ pats_n = rhs_n[z \leftarrow v_n] \end{array}} \end{pmatrix}$$

$$\Delta \begin{pmatrix} \dfrac{z :: stack}{\begin{array}{l} C_1 \ \overline{p_{11}} :: pats_{11} = rhs_{11}, \\ \dots \\ C_n \ \overline{p_{1k_1}} :: pats_{1k_1} = rhs_{1k_1} \\ C_n \ \overline{p_{n1}} :: pats_{n1} = rhs_{11}, \\ \dots \\ C_n \ \overline{p_{1k_n}} :: pats_{nk_n} = rhs_{nk_n} \end{array}} \end{pmatrix} = ty\_\mathsf{case} \ (\lambda v_1.M_1) \ \dots \ (\lambda v_n.M_n) \, z$$

$$\text{where } M_i = \Delta \begin{pmatrix} \dfrac{v_i :: stack}{\begin{array}{l} \overline{p_{i1}} \ @ \ pats_{k1} = rhs_{k1}, \\ \dots \\ \overline{p_{ik_i}} \ @ \ pats_{ik_i} = rhs_{ik_i} \end{array}} \end{pmatrix} \text{ for } i = 1, \dots, n$$

Fig. 3 Pattern matching

is defined. For example, the case expression for the natural numbers is defined as

$$(\mathsf{num\_case} \ b \ f \ 0 = b) \ \wedge \ (\mathsf{num\_case} \ b \ f \ (\mathsf{SUC} \ n) = f \ n) \, .$$

Case expressions form the target of the pattern-matching translation. The algorithm Δ shown in Fig. 3 converts a sequence of clauses of the form $[pat_i = rhs_i]_|$ into a nested case expression. Δ takes two arguments: a stack of subterms that are yet to be matched and a matrix whose rows correspond to the clauses in the pattern. All rows are of equal length, and the elements in a column should have the same type.

Conversion Δ proceeds from left to right, column by column. At each step the first column is examined. If each element in this column is a variable, then the head variable z in the stack is substituted for the corresponding v_i for the right-hand side of each clause. If each element in the column is the application of a constructor for type τ and τ contains constructor $c_1, \dots, c_n$, then the rows are partitioned into n groups of size $k_1, \dots, k_n$ according to the constructors. After partitioning, a row $(c(\bar{p}) :: pats; \ rhs)$ has its lead constructor discarded, resulting in a row expression $(\bar{p} \ @ \ pats; rhs)$. Here $::$ is the list constructor, and $@$ appends the second list to the first one. If constructor c_i has type $\tau_1 \to \cdots \to \tau_j \to \tau$, then new variables $v_i = v_1 : \tau_1, \dots, v_j : \tau_j$ are pushed onto the stack. Finally the results for all groups are combined into a case expression for the specified type.

Example 1 (Greatest Common Divisor).
The Greatest Common Divisor algorithm gcd can be specified by pattern matching as follows

```
Define
   `(gcd (0, y) = y) /\
    (gcd (x, 0) = x) /\
```

```
(gcd (x, y) =
   if y <= x then gcd (x - y, y)
                else gcd (x, y - x))`
```

After the pattern-matching transformation is used to define the function and termination is proved, we obtain the theorem (the irregularly named variables are an artefact of the pattern-matching translation)

```
|- gcd a =
    pair_case (\v v1.
       num_case
         (num_case 0 (\v7. SUC v7)  v1)
         (\v4. num_case (SUC v4)
                (\v9. if SUC v4 >= SUC v9
                        then gcd (SUC v4 - SUC v9,SUC v9)
                         else gcd (SUC v4,SUC v9 - SUC v4))
               v1)
        v) a
```

Then case expressions may be converted to conditional expressions based on the following theorem scheme mapping case expressions to their 'destructor' counterparts:

$$\vdash ty\_case \; f_1 \ldots f_n \; ob =$$
$$\text{if } \mathsf{is}C_1 \; ob \text{ then } f_1 \; (\mathsf{dest}C_{11} \; ob) \ldots (\mathsf{dest}C_{1k_1} \; ob) \text{ else}$$
$$\text{if } \mathsf{is}C_2 \; ob \text{ then } f_2 \; (\mathsf{dest}C_{21} \; ob) \ldots (\mathsf{dest}C_{2k_2} \; ob) \text{ else} \ldots$$

where operator $\mathsf{is}C_i$ tells whether a variable matches the ith constructor C_i, i.e. $\mathsf{is}C_i \; (C_j \; \vec{x}) = \mathsf{T}$ iff $i = j$; and operator $\mathsf{dest}C_{ij}$ is the jth projection function for constructor c_i. These operations are automatically defined when a datatype is declared. In addition, an optimization is performed to tuple variables: if an argument variable x has type $\tau_1\#\ldots\#\tau_n$, then it is replaced by a tuple of new variables $(x_1,\ldots,x_n)$. Superfluous branches and 'let' bindings are removed and some generally useful rewrites are applied. In this manner, the gcd equations are converted to

```
|- gcd (n1,n2) =
    if n1 = 0 then
       if n2 = 0 then 0 else n2
    else
       if n2 = 0 then n1
        else
          if n1 >= n2
            then gcd (n1 - n2, n2)
            else gcd (n1, n2 - n1)
```

4.2 Polymorphism

The *monomorphization* transformation eliminates polymorphism and produces a simply typed program that enables good data representations. The basic idea is to duplicate polymorphic datatype declarations at each ground type used and a function declaration at each type used, resulting in multiple monomorphic clones of polymorphic datatypes and functions. This step paves the way for subsequent conversions such as type-based defunctionalization. Although this approach would seem to lead to code explosion, it is manageable in practice. For example, MLton, a high-quality compiler for Standard ML, uses similar techniques and reports maximum increase of 30% in code size.

The first step is to build an *instantiation map* that enumerates, for each datatype and function declaration, the full set of instantiations for each polymorphic type. As mentioned above, a TFL program will be type checked by the HOL system and be decorated with polymorphic type variables such as $\alpha, \beta, \ldots$ when it is defined. In particular, type inference is done for (mutually) recursive function definitions. The remaining task is then to instantiate the generic types of a function with the actual types of arguments at its call sites, and this is also achieved by type inference.

The notation used in this section is as follows. A substitution rule $R = (t \hookrightarrow \{T\})$ maps a parameterized type t to a set of its type instantiations; an instantiation set $S = \{R\}$ is a set of substitution rules; and an instantiation map $M = \{z \hookrightarrow S\}$ maps a datatype or a function z to its instantiation set S. We write $M.y$ for the value at field y in the map M; if $y \notin \mathsf{Dom}\ M$ then $M.y$ returns an empty set. The union of two substitution sets $S_1 \cup_s S_2$ is $\{t \hookrightarrow (S_1.t \cup S_2.t) \mid t \in \mathsf{Dom}\ S_1 \cup \mathsf{Dom}\ S_2\}$. We write $\bigcup_s \{S\}$ for the union of a set of substitution rules. The union of two instantiation maps $M_1 \bigcup_m M_2$ is defined similarly. The composition of two instantiation sets S_1 and S_2, denoted as $S_1 \circ_r S_2$, is $\{z \hookrightarrow \bigcup \{S_2.t \mid t \in \mathsf{Dom}\ S_1\} \mid z \in Dom\ S_1\}$. Finally, the composition of an instantiation map M and a set S is defined as $M \circ_m S = \{z \hookrightarrow M.z \circ_r S \mid z \in \mathsf{Dom}\ M\}$.

The instantiation information of each occurrence of a polymorphic function and datatype is gathered into an instantiation map during a syntax-directed bottom-up traversal. The main conversion rules Γ and Δ shown in Fig. 4 build the instantiation map by investigating types and expressions respectively. The rule for a single variable/function declaration is trivial and omitted here: we just need to walk over the right-hand side of its definition. If a top-level function f is called in the body of another function g, then g must be visited first to generate an instantiation map M_g and then f is visited to generate M_f; finally these two maps are combined to a new one, i.e. $((M_f \circ M_g.f) \cup_m M_g)$. The clauses in mutually recursive functions can be visited in an arbitrary order.

This algorithm makes use of a couple of auxiliary functions provided by the HOL system. Function con2tp(c) maps a constructor c to the datatype to which it belongs; at_tp D returns σ if there is a datatype definition datatype $\sigma = D$ of $\ldots$; when x is either a function name or a constructor, and match_tp x τ matches the original type of x (i.e. the type when x is defined) with τ and returns a substitution set.

After the final instantiation map is obtained, we duplicate each polymorphic datatype and function declaration for all combinations of its type instantiations and replace each call of the polymorphic function with the call to its monomorphic clone with respect to the type. The automatic correctness proof for the transformation is

$$\begin{aligned}
\Gamma(\tau) &= \{\}, \quad \text{for } \tau \in \{T, t\} \\
\Gamma(\tau\ D) &= \{D \hookrightarrow \text{match_tp (at_tp } D)\ \tau\} \\
\Gamma(\tau_1\ op_t\ \tau_2) &= \Gamma(\tau_1)\ \cup_m\ \Gamma(\tau_2), \quad \text{for } op_t \in \{\#, \rightarrow\} \\
\Delta(i) &= \{\} \\
\Delta(v : \tau) &= \Gamma(\tau) \\
\Delta([e].) &= \bigcup_m \{\Gamma(e)\} \\
\Delta(p\ e) &= \Delta(e) \\
\Delta((c : \tau)\ e) &= \{\text{con2tp } c \hookrightarrow \text{match_tp (con2tp } c)\ \tau\} \\
&\quad \cup_m\ \Gamma(\tau)\ \cup_m\ \Delta(e) \\
\Delta((f : \tau)\ e) &= \{f_{id} \hookrightarrow \text{match_tp } f_{id}\ \tau\}\ \cup_m\ \Gamma(\tau)\ \cup_m\ \Delta(e) \\
\Delta(\text{if } e_1 \text{ then } e_2 \text{ else } e_3) &= \Delta(e_1)\ \cup_m\ \Delta(e_2)\ \cup_m\ \Delta(e_3) \\
\Delta(\text{case } e_1 \text{ of } [((c : \tau)\ e_2) \rightsquigarrow e_3]_|) &= \Delta(e_1)\ \cup_m\ \bigcup_m \{\{\text{con2tp } c \hookrightarrow \text{match_tp (con2tp } c)\ \tau\} \\
&\quad \cup_m\ \Delta(e_2)\ \cup_m\ \Delta(e_3)\} \\
\Delta(\text{let } v = e_1 \text{ in } e_2) &= (\Delta(e_1)\ \circ_m\ \Delta(e_2).v)\ \cup_m\ \Delta(e_2) \\
\Delta([\lambda\ v.]^* e) &= \Delta(e)
\end{aligned}$$

Fig. 4 Instantiation map construction for polymorphic components

trivial: each duplication of a polymorphic function computes the same function on the arguments of the instantiating types.

Example 2. Consider the following specification, in ML notation:

```
datatype (α, β) ty = c of α # β
fun f (x : α) = x
fun g (x : β, y : σ) = let h(z) = c(f(x), f(z)) in h(y)
j = (g (1 : num, F), g (F, T))
```

In this definition, h has the inferred type $\tau \rightarrow (\beta, \sigma)$ ty. The algorithm builds the following instantiation maps:

Investigate $j : M_j = \{g \hookrightarrow \{'c \hookrightarrow \{bool, num\}, 'd \hookrightarrow \{bool\}\}\}$
Investigate $g : M_g = \{f \hookrightarrow \{'a \hookrightarrow \{'c, 'd\}\}, \sigma \hookrightarrow \{'a \hookrightarrow \{'c\}, 'b \hookrightarrow \{'d\}\}\}$
Compose M_g and M_j: $M_{g \circ j} = M_g \circ M_j.g =$
$\quad \{f \hookrightarrow \{'a \hookrightarrow \{bool, num\}\}, \sigma \hookrightarrow \{'a \hookrightarrow \{bool, num\}, 'b \hookrightarrow \{bool\}\}\}$
Union M_g and $M_{g \circ j}$: $M_{\{g, j\}} = M_g \cup_m M_j =$
$\quad \{f \hookrightarrow \{'a \hookrightarrow \{bool, num\}\}, g \hookrightarrow \{'c \hookrightarrow \{bool, num\}, 'd \hookrightarrow \{bool\}\},$
$\quad\quad \sigma \hookrightarrow \{'a \hookrightarrow \{bool, num\}, 'b \hookrightarrow \{bool\}\}\}$
Investigate f : no changes, $M_{\{f, g, j\}} = M_{\{g, j\}}$

Then for datatype σ, function f, and function g, a monomorphic clone is created for each combination of instantiating types. Calls to the original functions are replaced with the appropriate copies of the right type. For example, expression j is converted to $(\mathsf{g}_{num\#bool}\ (1, \mathsf{F}), \mathsf{g}_{bool\#bool}\ (\mathsf{F}, \mathsf{T}))$, where $\mathsf{g}_{num\#bool}$ and $\mathsf{g}_{bool\#bool}$ are the two cloned definitions of g. The correctness of the conversion is proved based on the theorems showing that g's clones compute the same function as g

with respect to the instantiating types: $\vdash$ (g : *num#bool*) = $g_{num\#bool} \wedge$ (g : *bool#bool*) = $g_{bool\#bool}$.

4.3 *Sequentialization via CPS*

This transformation bridges the gap between the form of expressions and control flow structures in TFL and assembly. A TFL program is converted to a simpler form such that (1) the arguments to function and constructor applications are atoms like variables or constants; (2) discriminators in case expressions are also simple expressions; (3) compound expressions nested in an expression are lifted to make new 'let' bindings; and (4) curried functions are uncurried to a sequence of simple functions that take a single tupled argument.

To achieve this, a CPS transformation is performed. The effect is to sequentialize the computation of TFL expressions by introducing variables for intermediate results, and the control flow is pinned down into a sequence of elementary steps. It extends the one in our software compiler [27] by addressing higher level structures specific to TFL. Since there is no AST of programs in our approach, the CPS translation cannot be defined over a particular type; instead, it comprises a generic set of rewrite rules applicable to any HOL term. The basis is the following definition of the suspended application of a continuation parameter:

$$\vdash \mathsf{c}ek = ke \ .$$

To CPS an expression e we create the term $\mathsf{c}\ e\ (\lambda x.x)$ and then exhaustively apply the following rewrite rules, which are easy to prove since they are just rearrangements of simple facts about lambda calculus (Fig. 5).

Application of these rules pushes occurrences of c deeper into the term. After this phase of rewriting finishes, we rewrite with the theorem $\vdash \mathsf{c}\ e\ k = \mathsf{let}\ x = e\ \mathsf{in}\ k\ x$ and β-reduce to obtain a readable 'let'-based normal form.

$$
\begin{aligned}
&\vdash \mathsf{c}\ (e)\ k = k\ e, \quad \text{when } e \text{ is a primitive expression}\\
&\vdash \mathsf{c}\ (\lambda \overrightarrow{v}.e) = \lambda \overrightarrow{v}.\lambda k.\mathsf{c}\ (e)\ k\\
&\vdash \mathsf{c}\ (e_1, e_2)\ k = \mathsf{c}\ e_1\ (\lambda x.\mathsf{c}\ e_2\ (\lambda y.k\ (x, y)))\\
&\vdash \mathsf{c}\ (op\ e)\ k = \mathsf{c}\ (e)\ (\lambda x.k\ (op\ x)) \quad \text{when } op \in \{p, c, f_{id}\}\\
&\vdash \mathsf{c}\ (e_1\ e_2)\ k = \mathsf{c}\ (e_1)\ (\lambda x.\mathsf{c}\ (e_2)\ (\lambda y.k\ (x\ y)))\\
&\vdash \mathsf{c}\ (\mathsf{let}\ v = e_1\ \mathsf{in}\ e_2)\ k = \mathsf{c}\ (e_1)\ (\lambda x.\mathsf{c}\ (e_2)\ (\lambda y.k\ y))\\
&\vdash \mathsf{c}\ (\mathsf{if}\ e_1\ \mathsf{then}\ e_2\ \mathsf{else}\ e_3)\ k =\\
&\qquad \mathsf{c}\ (e_1)\ (\lambda x.k\ (\mathsf{if}\ x\ \mathsf{then}\ \mathsf{c}\ (e_2)\ (\lambda x.x)\ \mathsf{else}\ \mathsf{c}\ (e_3)\ (\lambda x.x)))\\
&\vdash \mathsf{c}\ (\mathsf{case}\ e_1\ \mathsf{of}\ c\ e_{2_1} \to e_{3_1}\ |\ c\ e_{2_2} \to e_{3_2}\ |\ \ldots)\ k =\\
&\qquad \mathsf{c}\ (e_1)\ (\lambda x.(\mathsf{c}\ (e_{2_1})\ (\lambda y_1.\mathsf{c}\ (e_{2_2})\ (\lambda y_2. \ldots,\\
&\qquad\quad k\ (\mathsf{case}\ x\ \mathsf{of}\ c\ y_1 \to \mathsf{c}\ (e_{3_1})\ (\lambda x.x)\ |\ c\ y_2 \to \mathsf{c}\ (e_{3_2})\ (\lambda x.x)\ |\ \ldots)))))
\end{aligned}
$$

Fig. 5 CPS conversion

Example 3. As a simple example with no control flow, consider the following operation found in the TEA block cipher [47]:

$$\mathsf{ShiftXor}\ (x, s, k_0, k_1) = (x \ll 4 + k_0) \oplus (x + s) \oplus (x \ll 5 + k_1)$$

All operations are on 32-bit machine words. In HOL4's ASCII representation this is

```
|- ShiftXor (x,s,k0,k1) =
     ((x << 4) + k0) ?? (x+s) ?? ((x >> 5) + k1)
```

The CPS rewriting pass flattens this to the equal form

```
|- ShiftXor (v1,v2,v3,v4) =
     let v5 = v1 << 4 in
     let v6 = v5 + v3 in
     let v7 = v1 + v2 in
     let v8 = v6 ?? v7 in
     let v9 = v1 << 5 in
     let v10 = v9 + v4 in
     let v11 = v8 ?? v10
     in v11
```

4.4 Defunctionalization

In the next phase of compilation, we convert higher order functions into equivalent first-order functions and hoist nested functions to the top level. This is achieved through a type-based closure conversion. After the conversion, no nested functions exist; and function call is made by dispatching on the closure tag followed by a top-level call.

Function closures are represented as algebraic data types in a way that, for each function definition, a constructor taking the free variables of this function is created. For each arrow type, we create a dispatch function, which converts the definition of a function of this arrow type into a closure constructor application. A nested function is hoisted to the top level with its free variables to be passed as extra arguments. After that, the calling to the original function is replaced by a calling to the relevant dispatch function passing a closure containing the values of this function's free variables. The dispatch function examines the closure tag and passes control to the appropriate hoisted function. Thus, higher order operations on functions are replaced by equivalent operations on first-order closure values.

As an optimization, we first run a pass to identify all 'targeted' functions which appear in the arguments or outputs of other functions and record them in a side effect variable `Targeted`. Non-targeted functions need not to be closure converted, and calls to them are made as usual. During this pass we also find out the functions to be defined at the top level and record them in `Hoisted`. Finally `Hoisted` contains all top-level functions and nested function to be hoisted.

The conversion works on simple typed functions obtained by monomorphisation. We create a closure datatype and a dispatch function for each of the arrow types that targeted functions may have. A function definition is replaced by a binding to an application of the corresponding closure constructor to this function's free variables. Suppose the set of targeted functions of type τ is $\{f_i\ x_i = e_i \mid i = 1, 2, \ldots\}$, then the following algebraic datatype and dispatch function are created, where tp_of and fv return the type and free variables of a term, respectively (and the type builder Γ will be described below):

$$clos_\tau = cons^\tau_{f_1}\ \text{of}\ \Gamma(\text{tp\_of}\ (\text{fv}\ f_1))\ |\ cons^\tau_{f_2}\ \text{of}\ \Gamma(\text{tp\_of}\ (\text{fv}\ f_2))\ |\ \ldots$$
$$(dispatch_\tau(cons^\tau_{f_1}, x_1, y_1) = (f_1 : \Gamma(\tau))\ (x_1, y_1))\quad \wedge$$
$$(dispatch_\tau(cons^\tau_{f_2}, x_2, y_2) = (f_2 : \Gamma(\tau))\ (x_2, y_2))\quad \wedge$$
$$\cdots$$

As shown in Fig. 6, the main translation algorithm inspects the references and applications of targeted functions and replaces them with the corresponding closures and dispatch functions. Function Γ returns the new types of variables. When walking over expressions, Δ replaces calls to unknown functions (i.e. those not presented in <u>Hoisted</u>) with calls to the appropriate dispatch function and calls to known functions with calls to hoisted functions. In this case, the values of free variables are passed as extra arguments. Function references are also replaced with appropriate closures. Finally <u>Redefn</u> contains all converted functions, which will be renamed and redefined in HOL at the top level.

Now we show the technique to prove the equivalence of a source function f to its converted form f'. We say that a variable $v' : \tau'$ corresponds $v : \tau$ iff (1) $v = v'$ if both τ and τ' are closure type or neither of them is. (2) $\forall x \forall x'.\ dispatch_{\tau'}(v', x') = v\ x$ if v' is a closure type and v is an arrow type, and x' corresponds to x; or vice

$$
\begin{array}{ll}
\Gamma(v : T) & = T \\
\Gamma(v : \tau_1 \to \tau_2) & = \text{if } v \in \underline{\text{Targeted}} \text{ then } clos_{\tau_1 \to \tau_2} \text{ else } \tau_1 \to \tau_2 \\
\Gamma(v : \tau\ D) & = \Gamma(\tau)\ D \\
\Gamma([v].) & = [\Gamma(v)]. \\
\Delta(v : \tau) & = \text{if } v \in \underline{\text{Targeted}} \text{ then } cons^\tau_v \text{ else } v : clos_\tau \\
\Delta([e].) & = [\Delta(e)]. \\
\Delta(p\ e) & = p\ (\Delta(e)) \\
\Delta(c\ e) & = c\ (\Delta(e)) \\
\Delta((f : \tau)\ e) & = \text{if } f \in \underline{\text{Hoisted}} \text{ then } (\text{new\_name\_of}\ f)\ (\Delta(e), \text{fv}\ f) \\
& \quad \text{else } dispatch_\tau\ (f : clos_\tau, \Delta(e)) \\
\Delta(\text{if } e_1 \text{ then } e_2 \text{ else } e_3) & = \text{if } \Delta(e_1) \text{ then } \Delta(e_2) \text{ else } \Delta(e_3) \\
\Delta(\text{case } e_1 \text{ of } [c\ e_2 \rightsquigarrow e_3]_|) & = \text{case } \Delta(e_1) \text{ of } [(\Delta(c\ e_2)) \rightsquigarrow \Delta(e_3)]_| \\
\Delta(\text{let } f = \lambda \vec{v}.e_1 \text{ in } e_2) & = (\Phi(f\ \vec{v} = e_1)\ ;\ \Delta(e_2)) \\
\Delta(\text{let } v = e_1 \text{ in } e_2) & = \text{let } v = \Delta(e_1) \text{ in } \Delta(e_2) \quad \text{when } e_1 \text{ is not a } \lambda \text{ expression}
\end{array}
$$

$\Phi(f_{id}\ (\vec{v} : \tau) = e) =$
    let $e' = \Delta(e)$ in
    <u>Redefn</u> := <u>Redefn</u> + $(f_{id} \hookrightarrow \underline{\text{Redefn}}.f_{id} \cup \{(f_{id} : \tau \to \Gamma(\text{tp\_of}\ e'))\ \vec{v} = e'\}$
$\Phi([f_{decl}]_\wedge) = [\Phi(f_{decl})].$

Fig. 6 Closure conversion

versa. Then f' is equivalent to f iff they correspond to each other. The proof process is simple, as it suffices to simply rewrite with the old and new definitions of the functions.

Example 4. The following higher order program

```
fun f (x : num) = x * 2 < x + 10
fun g (s : num → bool, x : num) =
    let h₁ = λy. y + x in if s x then h₁ else let h₂ = λy. h₁ y * x in h₂
fun k (x : num) = if x = 0 then 1 else g (f, x) (k (x − 1))
```

is closure converted to

$$\textsf{datatype } clos_{\tau_1} = cons_f^{\tau_1}$$

$$\textsf{datatype } clos_{\tau_2} = cons_{h_1}^{\tau_2} \textsf{ of } num \mid cons_{h_2}^{\tau_2} \textsf{ of } num$$

$$\textsf{fun } dispatch_{\tau_1} \ (cons_f^{\tau_1} : clos_{\tau_1}, x : num) = f' \ x \ \wedge \ f' \ x = x * 2 < x + 10$$

$$\textsf{fun } dispatch_{\tau_2} \ (cons_{h_1}^{\tau_2} \ y : clos_{\tau_2}, \ x : num) = h_1' \ (y, x)) \ \wedge$$

$$\textsf{fun } dispatch_{\tau_2} \ (cons_{h_2}^{\tau_2} \ y : clos_{\tau_2}, \ x : num) = h_2' \ (y, x)) \ \wedge$$

$$\textsf{fun } h_1' \ (y, x) = y + x \ \wedge \ \textsf{fun } h_2' \ (y, x) = h_1'(y, x) * x$$

$$\textsf{fun } g' \ (s : clos_{\tau_1}, x : num) = \textsf{if } dispatch_{\tau_1} \ (s, x) \textsf{ then } cons_{h_1}^{\tau_2} \ x \textsf{ else } cons_{h_2}^{\tau_2} \ x$$

$$\textsf{fun } k' \ (x : num) = \textsf{if } x = 0 \textsf{ then } 1 \textsf{ else } g \ (cons_f^{\tau_1}, x), (k' \ (x − 1))$$

where τ_1 and τ_2 stand for arrow types $num \rightarrow bool$ and $num \rightarrow num$ respectively. The following theorems (which are proved automatically) justify the correctness of this conversion:

$$\vdash f = f' \qquad \vdash k' = k$$
$$\vdash (\forall x. dispatch_{\tau_1} \ (s', x) = s \ x) \Rightarrow \forall x \forall y. dispatch_{\tau_2} \ (g' \ (s', x), y)$$
$$= (g \ (s, x)) \ y$$

4.5 Register Allocation

One of the most sophisticated algorithms in a compiler is register allocation, which supports the fiction of an unbounded number of variables with a fixed number of registers. Although many register allocation algorithms exist for imperative languages, we find them unnecessarily complicated for our purely functional language because variables are never destructively updated, obviating the standard notion of def-use chains. Operating over the SSA format, our algorithm is a simple greedy algorithm with backtracking for early spilling.

The basic policy of register allocation is to avoid registers already assigned to live variables. Variables live at the same time should not be allocated to the same register.

In this section, the naming convention is: variables yet to be allocated begin with v, variables spilled begin with m (memory variable), and those in registers begin with r (register variable). Notation '$\_$' matches a variable of any of these kinds. $\hat{v}$, $\hat{r}$, and $\hat{m}$ stand for a fresh variable, an unused register, and a new memory location, respectively. Predicate $r \leftarrow v$ specifies that variable v is assigned to register r; by definition $\forall r \in S_{\text{mach}}.\, r \leftarrow r$ and $\forall r \in_v S_{\text{mach}} \forall m.\, r \nleftarrow m$ (where S_{mach} is the set of machine registers). Notation avail e returns the set of available registers after allocating e, i.e. avail $e = S_{\text{mach}} - \{r \mid \forall w.\, w \in e \land r \leftarrow w\}$. Administrative terms app, save and restore are all defined as $\lambda x.x$; and loc $(v, l) = l$ indicates that variable v is allocated to location l (where $l = r$ or m). A function application is denoted by app.

When variable v in expression let $v = e_1$ in $e_2[v]$ is to be assigned a register, the live variables to be considered are just the free variables in e_2, excluding v. If live variables do not use up all the machine registers, then we pick an available register and assign v to it by applying rule assgn. Otherwise, we spill to the memory a variable consuming a register and assign this register to v. In some cases, we prefer to spill a variable as early as possible: in the early_spill rule variable w's value is spilled from r for future use; r may not be allocated to v in the subsequent allocation. When encountering a memory variable in later phases, we need to generate code that will restore its value from the memory to a register (the $\hat{v}$ in rule restore will be assigned a register by the subsequent application of rule assgn).

The allocation can be viewed as being implemented by rewriting with the following set of rules.

$$
\begin{array}{ll}
[\text{assign}] & \text{let } v = e_1 \text{ in } e_2[v] \;\longleftrightarrow\; \text{let } \hat{r} = e_1 \text{ in } e_2[\text{loc}(v, \hat{r})] \Leftarrow \text{avail } e_2 \neq \phi \\[4pt]
[\text{spill}] & \text{let } v = e_1 \text{ in } e_2[v, \text{loc}(w, r)] \;\longleftrightarrow\; \\
& \text{let } \hat{m} = \text{save } r \text{ let } r = e_1 \text{ in } e_2[\text{loc}(v, r), \text{loc}(w, \hat{m})] \Leftarrow \text{avail } e_2 = \phi \\[4pt]
[\text{early\_spill}] & \text{let } v = e_1 \text{ in } e_2[v, \text{loc}(w, r)] \;\longleftrightarrow\; \\
& \text{let } \hat{m} = \text{save } r \text{ in let } v = e_1 \text{ in } e_2[v, \text{loc}(w, \hat{m})] \Leftarrow \text{avail } e_2 = \phi \\[4pt]
[\text{restore}] & e[\text{loc}(v, m)] \;\longleftrightarrow\; \text{let } \hat{v} = \text{restore } m \text{ in } e[\hat{v}] \\[4pt]
[\text{caller\_save}] & \text{let } \_ = \text{app } f \text{ in } e[\_, \text{loc}(w, r)] \;\longleftrightarrow\; \\
& \text{let } \hat{m} = \text{save } r \text{ in let } \_ = \text{app } f \text{ in } e[\_, \text{loc}(w, \hat{m})] \\[4pt]
[\text{spill\_if}] & \text{let } \_ = \text{if } e_1 \text{ then } e_2[\text{loc}(w, r_1)] \text{ else } e_3[\text{loc}(w, r_2)] \text{ in } e_4[\text{loc}(w, r_0)] \;\longleftrightarrow\; \\
& \text{let } \hat{m} = \text{save } r_0 \text{ in let } \_ = \text{if } e_1 \text{ then } e_2[\text{loc}(w, \hat{m})] \text{ else } e_3[\text{loc}(w, \hat{m})] \text{ in} \\
& \quad e_4[\text{loc}(w, \hat{m})] \Leftarrow \neg(r_0 = r_1 = r_2)
\end{array}
$$

The format of a rule

$$[\text{name}]\; redex \;\longleftrightarrow\; contractum \;\Leftarrow\; P$$

specifies that an expression matching *redex* can be replaced with the instantiated *contractum* provided that side condition P over the *redex* holds. The declarative part of the rule, *redex* $\longleftrightarrow$ *contractum*, is a HOL theorem that characterizes the transformation to be performed; the control part, P, specifies in what cases the rewrite should be applied. Notation $e[v]$ stands for an expression that has free occurrences of expression v; and $e[v_1, \ldots, v_n] \longleftrightarrow e[w_1, \ldots, w_n]$ indicates that, for $\forall i.\ 1 \le i \le n$, all occurrences of v_i in e are replaced with w_i.

Saving is necessary not only when registers are spilled, but also when functions are called. Our compiler adopts the *caller-save* convention, so every function call is assumed to destroy the values of all registers. Therefore, we need to save the values of all registers that are live at that point, as implemented in the caller_save rule. In addition, as we allocate the two branches of a conditional expression separately, a variable may be assigned different registers by the branches. This will contradict the convention that a variable should be assigned only one register. In this case, we just need to early spill it through the spill_if rule.

In the final step, all save, store, and loc in an expression are eliminated. This results in an equivalent expression containing only register variables and memory variables.

$$[\text{elim\_save}] \quad \text{let } m = \text{save } r \text{ in } e[m] \longleftrightarrow e[r]$$
$$[\text{elim\_store}] \quad \text{let } r = \text{store } m \text{ in } e[r] \longleftrightarrow e[m]$$

In practice, in order to improve the performance we do not have to perform equivalence check for every rewrite step. Instead, after all the rewrites are done, by applying the following rules to the produced expression, we will obtain an expression that is α-equivalent to the original expression, thus validating that the register allocation on the entire expression is correct. This ease of verification of the correctness of the allocation is a definite 'win' for the translation validation approach over the verified compiler approach, especially when one considers the difficulty of verifying a register allocator. The technique here was discovered by Hickey and Nogin [20] and has also been exploited by Leroy [25].

Example 5. A round of encryption in TEA can be defined as follows

```
|- Round ((y,z),(k_0,k_1,k_2,k_3),s) =
    let s' = s + 2654435769w in
    let y' = y + ShiftXor (z,s',k_0,k_1)
    in ((y', z + ShiftXor (y',s',k_2,k_3)),
        (k_0,k_1,k_2,k_3),s')
```

Sequentialization yields the intermediate result

```
|- Round ((v1,v2),(v3,v4,v5,v6),v7) =
    let v8 = v7 + 2654435769w in
    let v9 = ShiftXor (v2,v8,v3,v4) in
    let v10 = v1 + v9 in
    let v11 = ShiftXor (v10,v8,v5,v6) in
    let v12 = v2 + v11
    in
        ((v10,v12),(v3,v4,v5,v6),v8)
```

and with four available registers, we obtain the allocation

```
|- Round ((r0,r1),(r2,r3,m1,m2),m3) =
      let m4 = r2 in
      let r2 = m3 in
      let r2 = r2 + 2654435769w in
      let m3 = r3 in
      let r3 = ShiftXor (r1,r2,m4,m3) in
      let r0 = r0 + r3 in
      let r3 = ShiftXor (r0,r2,m1,m2) in
      let r1 = r1 + r3
        in
          ((r0,r1),(m4,m3,m1,m2),r2)
```

5 Engaging with the Machine

Up to now, we have operated purely in logic, deriving ever-more assembly like
presentations of high-level functions, preserving equality all the while. We have
gone surprisingly far down the path to machine code. However, eventually the real
operational behaviour of the target machine and its limitations due to finite available
memory and word size have to be dealt with. We will not do so in this chapter.
However, Myreen has implemented a proof-producing compiler [30, 32] for a low-
level language which takes up approximately where our development here stops.
His compiler produces a theorem asserting that the logical function taken as input
is faithfully executed by the machine. The final gap between these two compilers is
one that we intend to bridge in future work. Among other things, we will have to
map to a uniform representation for datatypes and also deal with the allocation and
collection of garbage generated by our programs. The garbage collector verified by
Myreen in his thesis work will probably play a key role.

6 Related Work

Compiler verification is a venerable topic with many publications [11]. There
has also been a huge amount of research on translating functional languages;
one of the most influential on us has been the paper of Tolmach and Oliva [46]
which developed a translation from an SML-like functional language to Ada. Our
monomorphisation and closure conversion methods are similar to theirs, i.e. removing polymorphism by code specialization and higher order functions through closure
conversion. However, we target logic specification languages and perform correctness proofs on the transformations. Our work can be regarded as an extension of
theirs by now verifying the correctness of these two conversions in a translation-
validation style.

Hickey and Nogin [20] worked in MetaPRL to construct a compiler from a full
higher order, untyped, functional language to Intel x86 code, based almost entirely

on higher order rewrite rules. A set of unverified rewriting rules are used to convert a higher level program to a lower level program. They use higher order abstract syntax to represent programs and do not define the semantics of these programs. Thus, no formal verification of the rewriting rules is done.

Proof producing compilation for smaller subsets of logic has already been investigated in a prototype hardware compiler, which synthesizes Verilog netlists [14], and a software compiler [28], which produced low-level code from first-order HOL functions.

Hannan and Pfenning [17] constructed a verified compiler in LF for the untyped λ-calculus. The target machine is a variant of the CAM runtime and differs greatly from real machines. In their work, programs are associated with operational semantics; and both compiler transformation and verifications are modeled as deductive systems. Chlipala [8] further considered compiling a simply typed λ-calculus to assembly language. He proved semantics preservation based on denotational semantics assigned to the intermediate languages. Type preservation for each compiler pass was also verified. The source language in these works is the bare lambda calculus and is thus much simpler than TFL; thus, their compilers only begin to deal with the high-level issues, which we discuss in this paper.

Compared with Chlipala [8] who gives intermediate languages dependent types, Benton and Zarfaty [4] interpret types as binary relations. They proved semantic type soundness for a compiler from a simple imperative language with heap-allocated data into an idealized assembly language.

Leroy [7, 25] verified a compiler from a subset of C, i.e. Clight, to PowerPC assembly code in the Coq system. The semantics of Clight is completely deterministic and specified as a big-step operational semantics. Several intermediate languages are introduced, and translations between them are verified. The proof of semantics preservation for the translation proceeds by induction over the Clight evaluation derivation and case analysis on the last evaluation rule used; in contrast, our proofs proceed by verifying the rewriting steps.

A purely operational semantics-based development is that of Klein and Nipkow [22] which gives a thorough formalization of a Java-like language. A compiler from this language to a subset of Java Virtual Machine is verified using Isabelle/HOL. The Isabelle/HOL theorem prover is also used to verify the compilation from a type-safe subset of C to DLX assembly code [24], where a big step semantics and a small step semantics for this language are defined. Meyer and Wolff [29] derive in Isabelle/HOL a verified compilation of a lazy language called MiniHaskell to a strict language called MiniML based on the denotational semantics of these languages.

7 Conclusions and Future Work

We are in the midst of a change in perspective on compilers. In the past, a compiler was expected to generate correct and efficient code; in the future, a compiler will still be expected to generate good code, but it must also utilize and contribute to a range of program analyses, some of which may be done by the compiler itself, but some of which are external. Among the most important of these analyses is program verification. In a setting where program properties are important enough to be formally proved, the following are desirable:

- A source language with clear semantics and a good program logic.
- A compilation path, the result of which is machine code with a guarantee that executing that code on the target platform yields a result equal to that specified by the source program.

The approach taken in this paper satisfies the above criteria. Since the source language is a subset of HOL functions, the semantics are clear and the program logic is just HOL. We have also shown how compilation of a program can produce lower level programs as an integral part of proving the compilation run correct.

In future work, we want to scale up the compilation algorithms and investigate wider application of the technique. In particular, what applications formalize easily in pure functional programming and are important enough to warrant full functional verification and correctness of compilation?

Acknowledgement Thanks to David Hardin, Mike Gordon, Magnus Myreen, and Thomas Türk for help, encouragement, and advice.

References

1. Augustsson L (1985) Compiling pattern matching. In: Jouannnaud JP (ed) Conference on functional programming languages and computer architecture (LNCS 201) (Nancy, France), pp 368–381
2. Balakrishnan G, Reps TW, Melski D, Teitelbaum T (2005) Wysinwyx: what you see is not what you execute. In: Meyer B, Woodcock J (eds) VSTTE, Lecture notes in computer science, vol 4171. Springer, Berlin, pp 202–213
3. Barras B (2000) Proving and computing in HOL. In: Proceedings of TPHOLs 2000, LNCS, vol 1869. Springer, Berlin, pp 17–37
4. Benton N, Zarfaty U (2009) Formalizing and verifying semantic type soundness of a simple compiler. In: International conference on principles and practice of declarative programming (PPDP)
5. Bertot Y, Caste'ran P (2004) Interactive theorem proving and program development: Coq'art: the calculus of inductive constructions. Texts in theoretical computer science, An EATCS series. Springer, Berlin
6. Bishop S, Fairbairn M, Norrish M, Sewell P, Smith M, Wansbrough K (2005) Rigorous specification and conformance testing techniques for network protocols, as applied to TCP, UDP, and Sockets. In: Proceedings of SIGCOMM. ACM, New York, NY
7. Blazy S, Dargaye Z, Leroy X (2006) Formal verification of a C compiler front-end. In: 14th International symposium on formal methods (FM 2006), Hamilton, Canada
8. Chlipala A (2007) A certified type-preserving compiler from lambda calculus to assembly language. In: Conference on programming language design and implementation (PLDI'07)
9. Church A (1940) A formulation of the simple theory of types. J Symbolic Log 5:56–68
10. Conchon S, Filliatre J-C (2007) A persistent union-find data structure. In: 2007 ACM SIG-PLAN workshop on ML (Freiburg, Germany), October 2007

11. Dave MA (2003) Compiler verification: a bibliography. ACM SIGSOFT Softw Eng Notes 28(6):2
12. Gordon M, Melham T (1993) Introduction to HOL, a theorem proving environment for higher order logic. Cambridge University Press, Cambridge
13. Gordon M, Milner R, Wadsworth C (1979) Edinburgh LCF: a mechanised logic of computation, Lecture notes in computer science, vol 78. Springer, Berlin
14. Gordon M, Iyoda J, Owens S, Slind K (2005) Automatic formal synthesis of hardware from higher order logic. In: Proceedings of fifth international workshop on automated verification of critical systems (AVoCS), ENTCS, vol 145
15. Gordon MJC, Hunt WA, Kaufmann M, Reynolds J (2006a) An embedding of the ACL2 logic in HOL. In: Proceedings of ACL2 2006, ACM international conference proceeding series, vol 205. ACM, New York, NY, pp 40–46
16. Gordon MJC, Reynolds J, Hunt WA, Kaufmann M (2006b) An integration of HOL and ACL2. In: Proceedings of FMCAD 2006. IEEE Computer Society, Washington, DC, pp 153–160
17. Hannan J, Pfenning F (1992) Compiler verification in LF. In: Proceedings of the 7th symposium on logic in computer science
18. Harrison J (1995) Inductive definitions: automation and application. In: Schubert ET, Windley PJ, Alves-Foss J (eds) Proceedings of the 1995 international workshop on higher order logic theorem proving and its applications (Aspen Grove, Utah), LNCS, vol 971. Springer, Berlin, pp 200–213
19. Harrison J (1998) Theorem proving with the real numbers. CPHC/BCS distinguished dissertations, Springer, Berlin
20. Hickey J, Nogin A (2006) Formal compiler construction in a logical framework. High Order Symbolic Comput 19(2–3):197–230
21. Kaufmann M, Manolios P, Moore JS (2000) Computer-aided reasoning: an approach. Kluwer, Dordrecht
22. Klein G, Nipkow T (2006) A machine-checked model for a Java-like language, virtual machine and compiler. TOPLAS 28(4):619–695 619–695
23. Krauss K (2009) Automating recursive definitions and termination proofs in higher order logic. PhD thesis, Institut für Informatik, Technische Universität München
24. Leinenbach D, Paul W, Petrova E (2005) Towards the formal verification of a C0 compiler: code generation and implementation correctness. In: 4th IEEE international conference on software engineering and formal methods (SEFM 2006)
25. Leroy X (2006) Formal certification of a compiler backend, or: programming a compiler with a proof assistant. In: Proceedings of POPL 2006. ACM, New York, NY
26. Leroy X (2009) Formal verification of a realistic compiler. Commun ACM 52(7):107–115
27. Li G, Slind K (2007) Compilation as rewriting in higher order logic. In: Conference on automated deduction (CADE-21), July 2007
28. Li G, Owens S, Slind K (2007) Structure of a proof-producing compiler for a subset of higher order logic. In: 16th European symposium on programming (ESOP'07)
29. Meyer T, Wolff B (2004) Tactic-based optimized compilation of functional programs. In: Filliâtre J-C, Paulin-Mohring C, Werner B (eds) TYPES 2004. Springer, Heidelberg
30. Myreen M (2009) Formal verification of machine-code programs. PhD thesis, University of Cambridge
31. Myreen M, Gordon M (2007) Hoare logic for realistically modelled machine code. In: Proceedings of TACAS 2007, LNCS vol 4424. Springer, Berlin
32. Myreen M, Slind K, Gordon M (2009) Extensible proof-producing compilation. In: de Moor O, Schwartzbach M (eds) Compiler construction, LNCS, vol 5501. Springer, Heidelberg
33. Nipkow T, Paulson LC, Wenzel M (2002) Isabelle/HOL – a proof assistant for higher-order logic, LNCS, vol 2283. Springer, Berlin
34. Norrish M (2008) A formal semantics for C++. In: Informal proceedings of TTVSI
35. Norrish M, Slind K (2009) The HOL system: logic, 1998–2009. At http://hol.sourceforge.net/
36. Owens S (2008) A sound semantics for OCaml-Light, In: Proceedings of ESOP 2008, LNCS, vol 4960. Springer, Berlin

37. Owre S, Rushby JM, Shankar N, Stringer-Calvert DJ (1998) PVS system guide. SRI Computer Science Laboratory, Menlo Park, CA. Available at http://pvs.csl.sri.com/manuals.html
38. Paulson L (1983) A higher order implementation of rewriting. Sci Comput Program 3:119–149
39. Pfenning F, Elliot C (1988) Higher order abstract syntax. In: Proceedings of PLDI. ACM, New York, NY, pp 199–208
40. Pnueli A, Siegel M, Singerman E (1998) Translation validation. In: Proceedings of TACAS'98, Lecture notes in computer science, vol 1384. Springer, Berlin, pp 151–166
41. Rideau L, Serpette B, Leroy X (2008) Tilting at windmills with Coq: formal verification of a compilation algorithm for parallel moves. J Autom Reason 40(4):307–326
42. Ridge T (2009) Verifying distributed systems: the operational approach. In: Shao Z, Pierce BC (eds) Proceedings of the 36th ACM SIGPLAN-SIGACT symposium on principles of programming languages, POPL 2009, Savannah, GA, USA, January 21–23, 2009. ACM, New York, NY, pp 429–440
43. Sewell P, Nardelli F, Owens S, Peskine G, Ridge T, Sarkar S, Strnisa R (2007) Ott: effective tool support for the working semanticist. In: Proceedings of ICFP 2007. ACM, New York, NY
44. Slind K (1999) Reasoning about terminating functional programs. PhD thesis, Institut für Informatik, Technische Universität München
45. Slind K, Norrish M (2008) A brief overview of HOL4. In: Mohamed O, Muñoz C, Tahar S (eds) TPHOLs, Lecture notes in computer science, vol 5170. Springer, Heidelberg, pp 28–32
46. Tolmach A, Oliva DP (1998) From ML to Ada: strongly-typed language interoperability via source translation. J Funct Program 8(4):367–412
47. Wheeler D, Needham R (1999) TEA, a tiny encryption algorithm. In: Fast software encryption: second international workshop, Lecture notes in computer science, vol 1008. Springer, Berlin, pp 363–366

Specification and Verification of ARM Hardware and Software

Anthony C. J. Fox, Michael J. C. Gordon, and Magnus O. Myreen

1 Introduction and Overview

This introductory section provides a high-level summary of the history and evolving goals of the ARM verification project. Section 2, by Anthony Fox, is a more detailed look into the modelling and verification of ARM processors. Section 3, by Magnus Myreen, is more detailed than the others and introduces a new method for creating trustworthy software implementations directly on bare metal. This approach uses the Fox processor model for the semantics of a machine code programming logic that borrows some ideas from separation logic.

In the late 1990s, Graham Birtwistle, at the University of Leeds, was investigating the use of the Standard ML (SML) functional programming language for modelling ARM processors. He approached Mike Gordon, a longtime collaborator, about the possibility of a joint project to extend the Leeds modelling work to formal verification. Birtwistle and Gordon, together with contacts at ARM Ltd in Cambridge, submitted a research proposal to the UK Engineering and Physical Sciences Research Council (EPSRC) entitled "Formal Specification and Verification of ARM6". This application was initially turned down on the grounds that the ARM6 processor was obsolete. However, following a strong letter from ARM pointing out that they could not place more modern designs in the public domain, the project was funded on resubmission.

The EPSRC project supported two PhD students at Leeds: Dominic Pajak and Daniel Schostak and a postdoctoral researcher at Cambridge: Anthony Fox. Pajak and Schostak developed SML models of the ARMv3 ISA and ARM6 micro-architecture, respectively. They both had summer internships at ARM in Cambridge, and this enabled them to talk to ARM engineers to find out details, especially concerning the ARM6 micro-architecture, that were not easily available. Fox took details from Pajack and Schostak's models, and public ARM documentation, and developed formal specifications in higher order logic (HOL) suitable for formal

A.C.J. Fox (✉)
University of Cambridge, Cambridge, UK
e-mail: Anthony.Fox@cl.cam.ac.uk

D.S. Hardin (ed.), *Design and Verification of Microprocessor Systems
for High-Assurance Applications*, DOI 10.1007/978-1-4419-1539-9_8,
© Springer Science+Business Media, LLC 2010

verification. An overview of this work is in Sect. 2. Pajak and Schostak subsequently completed their PhDs at Leeds and took jobs at ARM in Cambridge.

The formal verification that a model of the ARM6 micro-architecture corresponded to the ARMv3 ISA was completed by Fox within a couple of years (much of which were spent developing general proof infrastructure for the HOL4 proof assistant, which was used for the verification). This demonstrated proof-of-concept for the verification of a simple commercial off-the-shelf (COTS) processor. More complex processors, like those implementing the x86 ISA, are widely considered to be too complex for complete formal verification, although very impressive work has been done by Intel, AMD and VIA (Centaur) on the formal verification of parts of implementations x86 processors and by Rockwell Collins on the AAMP7G specialised processor [15]. Many critical systems use simple processors comparable to ARM, and the Leeds-Cambridge project showed that the complete formal verification of these is within the current state-of-the-art.

Following the successful first project, Gordon and Fox applied for continued support and eventually got a new EPSRC grant entitled "Formal Specification and Verification of ARM-based Systems". The aim here was to go beyond the processor to surrounding system components and accurately model things like input/output, coprocessors, bus protocols, etc. with the goal of conducting case studies involving these. We also proposed to upgrade our formal ISA models to match more recent versions of ARM. It was decided not to upgrade the micro-architecture verification for two reasons (1) we would be unable to get access to more recent designs (processor implementations are confidential ARM IP, but ISA specifications are largely in the public domain) and (2) we felt that re-verifying new implementations would be a lot of detailed work without much research value. Current ARM ISAs are a lot more complex than ARM6, having, for example, instructions for floating point, vector processing, virtual addressing, etc. Current ARM micro-architecture implementations have complex pipelines that are much more complicated than that used in ARM6; this makes the relationship between micro-architecture and ISA computations harder to relate formally. The concepts needed for verifying complex (e.g. superscalar) implementations are reasonably well understood (it was the topic of Fox's PhD and several academic projects [3, 27]), but the actual verifications have significantly more work than those needed for the ARM6 three-stage pipeline. Despite this increased complexity, our feeling is that with modern theorem proving infrastructure (including that developed for the ARM6 verification), the complete formal verification of a modern ARM implementation would be similar in kind to the ARM6 proof. The ARM9 micro-architecture, which is still widely used in mobile devices, would be relatively straightforward, but the latest Cortex designs would be of very much more effort (e.g. at least 10× more). The academic research benefits of doing such micro-architecture verifications (e.g. the potential for publication) would not be commensurate with the effort required.

The second EPSRC project was significantly more challenging and the work is still continuing even though the end date of the project has passed. This is possible because the initial ARM research attracted some positive attention and we were offered additional funding from a US Government agency to continue the work and to extend it to explore high assurance cryptographic implementations. We were joined in this work by Konrad Slind and students at Utah, who concentrated on formal compilation of HOL specifications directly to a code representation close to ARM assembler. This is described in Slind's chapter in this book, so we will not say more here. At Cambridge, Joe Hurd joined the project to work on formalising the mathematics underlying elliptic curve cryptography (ECC). Our goal was to ensure, by machine checked formal proof, that ARM machine code, with a semantics provided by a high-fidelity processor model, correctly implemented ECC algorithms that were specified using the mathematical concepts of elliptic curves. To this end, Hurd developed HOL formalisations of the textbook level mathematical theory underlying ECC in the version of HOL supported by the HOL4 system [18, 19]. This lead to difficult proof challenges, such as mechanically proving the associativity of addition on elliptic curves [30].

In parallel with Hurd's investigation of elliptic curve mathematics, Magnus Myreen, then a PhD student at Cambridge, was developing a method of directly verifying ARM assembler. He verified example ARM code implementing some of the operations needed for ECC (e.g. Montgomery multiplication). The overall flow we envisaged was as follows:

1. Start with the textbook level mathematical specifications of ECC applications in HOL (Hurd)
2. Use a proof-producing compiler to translate the HOL specifications to ARM assembler (Slind)
3. Link compiled ARM code to verified runtime code (Myreen)

Although significant progress has been made on all these three steps, we have still (2009) to join everything up into a seamless flow.

One issue that arose as we upgraded to current ARM ISA specifications was the challenge of accurately modelling the communication between an ARM CPU and its environment, which might include a variety of memories, coprocessors, etc. This impacts especially on systems code (see Point 3 above). As described in Sect. 2, the current ARM model separates memory and coprocessors from the main CPU, reflecting how systems are configured. We thus represent an ARM system as a structure containing separately modelled CPUs, various memories and other hardware. An executable model – i.e. a next-state function – is derived by deduction from such a system structure, and it is this that provides the semantics of code. Since the ARM ISA model is complex, the first step in deriving verification infrastructure is to derive higher level rules for reasoning about code that hides the details of the derived next-state function from the verifier. The abstraction methods used for this are described in Myreen's section (Sect. 3). It turns out that these methods can also be used to validate synthesis from low-level HOL, which provides a way of linking the output of Slind's compiler to the Fox processor model. This is also outlined in Sect. 3.

2 Specification and Verification of ARM Architectures

2.1 *The Swansea Methodology*

Before coming to work at the University of Cambridge Computer Lab, Anthony Fox completed his PhD at the computer science department of the University of Wales, Swansea. His supervisor, Neal Harman, and the then head of the department, John Tucker, had written a series of papers examining algebraic correctness models for formally verifying computer hardware. For example, this included examining Mike Gordon's micro-programmed case study, see [12, 16]. Fox took Harman and Tucker's work further, adapting their approach to cover pipelined and superscalar micro-architectures. The key features of the Swansea approach are as follows:

- Modelling systems at identified *levels of abstraction*, with particular attention given to formally defining precise classes of *data* and *temporal* abstraction.[1] Correctness is expressed as a commutativity statement that formally relates two abstraction levels.
- When establishing the correctness of microprocessor designs, two key levels are considered: the *programmer's model* (PM) level and the *abstract circuit* (AC) level.
- Formal modelling is based on the use of equational specification, defining the *operational semantics* for a given system at established levels of abstraction. In particular, *primitive recursion* is the principle definition mechanism. This means that the formal specifications can be run or *symbolically evaluated*. Systems with and without I/O were considered.
- A verification approach based on the use of a series of single-step theorems was developed. This provides a way to verify systems by principally using case splitting and equational term rewriting. That is to say, without the need to carry out an explicit temporal induction or to define top-level invariant predicates. Instead, initialisation functions are used to specify the reachable state-space.
- The approach was designed to be tool neutral, enabling it to be implemented by a wide variety of proof assistants.

In Fox's thesis, a toy architecture with a pipelined implementation was defined and a pen-and-paper proof of correctness was also presented, see [4]. A superscalar implementation was also defined, together with a formal statement of correctness.[2]

---

[1] For example, the class of temporal abstraction maps required for superscalar designs is necessarily more general than that needed for conventional pipelined processors.

[2] At the time, a pen-and-paper proof of correctness was not feasible/attempted for the superscalar design. Since then some bugs have been identified.

2.2 Starting at Cambridge

In the latter half of 2000, Fox moved to Cambridge to start working on the ARM6 project. Work had already begun at Leeds; however, their ARM6 model had not been completed yet. This meant that Fox, who had no previous experience in using theorem provers, could gradually start learning HOL4 (then at version Taupo-4). Getting to know HOL can be challenging but fortunately he shared an office with Michael Norrish and there were various other HOL gurus around, including Konrad Slind. As an initial project, the Swansea approach was formalised in HOL. This involved defining predicates that characterised the various classes of state systems and abstraction maps (for example, state-dependent immersions); formalising the definition of correctness (one general enough to cover conventional pipelined processors); and proving the 1-step theorems. Then the framework was given a test run with the formal verification of a tiny micro-programmed CPU, see [5]. This was followed by the formal verification of the pipelined design from Fox's thesis.[3]

2.3 Modelling the ARM Instruction Set Architecture

Work on specifying the ARM instruction set architecture (ISA) in HOL began in 2001, see [6]. In this context, the ISA is taken to correspond with the assembly programmer's view of an architecture. In general, programmers have access to a fixed set of registers (contained in a CPU) and to a much larger main memory – this is usually connected to the processor via a memory bus.[4] To write code, the assembly programmer has at their disposal a set of low-level instructions – these all update the registers and memory in various precisely defined ways. For example, typically there will be a set of *data processing* instructions, which use an arithmetic logic unit (ALU) to perform primitive operations – such as addition, multiplication or bitwise logic – on registers.[5] There will also be a set of *memory access* instructions for loading data from memory to registers and for storing registers to memory. The overall set of instruction is often extended with the introduction of new architecture generations. The number and variety of registers and instructions can vary considerably across platforms but there will normally be *at least* a handful of registers and few dozen or soinstructions. Instructions are encoded as a sequence

---

[3] A minor bug was found in the pen-and-paper proof.

[4] In practice, memory may be implemented with a series of caches, firmware, RAM and sometimes with virtual memory e.g. a hard disk or a solid-state drive (SSD). However, memory details are invariably implementation dependent and are *mostly* hidden from the programmer. In some cases, the actual behaviour can be somewhat counter-intuitive, see [1].

[5] In CISC architectures, these instructions may address the memory as well as just registers.

of bits (machine code) to be stored in the main memory. In the x86 architecture, instructions have a variable length[6] but with the ARM architecture all instructions are 32-bits long.

The official descriptions of ISAs need to be relatively precise. This is invariably achieved through the use of pseudo-code and in some cases the descriptions are semi-formal. To define an operational semantics for the ARM architecture in HOL, Fox used the specifications produced by Birtwistle's group at Leeds, in conjunction with Steve Furber's book [10] and the official ARM610 data sheet. The objective was to accurately declare a type S, corresponding with the programmer's model state space (registers and memory), and to define a next state function $next : S \rightarrow S$ that specifies the operational semantics of the ARM instructions i.e. the effect of the instructions on the registers and memory. Fortunately, HOL provides excellent support for modelling systems in a functional style, thanks to its "type base" tools,[7] and by virtue of Slind's TFL environment, see [29]. The specification was structured according to instruction *classes* i.e. groups of similar instructions were specified as a whole. To begin with I/O was not considered, in particular, hardware interrupts were not modelled.

2.3.1 The State Space

The ARM architecture provides 16 user-accessible registers and a program status register, each 32-bit words – some of these are then shadowed with versions that are accessible only in *privileged* or *system* modes. These modes are used when running operating system and exception handling code. The main memory is effectively an array of bytes with a 32-bit address space. Thus, the overall state space is:

$$S = (RName \rightarrow word32) \times (word32 \rightarrow word8),$$

where *RName* represents the complete set of register names e.g. r8_usr and CPSR. Although HOL has good support for working with algebraic types, there was a slight problem with regards to modelling machine words. At the time, HOL had a theory of words developed by Wai Wong, see [31]; however, this theory was list based and made heavy use of *restricted quantifiers*, with predicates used to restrict the scope of universal quantifiers.[8] It was decided that this theory would be too cumbersome to use in the context of the ARM6 verification effort, particularly with regards to symbolic evaluation. This started the winding road to developing HOL's current

---

[6] This is mainly because the x86 architectures has its origins in 8-bit and 16-bit computing. Although this variable instruction length can greatly complicate the hardware needed to decode instructions, it can give excellent code density. ARM added a set of 16-bit (Thumb) instructions in order to improve code density.

[7] It is possible to define and work with types in HOL that correspond with algebraic data types.

[8] HOL is based on simple type theory and does not directly support predicate sub-typing.

theory of n-bit words, with the latest version using an idea from John Harrison (see [17]) in order to get around the perceived need for restricted quantifiers.

2.4 Modelling the ARM6 Micro-Architecture

In 2002, the ARM6 micro-architecture was modelled in HOL, see [7], and in 2003 a formal verification was completed, see [8]. The ARM6 microprocessor dates from around 1994 and was widely deployed in a number of low-powered devices, such as the Apple Newton PDA. The processor's micro-architecture is relatively simple, employing a three-stage pipeline with *fetch, decode* and *execute* stages. As with other commercial designs, details of the processor's implementation are not in the public domain. It was only through collaboration with ARM Ltd, and Daniel Schostak's internship there, that it was possible to develop the formal model. The ARM6 processor is no longer in production, which was a factor in us gaining permission to carry out this research. However, it is worth noting that the ARM9 (circa 2004 and used in the Nokia N-Gage) is *not* a superscalar design and has a five-stage pipeline. It can be argued therefore that the verification of the ARM6 is still pertinent with respect to some more modern designs.

Daniel Schostak produced a very detailed model of the ARM6 for his thesis, see [28]. He only introduced a limited amount of abstraction, modelling the RTL (register transfer level) with a two-phase clock model. A limited amount of data abstraction was applied when producing the cycle accurate HOL model. One of the most useful resources in achieving this was Schostak's tabular style paper specification.[9] For example, his tabular description of the DIN latch (which stores input from the data bus) is shown below.

DIN

IC	IS	
*	*	
data_proc	t_2	*IREG*
mrs_msr	t_2	*IREG*
ldr	t_2	*IREG*
ldr	t_4	*DATA*
str	t_2	*IREG*
ldm	t_4	*DATA*
ldm	t_n	*DATA*
swp	t_4	*DATA*
br	t_2	*IREG*
mrc	t_4	*DATA*
ldc	t_2	*IREG*
stc	t_2	*IREG*
X	X	X

---

[9] Schostak produced extensive paper specifications of the ARM6 using various styles. He also produced a high-fidelity implementation in ML and now works full time at ARM Ltd.

This was translated into the following HOL definition:

```
⊢ ∀ ic is ireg data.
    DIN ic is ireg data =
    if
      ((ic = ldr) ∨ (ic = ldm) ∨ (ic = swp) ∨ (ic = mrc)) ∧
      (is = t4) ∨ (ic = ldm) ∧ (is = tn)
    then
      data
    else
      ireg
```

Here `ic` represents the instruction class and `is` is the instruction *step*, for example, `t3` is the first cycle of the pipeline execute stage and `tn` represents an iterated phase. By defining the next-state behaviour of all of the processor's latches and buses, it was possible to define a next-state function for the entire ARM6 core. Formal verification was proceeded by case splitting over the instruction class – the final version had 17 such classes. Inevitably, a small number of bugs were found in all of the specifications. Ultimately the ARM6 can be regarded as a reference implementation and so the formal verification can be seen as an exercise in developing an ISA model that is a verified abstraction of the processor.

2.4.1 Coverage

Somewhat confusingly, the ARM6 processor implements version three of the ARM architecture, written as ARMv3. To begin with, all of the ARMv3 instructions were modelled at the ISA level but some "hard" features were not included in the first ARM6 model – accordingly they were dropped from the ISA model prior to carrying out the initial verification attempt. The omissions included the `mul`, `ldm` and `stm` instructions, which all have relatively complex low-level behaviour (an iterated phase).[10] To complete, the formal proof invariants were constructed for these cases. The coprocessor instructions and hardware interrupts required models with input and output and this is discussed below. A feature complete formal verification was finished in 2005, see [9].

2.4.2 Input and Output

To accommodate input and output (I/O) features, the HOL formalisation of the Swansea approach was extended. It was also necessary to make significant changes to the ISA and micro-architecture models, and the formal verification required a fair amount of reworking. More sophisticated reasoning is required when verifying the correctness of coprocessor instructions and hardware interrupts. For example, the

---

[10] The ARM6 ALU does not contain a multiplier, so instead the processor's adder and shifter are used to implement Booth's algorithm over a number of clock cycles.

communication between the ARM core and coprocessor happens through a busy-wait loop, which has to be assumed to terminate after some indeterminate interval. It is also necessary to reason about the priority and timing of interrupts and, added to this mix, a reset signal can abort instructions at any cycle.

In the process of adding I/O, the memory was removed from the state-space of the ISA and micro-architecture specifications. At the ISA level, this meant that the state-space consisted of just the programmer's model registers, together with an instruction register (the op-code of the instruction to be run) and a exception status field, that is:

$$S = (RName \rightarrow word32) \times word32 \times Exception.$$

The next-state function is then of the form $next : S \times I \rightarrow S$, where I represents a set of input values i.e. data from memory and coprocessors, together with hardware interrupts. There is also an output function $out : S \rightarrow O$ that models data being passed from the processor to the memory and coprocessors. One consequence of these changes is that the resulting next-state functions no longer provide a *direct* means to run programs i.e. there is no longer a prescriptive model of memory, just an interface.

2.5 Beyond the ARM6

Following the formal verification of 2005, it was decided to extend the ISA model and focus on machine code verification, forgoing the considerable overhead associated with further extending and re-verifying the ARM6 model. It was at about this time that Magnus Myreen started his PhD at Cambridge. To begin with ARMv3M was supported (with the inclusion of long [64-bit] multiplies) and then ARMv4 was covered (through the addition of half-word and signed load and store instructions). At the time of writing this article, the ARMv4 architecture is still very much in use – it is implemented by a selection of processors in the ARM7, ARM8 and ARM9 families (as used in the Nintendo DS and Apple iPod).

After making these extensions, the next step was to provide support for reasoning about assembly code. In particular, it is not especially practical to work directly with 32-bit machine-code values. To this end, a HOL type was added to represent decoded ARM instructions; a parser/assembler was written;[11]; there was also support for pretty-printing instructions i.e. providing disassembly of machine code.

A new top-level next-state function was defined (using the existing definitions as sub-functions) and this reintroduced the main memory as part of the state-space. Consequently it was again possible to run code using the model and one could

---

[11] This was originally done using `mosmllex` and `mosmlyac` and later ported to `mllex` and `mlyacc`, so as to generate Standard ML.

also start reasoning about the semantics of programs. A pure memory model was assumed, that is to say, the memory was treated as a simple array with read and write accesses that never fail. A fast method for running code (useful in testing the model) was provided through the use of Konrad Slind's EmitML tool – this converted the HOL definitions into Standard ML. This ML code was compiled with MLton, resulting in an instruction throughput performance of approximately 10,000 instructions-per-second (10 kips).

It was then necessary to address the problem that the formal model somewhat obfuscates the behaviour of particular instruction instances. For example, one cannot read the specification and immediately see the effect of the instruction add r1,r2,r3. The reasons for this are: the underlying model is based on machine code; the specification is structuring according to instructions classes (not instruction instances); the overall semantics is expressed through one monolithic, top-level next-state function. To address this, a collection of *single-step* theorems of the following form are generated:

$$P(s) \Rightarrow (next(s) = s') \, .$$

Here the antecedent predicate P represents the context (showing exactly whichinstruction instance is to be run) and s' is the result of symbolically evaluating the model in this context. These theorems are generated using *forward-proof* (as opposed to goal-directed proof) and simplifications are applied to make the results as user-friendly as possible. The resulting theorems make the specification more accessible and usable. The term representing s' can be examined to see which registers and memory locations have been read and/or updated, and this is pertinent to Magnus Myreen's code verification work.

2.5.1 Further Refinements

With the addition of the 16-bit Thumb instructions, the ARMv4 model was later extended to ARMv4T. A more advanced mechanism for constructing a complete system was also examined i.e. building a system composed of ISA, memory and coprocessors models. A compositional, circuit-based style was adopted, wherein the output from one unit is connected to the input of another. This means that one can more easily consider different system configurations, for example, "plugging-in" different memory models. This contrasts with the previous approach wherein the memory was more hard-wired into the ISA specification.

2.6 Going Monadic

In addition to his work with the ARM architecture, Myreen has also worked with formal models of the x86 and PowerPC architectures. The x86 model initially came from collaborating with Susmit Sarkar, who has been working with Peter Sewell and others in the field of relaxed memory models, see [1]. This group made enquiries as to the suitability of the ARM model with respect to their research. However, a single-step operational semantics was not what they are after – they needed to know the precise order of all memory and register accesses. In collaboration with Myreen, they had developed a monadic approach to ISA specification and, inspired by this, Fox agreed to completely re-specify the entire ARM ISA using this approach. This would provide an event-based semantics for work on relaxed memory models and an operational semantics for work on code verification.

In their monadic approach, three principle operators are used: sequencing (`seqT` or `>>=`), parallel composition (`parT` or `|||`) and returning a constant value (`constT` or `return`). For example, in

```
(f ||| g) >>= (λ (x,y) . return (x + y + 1))
```

the operations `f` and `g` are performed in parallel and the results are then combined in a summation and returned. The overall type of this term is `num M` and the precise details of this type are hidden underneath a HOL-type abbreviation on `M`. For example, in the standard ARM operational semantics, we have:

```
'a M = arm_state → ('a, arm_state) error_option
```

Here `arm_state` is the state-space and `error_option` is just like the standard functional *option* type, except that the "none" case is tagged with a string, which provides a useful mechanism for reporting erroneous behaviour. In this *sequential* operational semantics, the `parT` operator is evaluated sequentially with a left-to-right ordering e.g. `f` is applied before `g` in the example above.

There are many advantages to working in this monadic style, these include:

- The ability to modify the underlying semantics by simply changing the monad's type and the definitions of the monadic operators.
- The ability to avoid excessive parameter passing and to hide details of the state-space. In some cases, there might not even be a state-space.
- It provides a clean way to handle erroneous cases. In particular, it is easy to model behaviour that the ARM architecture classifies as *Unpredictable*.
- With some pretty-printing support, the definitions look more like imperative code. This makes the specifications more readable to those unfamiliar with functional programming and it also provides a more visible link with pseudo-code from reference manuals.

For example, consider the following pseudo-code from the ARM architectural reference manual:

```
// BranchWritePC()
// ===============

BranchWritePC(bits(32) address)
    if CurrentInstrSet() == InstrSet_ARM then
        if ArchVersion() < 6 && address<1:0> != '00' then UNPREDICTABLE;
        BranchTo(address<31:2>:'00');
    else
        BranchTo(address<31:1>:'0');
```

With pretty-printing turned on this corresponds with the following HOL code:

```
⊢ ∀ ii address.
    branch_write_pc ii address =
    do
      iset ← current_instr_set ii;
      if iset = InstrSet_ARM then
        do
          version ← arch_version ii;
          if version < 6 ∧ (1 >< 0) address <> 0w then
            errorT "branch_write_pc: unpredictable"
          else
            branch_to ii ((31 <> 2) address)
        od
      else
        branch_to ii ((31 <> 1) address)
    od
```

Although the translation is not literal, there is clearly a connection between the two specifications. The function `branch_write_pc` has return type `unit M`, that is to say, it is similar to a procedure (or `void` function in C). The HOL model introduces a variable `ii`, which is used to uniquely identify the source of all read and write operations – this becomes significant in multi-core systems with shared memory. The operator `errorT` is used to handle the unpredictable case. The word *extract* and *slice* operations (`><` and `<>`) are used to implement the bit operations shown in the ARM reference. Inequality is overloaded to be `<>`, which corresponds with `!=` in the pseudo-code. Observe that the HOL specification does not explicitly refer to state components; such details are hidden by the monad, and the operations `arch_version` and `current_instr_set` automatically have access to all the data that they need. In the sequential model, the state actually contains a component that identifies the specific version of the architecture being modelled e.g. `ARMv4` or `ARMv4T`, both of which give a version number of four. This makes it possible to simultaneously support multiple architecture versions. Further refinement has also been made in the process of producing the new specification, especially with regard to instruction decoding and the representation of instructions.

2.6.1   Coverage

The monadic specification covers all of the currently supported ARM architecture versions, that is to say: ARMv4, ARMv4T, ARMv5T, ARMv5TE, ARMv6, ARMv6K, ARMv6T2, ARMv7-A, ARMv7-R and ARMv7-M. A significant number of new instruction were introduced with ARMv6, which was introduced with the ARM11 family of processors. The latest generation (ARMv7) has only a small number of extra ARM instructions but these versions do all support Thumb2 – this provides a large number of double-width Thumb instructions, which cover nearly all of the functionality of the standard ARM instructions. In fact, the latest Cortex-M processors only run Thumb2 instructions and are designed to be used as microcontrollers. The Cortex-A8 processor (as found in the Apple iPhone 3GS and Palm Pre) implements ARMv7-A.

The HOL model also covers the Security and Multiprocessor extensions. It does not support Jazelle, which provides hardware support for running Java bytecode. Technical details about Jazelle and its implementations are restricted to ARM licensees only, see [2]. Consequently, a HOL specification of Jazelle is very unlikely. Documentation is available for the ThumbEE, VFP (vector floating-point) and Advanced SIMD extensions, but they have not been specified yet – the SIMD extensions were introduced with ARMv7 and the associated infrastructure is referred to as NEON™ technology.

2.6.2 Single-Step Theorems

Recently a tool for generating single-step theorems for the monadic model has been developed. These theorems are now generated entirely *on-the-fly* for specific opcodes.[12] This contrasts with the previous approach whereby a collection of pre-generated theorems (effectively templates) are stored and then specialised prior to use. The old approach is not practical in the context of the much larger number of instructions and range of contexts. The single-step theorems are generated entirely through forward proof and so the process is not especially fast. Consequently, it may prove necessary to store some of the resulting theorems in order to improve runtimes further down the line.

The function call

```
arm_stepLib.arm_step "v6T2,be,thumb,sys" "FB02F103";
```

produces the following theorem

```
⊢ ∀ state.
    (ARM_ARCH state = ARMv6T2) ∧ (ARM_EXTENSIONS state = {}) ∧...∧
    (ARM_MODE state = 31w) ∧ aligned (ARM_READ_REG 15w state,2) ∧
    (ARM_READ_MEM (ARM_READ_REG 15w state + 3w) state = SOME 3w) ∧
    (ARM_READ_MEM (ARM_READ_REG 15w state + 2w) state = SOME 241w) ∧
    (ARM_READ_MEM (ARM_READ_REG 15w state + 1w) state = SOME 2w) ∧
    (ARM_READ_MEM (ARM_READ_REG 15w state) state = SOME 251w) ⇒
    (ARM_NEXT state =
     SOME
       (ARM_WRITE_REG 1w
          (ARM_READ_REG 2w state * ARM_READ_REG 3w state)
          (ARM_WRITE_REG 15w (ARM_READ_REG 15w state + 4w) state)))
```

For brevity/clarity, some parts of the antecedent have been omitted. The first argument to `arm_step` is a string containing configuration options e.g. the architecture version and the byte ordering. The second string is the instruction op-code. In the example above, 0xFB02F103 is the machine code for the Thumb2 instruction `mul r1,r2,r3`.[13] The four instruction bytes are read from memory using the program-counter, which is register 15.

---

[12] This tool makes heavy use of the HOL conversion EVAL.

[13] At the moment op-codes are being generated using GNU's `binutils` tools. FB02F103 breaks up into 251w, 2w, 241w and 3w, which are used in the theorem.

The next-step theorem shown above bears little or no *visible* resemblance to the underlying monadic specification. The functions in uppercase are defined in post hoc manner, so as to present a more conventional state-oriented semantics. The top-level next-state function ARM_NEXT returns an option type – if an error occurred (e.g. with an unpredictable case), then the result would be NONE, but in practice the tool raises an ML exception for such cases.

2.6.3 Active and Future Work

Current work includes focusing on updating the parser, assembler and pretty-printing support. The instruction parser has been completely rewritten in ML, abandoning the use of mllex and mlyacc. This provides greater flexibility in supporting multiple architectures and encoding schemes. It should also facilitate better code portability. It would have been possible to avoid writing an assembler and instead interface with GNU's binutils but this would require users to specifically install these tools, configuring them as an ARM cross-compiler. Also, only the very latest version of the GNU assembler supports Thumb2, and it does appear to have a small number of teething problems (bugs) in that area. The parser is nearly complete – generating op-codes is the next stage.

Another future area of work will be in handling I/O. It should be relatively straightforward to add hardware interrupts. Readers may have observed that the sequential version of the monadic model again includes the memory as part of the state-space. This could be considered to be a regressive step in comparison with the approach discussed at the end of Sect. 2.5. However, the monadic approach does make it easier to modify the underlying memory model.[14] It is expected that memory-mapped I/O (MMIO) can be supported by interleaving calls to next-state functions i.e. the ISA next-state function would be interleaved with a MMIO next-state function.

3 High-Assurance Software Engineering

In 2005, Myreen started his PhD which came to focus on theories and tools for proving ARM machine code correct on top of Fox's formal specification of the ARM ISA. This section presents the current state of the resulting framework which has come to support both formal verification of existing ARM machine code and synthesis of new ARM code from functional specifications. The three layers of this framework are as follows:

1. *Hoare logic for machine code* is used for making concise and composable formal specifications about ARM code (Sect. 3.1).

---

[14] Although at the moment, the arm_step tool does make some assumptions about the memory model.

2. A *decompiler* aids verification by automatically extracting functional descriptions of ARM code from Fox's detailed ISA specification (Sect. 3.2).
3. A *compiler* is used for synthesis of new ARM code from, possibly partial, functional specifications (Sect. 3.3).

Our to-date largest case study, synthesis of a formally verified LISP interpreter, is outlined in Sect. 3.4.

3.1 Machine-Code Hoare Logic

Machine-code programs operate over a heterogeneous state consisting of register, memory locations and various status bits. As a result, keeping track of which resources might have been altered by some ARM code can easily become tedious. In order to avoid always explicitly stating "... and nothing else changed" (a frame property), we write our theorems in terms of a machine-code Hoare triple $\{p\}\ c\ \{q\}$ which implicitly formalise a frame property (from separation logic [26]):

> $\{p\}\ c\ \{q\}$ is true if any state of an ARM processor s which satisfies precondition p, can through execution of code c on the ARM ISA reach a state s' which satisfies postcondition q; and, furthermore, all resources not mentioned in p will remain unchanged in the transition from s to s'.

A formal definition of this Hoare triple will be given later, also see [22].

The frame property manifests itself in practice as a proof rule called *the frame rule* (again borrowed from separation logic). The frame rules allows an arbitrary assertion r to be added to any Hoare triple $\{p\}\ c\ \{q\}$ using the separating conjunction $*$ (defined later):

$$\{p\}\,c\,\{q\}\ \Rightarrow\ \forall r.\ \{p * r\}\,c\,\{q * r\}$$

The frame property of our Hoare triples allows us to only mention locally relevant resources, e.g. a theorem describing the ARM instruction, add r4,r3,r4 (encoded as 0xE0834004), need only mention resources register 3, 4 and 15 (the program counter). For example the following Hoare-triple theorem states, if register 3 has value a, register 4 has value b and the program counter is p, then the code E0834004 at location p will reach a state where register 3 has value a, register 4 holds $a + b$ and the program counter is set to $p + 4$:

$$\{\,\mathsf{r3}\ a * \mathsf{r4}\ b * \mathsf{pc}\ p\,\}$$
$$p : \mathtt{E0834004}$$
$$\{\,\mathsf{r3}\ a * \mathsf{r4}\ (a + b) * \mathsf{pc}\ (p + 4)\,\}$$

The frame rule allows us to infer that the value of register 5 is left unchanged by the above ARM instruction, since we can instantiate r in the frame rule above with an assertion saying that register 5 holds value c, i.e. $\mathsf{r5}\ c$.

$$\{\,\mathsf{r3}\ a\ *\ \mathsf{r4}\ b\ *\ \mathsf{pc}\ p\ *\ \mathsf{r5}\ c\,\}$$
$$p : \texttt{E0834004}$$
$$\{\,\mathsf{r3}\ a\ *\ \mathsf{r4}\ (a+b)\ *\ \mathsf{pc}\ (p+4)\ *\ \mathsf{r5}\ c\,\}$$

All user-level ARM instructions satisfy specification in this style. Memory reads and writes are not much different, e.g. Hoare-triple theorem describing the instruction `swp r4,r4, [r3]` (E1034094) for swapping the content of memory location a given in register 3 with that of register 4 is given as follows. Here $\mathsf{m}\ m$ states that a function m, a partial mapping from addresses (32-bit words) to values (32-bit words), correctly represents a portion of memory (addresses $\mathsf{domain}\ m$), address a must be in the memory portion covered by m and for tidiness needs to be word-aligned, i.e. $a\ \&\ 3 = 0$; we write $m[a \mapsto b]$ for m updated to map a to b.

$$a\ \&\ 3 = 0 \wedge a \in \mathsf{domain}\ m \Rightarrow$$
$$\{\,\mathsf{r3}\ a\ *\ \mathsf{r4}\ b\ *\ \mathsf{m}\ m\ *\ \mathsf{pc}\ p\,\}$$
$$p : \texttt{E1034094}$$
$$\{\,\mathsf{r3}\ a\ *\ \mathsf{r4}\ (m(a))\ *\ \mathsf{m}\ (m[a \mapsto b])\ *\ \mathsf{pc}\ (p+4)\,\}$$

The following subsections will present the definition of our machine-code Hoare triple and some proof rules (HOL theorems) that have been derived from the definition of the Hoare triple and are hence sound.

3.1.1 Set-Based Separating Conjunction

The definition of our machine-code Hoare triple uses the separating conjunction, which we define unconventionally to split sets rather than partial functions. Our definition of the set-based separating conjunction states that $(p * q)\ s$ whenever s can be split into two disjoint sets u and v such that p holds for u and q holds for v:

$$(p * q)\ s\ =\ \exists u\ v.\ p\ u \wedge q\ v \wedge (u \cup v = s) \wedge (u \cap v = \{\})$$

In order to make use of this set-based separating conjunction, we need to translate ARM states into sets of state components. We define the type of an ARM state elements as a data-type with constructors:

```
Reg     : word₄  → word₃₂ → arm_state_element
Status  : status_names → boolean → arm_state_element
Memory  : word₃₀ → word₃₂ → arm_state_element
Undef   : bool → arm_state_element
```

We define a function `arm2set` for translating states representation used in the ARM ISA specification into sets of ARM state elements, using read functions `arm_read_reg`, `arm_read_mem`, `arm_read_status` and `arm_read_undefined`, which, respectively, read a register, memory location, status bit and undefined flag.

Here range $f = \{\, y \mid \exists x.\ f\ x = y\, \}$.

$$
\begin{aligned}
&\textsf{arm2set}\ \textit{state} = \\
&\quad \textsf{range}\ (\lambda r.\ \textsf{Reg}\ r\ (\textsf{arm\_read\_reg}\ r\ \textit{state}))\ \cup \\
&\quad \textsf{range}\ (\lambda a.\ \textsf{Mem}\ a\ (\textsf{arm\_read\_mem}\ a\ \textit{state}))\ \cup \\
&\quad \textsf{range}\ (\lambda s.\ \textsf{Status}\ s\ (\textsf{arm\_read\_status}\ s\ \textit{state}))\ \cup \\
&\quad \{\, \textsf{Undef}\ (\textsf{arm\_read\_undefined}\ \textit{state})\, \}
\end{aligned}
$$

Some basic assertions are defined over sets of ARM state elements as follows. We often write r1 a, r2 b, etc. as abbreviations for reg 1 a, reg 2 b, etc.

$$
\begin{aligned}
(\textsf{reg}\ i\ a)\ s &= (s = \{\, \textsf{Reg}\ i\ a\, \}) \\
(\textsf{mem}\ a\ w)\ s &= (s = \{\, \textsf{Mem}\ a\ w\, \})
\end{aligned}
$$

These assertions have their intended meaning when used with arm2set:

$$
\begin{aligned}
\forall p\ s.\ (\textsf{mem}\ a\ w * p)\ (\textsf{arm2set}\ s)\ &\Rightarrow\ (\textsf{arm\_read\_mem}\ a\ s = w) \\
\forall p\ s.\ (\textsf{reg}\ i\ v * p)\ (\textsf{arm2set}\ s)\ &\Rightarrow\ (\textsf{arm\_read\_reg}\ i\ s = v)
\end{aligned}
$$

The separating conjunction separates assertions:

$$
\forall p\ s.\ (\textsf{mem}\ a\ x * \textsf{mem}\ b\ y * \textsf{reg}\ i\ u * \textsf{reg}\ j\ v * p)\ (\textsf{arm2set}\ s)\ \Rightarrow\ a \neq b \wedge i \neq j
$$

Other assertions used in this text are:

$$
\begin{aligned}
\textsf{aligned}\ a &= (a\ \&\ 3 = 0) \\
\textsf{emp}\ s &= (s = \{\}) \\
\langle b \rangle\ s &= (s = \{\}) \wedge b \\
(\textsf{code}\ c)\ s &= (s = \{\, \textsf{Mem}\ (a[31\text{---}2])\ i \mid (a,i) \in c\, \}) \\
(\textsf{m}\ m)\ s &= (s = \{\, \textsf{Mem}\ (a[31\text{---}2])\ (m\ a) \mid a \in \textsf{domain}\ m \wedge \textsf{aligned}\ a\, \}) \\
(\textsf{pc}\ p)\ s &= (s = \{\textsf{Reg}\ 15\ p, \textsf{Undef}\ \textsf{F}\}) \wedge \textsf{aligned}\ p \\
(\textsf{s}\ (n,z,c,v))\ s &= (s = \{\, \textsf{Status}\ \textsf{N}\ n, \textsf{Status}\ \textsf{Z}\ z, \textsf{Status}\ \textsf{C}\ c, \textsf{Status}\ \textsf{V}\ v\, \})
\end{aligned}
$$

3.1.2 Definition of Hoare Triple

Let $\textsf{run}(n,s)$ be a function which applies the next-state function from our ARM ISA specification n times to ARM state s.

$$
\begin{aligned}
\textsf{run}(0,s) &= s \\
\textsf{run}(n{+}1,s) &= \textsf{run}(n, \textsf{arm\_next\_state}(s))
\end{aligned}
$$

Our machine-code Hoare triple has the following definition: any state s which satisfies p separately from code c and some frame r (written $p * \textsf{code}\ c * r$) will

$$\{\,\mathsf{r7}\ x * \mathsf{r8}\ y * \mathsf{pc}\ p\,\}\ \ p : \mathtt{E2878001}\ \ \{\,\mathsf{r7}\ x * \mathsf{r8}\ (x{+}1) * \mathsf{pc}\ (p{+}4)\,\}$$

$$= $$

$$
\begin{aligned}
\forall state.\ &(\mathsf{arm\_read\_reg}\ 7\ state = x) \land\\
&(\mathsf{arm\_read\_reg}\ 8\ state = y) \land\\
&(\mathsf{arm\_read\_reg}\ 15\ state = p) \land \mathsf{aligned}\ p \land\\
&(\mathsf{arm\_read\_undefined}\ 15\ state = \mathsf{F}) \land\\
&(\mathsf{arm\_read\_mem}\ p\ state = \mathtt{E2878001}) \Rightarrow\\
&\exists n\ state'.\ (state' = run(n, state)) \land\\
&\qquad (\mathsf{arm\_read\_reg}\ 7\ state' = x) \land\\
&\qquad (\mathsf{arm\_read\_reg}\ 8\ state' = x{+}1) \land\\
&\qquad (\mathsf{arm\_read\_reg}\ 15\ state' = p{+}4) \land \mathsf{aligned}\ (p{+}4) \land\\
&\qquad (\mathsf{arm\_read\_undefined}\ 15\ state' = \mathsf{F}) \land\\
&\qquad (\mathsf{arm\_read\_mem}\ p\ state' = \mathtt{E2878001}) \land\\
&\qquad (\mathsf{arm2set}\ state - Frame = \mathsf{arm2set}\ state' - Frame)
\end{aligned}
$$

where $Frame = \mathsf{range}\ (\lambda w.\ \mathsf{Reg}\ 7\ w)\ \cup\ \mathsf{range}\ (\lambda w.\ \mathsf{Reg}\ 8\ w)\ \cup\ \mathsf{range}\ (\lambda w.\ \mathsf{Reg}\ 15\ w)$

Fig. 1 A machine-code Hoare triple expanded

reach (after some k applications of the next-state function) a state which satisfies q separately from the code c and frame r (written $q * \mathsf{code}\ c * r$).

$$
\begin{aligned}
\{p\}\ c\ \{q\} = \forall s\ r.\ &(p * \mathsf{code}\ c * r)\ (\mathsf{arm2set}(s)) \Rightarrow\\
&\exists k.\ (q * \mathsf{code}\ c * r)\ (\mathsf{arm2set}(run(k, s)))
\end{aligned}
$$

As a convention, we write concrete code sets $c = \{(p, i), (q, j), \ldots\}$ as "$p : i, q : j, \ldots$" in order to avoid confusion with the curly brackets used when writing Hoare triples.

An example of what a machine-code Hoare triple means in terms of the basic read functions is shown in Fig. 1. The last line which relates $\mathsf{arm2set}\ state$ to $\mathsf{arm2set}\ state'$ states that nothing (observable through the read functions) changed other than registers 7, 8 and 15. This fact that nothing outside of the foot-print of the specification was affected, comes from the universally quantified frame r in the definition of the machine-code Hoare triple.

3.1.3 Proof Rules

Below we list some theorems proved from the definition of our Hoare triple. These theorems are cumbersome to use in manual proofs, but easy to use in building proof automation, which is the topic of the next two sections.

Frame: $\{p\}\ c\ \{q\} \Rightarrow \forall r.\ \{p * r\}\ c\ \{q * r\}$

The frame rule allows any assertions to be added to the pre- and postconditions of a Hoare triple, often applied before composition.

Composition: $\{p\}\ c_1\ \{q\} \wedge \{q\}\ c_2\ \{r\} \ \Rightarrow\ \{p\}\ c_1 \cup c_2\ \{r\}$

The composition rule composes two specifications and takes the union of the two code sets (the code sets may overlap, happens for loops).

Precondition strengthening: $\{p\}\ c\ \{q\} \wedge (\forall s.r\,s \Rightarrow p\,s) \Rightarrow \{r\}\ c\ \{q\}$
Postcondition weakening: $\{p\}\ c\ \{q\} \wedge (\forall s.q\,s \Rightarrow r\,s) \Rightarrow \{p\}\ c\ \{r\}$

Preconditions can be strengthened, postconditions can be weakened.

Precondition exists: $\{\exists x.\ p\,x\}\ c\ \{q\} \ \Longleftrightarrow\ \forall x.\ \{p\,x\}\ c\ \{q\}$

Existential quantifiers in the precondition are equivalent to universal quantifiers outside of the specification.

Move pure condition: $\{p * \langle b \rangle\}\ c\ \{q\} \ \Longleftrightarrow\ (b \Rightarrow \{p\}\ c\ \{q\})$

Pure stateless assertions, written $\langle b \rangle$, can be pulled out of the precondition. Here b has type bool.

Code extension: $\{p\}\ c\ \{q\} \ \Rightarrow\ \forall e.\ \{p\}\ c \cup e\ \{q\}$

The code can be extended arbitrarily. This rule highlights that $\{p\}\ c\ \{q\}$ states that c is *sufficient* to transform any state satisfying p into a state satisfying q. Thus any larger set $c \cup e$ is also sufficient.

3.2   Decompilation of ARM Code

To aid verification of machine code, we have developed a novel verification technique [24] which is based on decompiling machine code into functions in the logic of a theorem prover, in this case the HOL4 theorem prover.

3.2.1   Example

Given some ARM code, which calculates the length of a linked list,

```
 0: E3A00000        mov r0, #0        ; set reg 0 to 0
 4: E3510000     L: cmp r1, #0        ; compare reg 1 with 0
 8: 12800001        addne r0, r0, #1  ; if not equal: add 1 to reg 1
12: 15911000        ldrne r1, [r1]    ;         load mem[reg 1] into reg 1
16: 1AFFFFFB        bne L             ;         jump to compare
```

the decompiler reads the hexadecimal numbers, extracts a function f and a safety condition f_{pre} which describe the data-update performed by the ARM code:

$f(r_0, r_1, m) =$
$\quad$ let $r_0 = 0$ in $g(r_0, r_1, m)$

$g(r_0, r_1, m) =$
$\quad$ if $r_1 = 0$ then (r_0, r_1, m) else
$\quad\quad$ let $r_0 = r_0 + 1$ in
$\quad\quad$ let $r_1 = m(r_1)$ in
$\quad\quad\quad g(r_0, r_1, m)$

$f_{pre}(r_0, r_1, m) =$
$\quad$ let $r_0 = 0$ in $g_{pre}(r_0, r_1, m)$

$g_{pre}(r_0, r_1, m) =$
$\quad$ if $r_1 = 0$ then T else
$\quad\quad$ let $r_0 = r_0 + 1$ in
$\quad\quad$ let $cond = r_1 \in$ domain $m \wedge$ aligned r_1 in
$\quad\quad$ let $r_1 = m(r_1)$ in
$\quad\quad\quad g_{pre}(r_0, r_1, m) \wedge cond$

The decompiler also proves the following theorem which state that f is accurate with respect to the ARM model, for input values that satisfy f_{pre}. Here $(k_1, k_2, \ldots, k_n)$ is $(x_1, x_2, \ldots, x_n)$ abbreviates $k_1\ x_1 * k_2\ x_2 * \cdots * k_n\ x_n$, i.e. expression $(r0, r1, m)$ is (r_0, r_1, m) states that register 0 has value r_0, register 1 is r_1 and part of memory is described by m.

$$\{ (r0, r1, m)\text{ is }(r_0, r_1, m) * s * pc\ p * \langle f_{pre}(r_0, r_1, m)\rangle \}$$

$$p : \texttt{E3A00000, E3510000, 12800001, 15911000, 1AFFFFFB} \qquad (1)$$

$$\{ (r0, r1, m)\text{ is }(f\ (r_0, r_1, m)) * s * pc\ (p + 20) \}$$

The user can then prove that the original machine code indeed calculates the length of a linked-list by simply proving that the extracted function f does that. Let list state that abstract list l is stored in memory m from address a onwards.

$$\text{list }(\text{nil}, a, m) \;=\; a = 0$$
$$\text{list }(x::l, a, m) \;=\; \exists a'.\ m(a) = a' \wedge m(a+4) = x \wedge a \neq 0 \wedge$$
$$\text{list }(l, a', m) \wedge \text{aligned } a$$

Let length l be the length of an abstract list l, e.g. length $(4::5::\text{nil}) = 2$. It is easy (15 lines of HOL4) to prove, by induction on the abstract list l, that the function f, from above, calculates the length of a linked list and also that list implies the precondition f_{pre}.

$$\forall x\ l\ a\ m.\quad \text{list }(l, a, m) \;\Rightarrow\; f(x, a, m) = (\text{length } l, 0, m) \qquad (2)$$
$$\forall x\ l\ a\ m.\quad \text{list }(l, a, m) \;\Rightarrow\; f_{pre}(x, a, m) \qquad (3)$$

Given (2) and (3), it follows immediately from (1) that the ARM code calculates the length of a linked-list correctly:

$$\{ (r0, r1, m)\text{ is }(r_0, r_1, m) * s * pc\ p * \langle \text{list }(l, r_1, m)\rangle \}$$

$$p : \texttt{E3A00000, E3510000, 12800001, 15911000, 1AFFFFFB}$$

$$\{ (r0, r1, m)\text{ is }(\text{length } l, 0, m) * s * pc\ (p + 20) \}$$

3.2.2 Implementation

The following loop-introduction rule is the key idea behind our decompiler implementation. This rule can introduce any tail-recursive function tailrec, with safety condition tailrec_pre, of the form:

$$\text{tailrec } x \;=\; \text{if } G\ x \text{ then tailrec } (F\ x) \text{ else } (D\ x)$$
$$\text{tailrec\_pre } x \;=\; Q\ x \wedge (G\ x \Rightarrow \text{tailrec\_pre } (F\ x))$$

Given a theorem for the step case, $\{r(x)\}\ c\ \{r(F\ x)\}$, and one for the base case, $\{r(x)\}\ c\ \{r'(D\ x)\}$, the loop rule can introduce tailrec:

$$\forall r\ r'\ c.\ (\forall x.\ Q\ x \wedge G\ x \Rightarrow \{r(x)\}\ c\ \{r(F\ x)\}) \wedge$$
$$(\forall x.\ Q\ x \wedge \neg G\ x \Rightarrow \{r(x)\}\ c\ \{r'(D\ x)\})$$
$$\Rightarrow (\forall x.\ \mathsf{tailrec\_pre}\ x \Rightarrow \{r(x)\}\ c\ \{r'(\mathsf{tailrec}\ x)\})$$

Parameters F, D, G, Q, r, r' were instantiated as follows for introduction of g in our example above.

$$
\begin{aligned}
F = D &= \lambda(r_0, r_1, m).\ \text{if}\ r_1 = 0\ \text{then}\ (r_0, r_1, m)\ \text{else}\ (r_0{+}1, m(r_1), m)\\
G &= \lambda(r_0, r_1, m).\ \text{if}\ r_1 = 0\ \text{then}\ \mathsf{F}\ \text{else}\ \mathsf{T}\\
Q &= \lambda(r_0, r_1, m).\ \text{if}\ r_1 = 0\ \text{then}\ \mathsf{T}\ \text{else}\ (r_1 \in \text{domain}\ m \wedge \text{aligned}\ r_1)\\
r &= \lambda(r_0, r_1, m).\ (\mathsf{r0}, \mathsf{r1}, \mathsf{m})\ \text{is}\ (r_0, r_1, m) * \mathsf{s} * \mathsf{pc}\ p\\
r' &= \lambda(r_0, r_1, m).\ (\mathsf{r0}, \mathsf{r1}, \mathsf{m})\ \text{is}\ (r_0, r_1, m) * \mathsf{s} * \mathsf{pc}\ (p + 20)
\end{aligned}
$$

The loop rule can be derived from the rule for composition of Hoare triples given in Sect. 3.1. For details of decompilation, see [22, 24].

3.3 Extensible Proof-Producing Compilation

It is often the case that we prefer to synthesise ARM code from specifications rather than apply post hoc verification to existing ARM code. For this purpose, we have developed a proof-producing compiler [25] which maps tail-recursive functions in HOL4, i.e. functional specifications, to ARM machine code and proves that the ARM code is a valid implementation of the original HOL4 functions.

3.3.1 Example 1

Given a function f,

$$\mathsf{f}(r_1) = \text{if}\ r_1 < 10\ \text{then}\ r_1\ \text{else}\ \text{let}\ r_1 = r_1 - 10\ \text{in}\ \mathsf{f}(r_1)$$

the compiler produces machine code,

```
0:  E351000A    L:  cmp  r1,#10        ;  compare reg 1 with 10
4:  2241100A        subcs r1,r1,#10    ;  if less: subtract 10 from reg 1
8:  2AFFFFFC        bcs L              ;       jump to compare
```

and proves that the generated code calculates f:

$$\{\mathsf{r1}\ r_1 * \mathsf{pc}\ p * \mathsf{s}\}\ \ p : \mathtt{E351000A, 2241100A, 2AFFFFFC}\ \{\mathsf{r1}\ \mathsf{f}(r_1) * \mathsf{pc}\ (p{+}12) * \mathsf{s}\}$$

In case we have manually proved that f calculates unsigned-word modulus of 10, i.e. $\forall x.\ f(x) = x \bmod 10$, then we immediately know that the ARM code calculates modulus of 10:

$$\{\text{r1 } r_1 * \text{pc } p * \text{s}\}\quad p : \texttt{E351000A, 2241100A, 2AFFFFFC}\quad \{\text{r1 } (r_1 \bmod 10)\ *\text{pc } (p{+}12) * \text{s}\}$$

3.3.2 Example 2

An important feature of this compiler is its support for extensions. If the compiler is supplied with the above theorem which states that ARM code `E351000A 2241100A 2AFFFFFC` assigns $r_1 \bmod 10$ to r_1. Subsequent compilations can make use of this verified code. For example,

$$\begin{aligned}
f(r_1, r_2, r_3) = {} &\text{let } r_1 = r_1 + r_2 \text{ in}\\
&\text{let } r_1 = r_1 + r_3 \text{ in}\\
&\text{let } r_1 = r_1 \bmod 10 \text{ in}\\
&r_1
\end{aligned}$$

will compile successfully into a theorem which makes use of the previously verified code (the last three instructions in the code below):

$$\begin{aligned}
&\{\text{r1 } r_1 * \text{r2 } r_2 * \text{r3 } r_3 * \text{pc } p * \text{s}\}\\
&\quad p : \texttt{E0811002, E0811003, E351000A, 2241100A, 2AFFFFFC}\\
&\{\text{r1 } (f(r_1, r_2, r_3)) * \text{r2 } r_2 * \text{r3 } r_3 * \text{pc } (p{+}20) * \text{s}\}
\end{aligned}$$

3.3.3 Implementation

The compiler is implemented using translation validation based on the decompiler from above; for each function f, the compiler will:

1. Generate machine code for input function f with an unverified algorithm
2. Decompile the generated code into a function f′
3. Automatically prove f = f′

The compiler returns to the user, the theorem certificate produced in step 2, but with f′ replaced by f using the rewrite theorem produced in step 3.

The compiler's separation between code generation (step 1) and certification (steps 2 and 3) has two useful consequences: code generation need not be proof-producing, and multiple lightweight optimisations can be made in step 1 with practically no added proof burden for steps 2 and 3. Step 1 is allowed to produce any code for step 3 will be able to prove f = f′. For example, just doing expansion of let expressions in step 3 immediately makes optimisation such as register renaming, some instruction reordering and dead-code removal unobservable.

Extensions are implemented by making the decompiler use the theorems provided when constructing the step- and base-theorems for instantiating its loop rule, as explained in Sect. 3.2.

3.4 Case Study: Verified LISP Interpreter

The construction of a verified LISP interpreter [23] is the, to date, largest case study conducted on top of the ARM model. This case study included producing and verifying implementations for:

- A copying garbage collector
- An implementation of basic LISP evaluation
- A parser and printer for s-expressions

These components were combined to produce an end-to-end implementation of a LISP-like language, similar to the core of the original LISP 1.5 by McCarthy [20].

3.4.1 Example

For a flavour of what we have implemented and proved, consider an example: if our implementation is supplied with the following call to function `pascal-triangle`,

```
(pascal-triangle '((1)) '6)
```

it parses the string, evaluates the expression and prints a string,

```
((1 6 15 20 15 6 1)
 (1 5 10 10 5 1)
 (1 4 6 4 1)
 (1 3 3 1)
 (1 2 1)
 (1 1)
 (1))
```

where `pascal-triangle` had been supplied to it as

```
(label pascal-triangle
  (lambda (rest n)
    (cond ((equal n '0) rest)
          ('t (pascal-triangle
                (cons (pascal-next '0 (car rest)) rest)
                (- n '1))))))
```

with auxiliary function:

```
(label pascal-next
  (lambda (p xs)
    (cond ((atomp xs) (cons p 'nil))
          ('t (cons (+ p (car xs)) (pascal-next (car xs)
              (cdr xs)))))))
```

The theorem we have proved about our LISP implementation can be used to show e.g. that running `pascal-triangle` will terminate and print the first $n + 1$ rows of Pascal's triangle, without a premature exit due to lack of heap space. One can use our theorem to derive sufficient conditions on the inputs to guarantee that there will be enough heap space.

3.4.2 LISP Evaluation

The most interesting part of this case study is possibly the construction of verified code for LISP evaluation. For this we used our extensible compiler, described above.

First, the compiler's input language was extended with theorems that provide ARM code that performs LISP primitives, car, cdr, cons, equal, etc. These theorems make use of an assertion lisp, which states that a heap of s-expressions $v_1 \ldots v_6$ is present in memory. For car of s-expressions v_1, we have the theorem:

$$\text{is\_pair } v_1 \Rightarrow$$
$$\{ \text{ lisp } (v_1, v_2, v_3, v_4, v_5, v_6, l) * \text{pc } p \}$$
$$p : \texttt{E5933000}$$
$$\{ \text{ lisp } (\text{car } v_1, v_2, v_3, v_4, v_5, v_6, l) * \text{pc } (p + 4) \}$$

The cons primitive was the hardest one to construct and prove correct, since the implementation of cons contains the garbage collector: cons is guaranteed to succeed whenever the size of all live s-expressions is less than the heap limit l.

$$\text{size } v_1 + \text{size } v_2 + \text{size } v_3 + \text{size } v_4 + \text{size } v_5 + \text{size } v_6 < l \Rightarrow$$
$$\{ \text{ lisp } (v_1, v_2, v_3, v_4, v_5, v_6, l) * \text{pc } p \}$$
$$p : \texttt{E50A3018 E50A4014 E50A5010 E50A600C ... E51A8004 E51A7008}$$
$$\{ \text{ lisp } (\text{cons } v_1 \ v_2, v_2, v_3, v_4, v_5, v_6, l) * \text{pc } (p + 332) \}$$

The above-mentioned theorems extend the compiler input language with:

$$\text{let } v_1 = \text{car } v_1 \text{ in } \quad \text{and} \quad \text{let } v_1 = \text{cons } v_1 \ v_2 \text{ in}$$

Once the compiler understood enough LISP primitives, we defined lisp_eval as a lengthy tail-recursive function and used the compiler to synthesise ARM code for implementing lisp_eval.

In order to verify the correctness of lisp_eval, we proved that lisp_eval will always evaluate s to r in environment ρ whenever a clean relation semantics for LISP evaluation, which had been developed in unrelated previous work [11], evaluates s to r in environment ρ, written $(s, \rho) \rightarrow_{eval} r$. Here s-expression nil initialises variables v_2, v_3, v_4 and v_6; functions t and u are translation functions from one form of s-expression into another.

$$\forall s \, \rho \, r. \ (s, \rho) \rightarrow_{eval} r \ \Rightarrow \ \text{fst } (\text{lisp\_eval } (t \ s, \text{nil}, \text{nil}, \text{nil}, u \ \rho, \text{nil}, l)) = t \ r$$

3.4.3 Parsing and Printing

The heap of s-expressions defined within the lisp assertion used above is non-trivial to set up. Therefore we constructed verified code for setting up and tearing down a heap of s-expressions. The set-up code also parses s-expressions stored as a string in memory and sets up a heap containing that s-expression. The tear down code prints into a buffer in memory, the string representation of an s-expression from the heap. The code for set-up/tear-down, parsing/printing, was again synthesised from functions in the HOL4 logic.

3.4.4 Final Correctness Theorem

By composing theorems for parsing, evaluation and printing, we get the final correctness theorem: if $\rightarrow_{eval}$ relates s with r under the empty environment (i.e. $(s, [])$ $\rightarrow_{eval}$ r), no illegal symbols are used (i.e. sexp_ok $(t\ s)$), running lisp_eval on $t\ s$ will not run out of memory (i.e. lisp_eval_pre$(t\ s, \text{nil}, \text{nil}, \text{nil}, \text{nil}, \text{nil}, l)$), the string representation of $t\ s$ is in memory (i.e. string a (sexp2string $(t\ s)$)) and there is enough space to parse $t\ s$ and set up a heap of size l (i.e. enough_space $(t\ s)\ l$), then the code will execute successfully and terminate with the string representation of $t\ r$ stored in memory (i.e. string a (sexp2string $(t\ r)$)). The ARM code expects the address of the input string to be in register 3, i.e. r3 a.

$$\forall s\ r\ l\ p.$$
$$(s, []) \rightarrow_{eval} r \wedge \text{sexp\_ok}\ (t\ s) \wedge \text{lisp\_eval\_pre}(t\ s, \text{nil}, \text{nil}, \text{nil}, \text{nil}, \text{nil}, l) \Rightarrow$$
$$\{\ \exists a.\ \text{r3}\ a * \text{string}\ a\ (\text{sexp2string}\ (t\ s)) * \text{enough\_space}\ (t\ s)\ l * \text{pc}\ p\ \}$$
$$p : ...\ \text{code not shown} ...$$
$$\{\ \exists a.\ \text{r3}\ a * \text{string}\ a\ (\text{sexp2string}\ (t\ r)) * \text{enough\_space'}\ (t\ s)\ l$$
$$*\text{pc}\ (p+10404)\ \}$$

We have also proved this result for similar x86 and PowerPC code. Our verified LISP implementations run can be run on ARM, x86 and PowerPC hardware.

4 Conclusions and Future Research

The ARM verification project has been a fairly modest scale effort: one person full-time specifying and verifying the hardware (Fox) and one to two part time researchers looking at software and the background mathematical theories (Hurd, Myreen). In addition, some students have spent time assisting the research, namely Scott Owens, Guodong Li and Thomas Tuerk.

   The project aims to verify systems built out of COTS components where everything – micro-architecture up to abstract mathematics – is formalised within a single framework. The research is still in progress and, unlike the celebrated CLI

Stack [21], we have not yet completely joined up the various levels of modelling, but this remains our goal. Unlike most other work, we have used a COTS processor and have tried (and are still trying) to formally specify as much as possible, including difficult features like input/output and interrupts. The closest work we know of is the verification of security properties of the Rockwell Collins AAMP7G processor [13, 14]. More on AAMP7G can be found in other chapters of this book.

Even though the ARM ISA is relatively simple, the low-level details can overwhelm verification attempts. During the project we have found that it is important to abstract as much as possible so that proofs are not cluttered with such details. A key tool for this has been the derivation of a next-state function for CPU–memory combinations which then can be used to derive clean semantic specifications for instruction uses-cases and then support a further abstraction to Hoare-like rules for machine code segments, with the frame problem managed via a separating conjunction. Some of the technical details pertaining to this abstraction methodology are sketched in the preceding two sections.

Although our formal specifications include input/output, interrupts and facilities for modelling complex memory models, we have yet (2009) to demonstrate significant verification case studies that utilize them. Our current work aims to create a complete functional programming platform on bare metal, with high-fidelity modelling of system level timing and communication with the environment. We expect that achieving this will take several more years of research at the current level of effort.

References

1. Alglave J, Fox A, Ishtiaq S, Myreen M, Sarkar S, Sewell P, Nardelli FZ (2009) The semantics of Power and ARM multiprocessor machine code. In: Basin D, Wolff B (eds) Proceedings of the 4th ACM SIGPLAN workshop on declarative aspects of multicore programming. Association for Computing Machinery, New York, NY, pp 13–24
2. ARM Ltd. (2009) Jazelle technology. http://www.arm.com/products/multimedia/java/jazelle.html (accessed in July 2009)
3. Burch J, Dill D (1994) Automatic verification of pipelined microprocessor control. Springer, Berlin, pp 68–80
4. Fox ACJ (1998) Algebraic models for advanced microprocessors. PhD thesis, University of Wales, Swansea
5. Fox ACJ (2001a) An algebraic framework for modelling and verifying microprocessors using HOL. In: Technical report 512, University of Cambridge Computer Laboratory, April 2001
6. Fox ACJ (2001b). A HOL specification of the ARM instruction set architecture. In: Technical report 545, University of Cambridge Computer Laboratory, June 2001
7. Fox ACJ (2002) Formal verification of the ARM6 micro-architecture. In: Technical report 548, University of Cambridge, Computer Laboratory, 2002
8. Fox ACJ (2003) Formal specification and verification of ARM6. In: Basin D, Wolff B (eds) Theorem proving in higher order logics, vol 2758 of Lecture notes in computer science. Springer, Berlin, pp 25–40
9. Fox ACJ (2005) An algebraic framework for verifying the correctness of hardware with input and output: a formalization in HOL. In: Fiadeiro J, Harman N, Roggenbach M, Rutten

JJMM (eds) CALCO 2005, vol 3629 of Lecture notes in computer science. Springer, Berlin, pp 157–174
10. Furber S (2000) ARM: system-on-chip architecture, 2nd edn. Addison-Wesley, Reading, MA
11. Gordon M (2007) Defining a LISP interpreter in a logic of total functions. In: The ACL2 theorem prover and its applications (ACL2)
12. Gordon MJC (1983) Proving a computer correct with the LCF-LSM hardware verification system. In: Technical report 42, University of Cambridge Computer Laboratory, 1983
13. Greve D, Wilding M, Vanfleet WM (2003) A separation kernel formal security policy. In: ACL2 workshop 2003, June 2003
14. Greve D, Richards R, Wilding M (2004) A summary of intrinsic partitioning verification. In: ACL2 Workshop 2004, November 2004
15. Hardin D (2008) Invited tutorial: considerations in the design and verification of micropro- cessors for safety-critical and security-critical applications. In: Proceedings of FMCAD 2008, November 2008
16. Harman NA, Tucker JV (1997) Algebraic models of microprocessors: the verification of a sim- ple computer. In: Stavridou V (ed) Mathematics of dependable systems II. Oxford University Press, Oxford, pp 135–170
17. Harrison JR (2005) A HOL theory of Euclidean space. In: Hurd J, Melham T (eds) Theo- rem proving in higher order logics, 18th International conference, TPHOLs 2005, vol 3603 of Lecture notes in computer science, Oxford, UK. Springer, Berlin, pp 114–129
18. Hurd J (2005) Formalizing elliptic curve cryptography in higher order logic. Available from the author's Web site, October 2005
19. Hurd J, Gordon M, Fox A (2006) Formalized elliptic curve cryptography. In: High confidence software and systems: HCSS 2006, April 2006
20. McCarthy J, Abrahams PW, Edwards DJ, Hart TP, Levin MI (1966) LISP 1.5 programmer's manual. MIT, Cambridge, MA
21. Moore JS (foreword) (1989) Special issue on systems verification. J Autom Reason 5(4): 461–492
22. Myreen MO (2009a) Formal verification of machine-code programs. PhD thesis, University of Cambridge
23. Myreen MO (2009b) Verified implementation of LISP on ARM, x86 and PowerPC. In: Theorem proving in higher-order logics (TPHOLs). Springer, Berlin
24. Myreen MO, Slind K, Gordon MJC (2008) Machine-code verification for multiple architec- tures – an application of decompilation into logic. In: Formal methods in computer aided design (FMCAD). IEEE, New York, NY
25. Myreen MO, Slind K, Gordon MJC (2009) Extensible proof-producing compilation. In: Com- piler construction (CC). Springer, Heidelberg
26. Reynolds J (2002) Separation logic: a logic for shared mutable data structures. In: Proceedings of logic in computer science (LICS). IEEE Computer Society, Washington, DC
27. Sawada J, Hunt WA Jr (2002) Verification of fm9801: an out-of-order microprocessor model with speculative execution, exceptions, and program-modifying capability. Formal Methods Syst Des 20(2):187–222
28. Schostak D (2003) Methodology for the formal specification of RTL RISC processor designs (with particular reference to the ARM6). PhD thesis, University of Leeds
29. Slind K (2009) TFL: an environment for terminating functional programs. http://www.cl.cam. ac.uk/~ks121/tfl.html (accessed in July 2009)
30. Thery L (2007) Proving the group law for elliptic curves formally. In: Technical report RT- 0330, INRIA, 2007
31. Wong W (1983) Formal verification of VIPER's ALU. In: Technical report 300, University of Cambridge Computer Laboratory, April 1983

Information Security Modeling and Analysis

David A. Greve

1 Introduction

The focus of our information security research is on how information is protected within the context of computing systems. In particular, our concern is for how information may or may not be communicated within computing systems. The objective of our research is to develop formal information security specifications for computing system components that can be mathematically verified against the implementation of those components and then used to reason about the information security properties of systems that use those components. The goal is to enable formal proofs that our high-assurance secure information processing systems satisfy their system-level security policies. Developing formal security specifications requires reasonable formal models of computing system components, information, and communication.

1.1 Formal PVS Specifications

Throughout this chapter we will present concrete formalizations of selected topics in the language of PVS [14]. PVS is a mechanized reasoning environment for formal specification and verification. The specification language of PVS is based on classical, typed higher order logic. Specifications are organized into (potentially parameterized) theories, which are collections of type and function definitions, assumptions, axioms, and theorems. The proof language of PVS is composed of a variety of primitive inference procedures that may be combined to construct more powerful proof strategies. Proofs in PVS are normally associated with theories but are hidden from the casual observer. In some of our examples, an extension to PVS called ProofLite [9] has been used that enables a literate-proof style, reminiscent of

D.A. Greve (✉)
Rockwell Collins, Inc., Cedar Rapids, IA, USA
e-mail: dagreve@rockwellcollins.com

D.S. Hardin (ed.), *Design and Verification of Microprocessor Systems for High-Assurance Applications*, DOI 10.1007/978-1-4419-1539-9_9,
© Springer Science+Business Media, LLC 2010

Coq, in which proof scripts are stored as special comments within PVS theories. Many of the theories we present here have been edited for the sake of brevity, but a Web site containing the complete proof scripts can be found by visiting http://extras.springer.com and entering the ISBN for this book.

While many of the concepts in this chapter were first formalized in the logic of ACL2, PVS provides a convenient, generally accessible mathematical framework for presenting high-level concepts involving sets, quantifiers, and first-class functions not available in ACL2. These formalizations also act as a sanity check, helping to ensure that our understanding of the concepts is consistent and portable across different formalizations. The theories presented here were developed using PVS 4.2 and ProofLite 4.0.

2 A Formal Model of Computing Systems

In our framework, sequential computing systems that interact with the external environment are generally modeled as state machines. A state machine model suggests a state transition (or step) function that operates over a set of inputs and an initial state to produce a set of outputs and a next state. This function can be applied iteratively to successive inputs and states to simulate the evolution of the system state and its outputs over time.

State machine models are significant in our analysis because they allow us to decompose our analysis into both the single-step and the multistep (trace) behavior of the system. Many security properties are best stated as single-step properties. Some properties, however, must be analyzed over an entire execution trace. State machine models support the analysis of both.

State transition operations in our framework are modeled functionally. A functional model is one in which an output is computed only from the inputs with no hidden side-effects. State transition functions must therefore accept as input an initial state plus inputs, all state changes and outputs must be computed relative to the initial state and the inputs, and the updated state plus outputs must be returned by the function. By eliminating side-effects, functional models require that all computations be explicit. This is important in a security context, where it is essential to be able to account for all system behavior.

2.1 A Formal Model of Information

There are many obvious and effective techniques for formally modeling computing systems. This seems not to be the case for the information processed by computing systems. This is not to say that there are no useful mathematical models of information. Information entropy (or Shannon entropy) is a very useful model of information used in information theory that has yielded many remarkable results, both in theory and in practice [13]. However, the formal analysis of a computing system in which information is modeled as a property of its *value* (or expected value) seems overwhelming. While such an undertaking might be appropriate in such specialized fields as cryptoanalysis, it is overkill for many of the properties that one typically encounters in the broader context of information security.

Our approach to modeling information in computing systems, while somewhat indirect, is one that focuses the *location* of information in the system, rather than the value of that information. In our model, the concept of information is defined relative to the system processing the information. We say things such as, "these inputs (the inputs located here) are classified as SECRET while these other inputs (located over there) are not." Much of the focus of our research is in formalizing what we mean by "locations" and then connecting that model to our formal model of computation.

2.1.1 The Calculus of Indices

In our model, the location of a piece of information is represented as an *index*. An index is a value in the logic that is associated with a specific portion of the system state. Implicitly associated with an index is a means of extracting (or inserting) the portion of system state associated with that index. A *basis* is a collection of distinct index values, each of which is associated with a portion of the system state. Index values and the functions used to manipulate and reason about them are all part of what we call the *calculus of indices*.

The calculus of indices is characterized within the logic, not the metalogic. Indices can be expressed, manipulated, quantified over, and computed[1] (possibly from other portions of the system state), all within the logic. For example, in the Lisp expression `(let* ((x 3) (y 3))..)`, "x" and "y" (as they appear in the expression) are in the metalogic (of Lisp) and thus are not a part of our calculus. It is, for example, not possible to quantify over "x" and "y" or to express the fact that "x" and "y" are different (their values in this fragment are, after all, the same!) in the sense that the act of binding a value to "y" does not interfere with the value bound to "x." The fact that "x" and "y" are independent follows from the fact that "x" and "y" are syntactically distinct symbols in the metalogic.[2] Similarly, in the Hoare-style statement `{P[x/E]} x: = E {P}`, the syntactic expression "x" is in the metalogic and thus would not be considered part of a calculus.

---

[1] Because index values are computed, analysis in our model is, in general, undecidable.

[2] Note that such metalogical conclusions often disguise pointer aliasing issues: the faulty intuition being that syntactically unique symbols must point to unique address locations.

Every indexing scheme is associated with a specific type of object, though each type of object may have many indexing schemes. Most common data structures suggest obvious indexing schemes. The set of field names is a natural choice for records. Tuples might be well served by a natural number basis that maps to each tuple position. The contents of a list could be described by their position in the list and the contents of an association by the keys of the association. Arrays suggest a very natural indexing scheme: each index value is simply an index into the array. The fact that array indices are generally computable within the logic highlights the need to support computation within the calculus. Arguments to functions may also be indexed, either by name or position, as they may return values.

Sets of index values are important in our formalizations. The PVS theory we use to model index sets introduces many common set operations as well as short-hand (infix) notations set insertion, deletion, and union.

```
IndexSet [index: TYPE]: THEORY
BEGIN

  IMPORTING sets_lemmas [index]

  set: TYPE = [ index -> bool ]

  Empty: set = (lambda (i: index): false)

  insert (a: index, x: set): set =
    (lambda (i: index): (i = a) OR x (i))

  union (x, y: set): set =
    (lambda (i: index): x (i) OR y (i));

  remove (s1, s2: set): set =
    (LAMBDA (i: index): s1 (i) & not (s2 (i)));

  intersection (s1, s2: set): set =
    (LAMBDA (i: index): s1 (i) & s2 (i));

  not (s: set): set = (LAMBDA (i: index): (not (s (i))));

  + (s: set, a: index): set = insert (a, s);

  + (a, b: index): set = insert (a, singleton (b));

  + (s1, s2: set): set = union (s1, s2);

  - (s1, s2: set): set = remove (s1, s2);

  - (s1: set, a: index): set = remove (s1, singleton (a));
```

```
   &(s1,s2: set): set = intersection(s1,s2);

   intersects(s1,s2: set): bool =
     (EXISTS (i: index): member(i,s1) AND member(i,s2))

   disjoint(s1,s2: set): bool = not(intersects(s1,s2))

END IndexSet
```

2.1.2 Paths

A hierarchical indexing scheme is a useful generalization providing intuitive representations for locations within hierarchical data structures. Paths, as commonly used in operating systems to describe specific locations in hierarchical file systems, are a sequence of directory names. We define a path as a sequence of index values used to identify a specific location in a hierarchical data structure. For example, if the third argument to a function is a record structure that contains an array in a field named "x," the path specifying the location of the 5th element of that array might be represented as the sequence (3 "x" 5).

While paths are exceptionally useful in modeling computing systems, the concepts we wish to convey are independent of the index representation. To simplify our presentation, we consider only simple (nonhierarchical) indexing systems. Suffice it to say that a useful calculus that employs paths must consider a rich set of binary relations between paths (including subsumption) that will impact the formulation of the theory.

2.1.3 A Simple Calculus in PVS

An indexing scheme is always formalized relative to the type of the underlying data. Of particular importance is the function that projects from the state the value associated with each index value. This function *is* the interpretation of the basis set relative to the state. Here we formalize an indexing scheme for a specific record type in PVS. The projection function in this example is called get. Note the hoops required to work around the strong type system.

```
IndexingExample : THEORY

  BEGIN

  enum: TYPE = {red, green, blue}

%% An Example Record State
```

```
   state: TYPE = [# a: int, b: bool, c: enum #]

 %% An Example Index Set

   index: TYPE = { a, b, c }

   polyvalue: TYPE = [ int + bool + enum ]

   index_predicate(i: index, v: polyvalue): bool =
     COND
       a?(i) -> IN?_1(v),
       b?(i) -> IN?_2(v),
       c?(i) -> IN?_3(v)
     ENDCOND

   valuetype(i: index): TYPE =
     {v: polyvalue | index_predicate(i,v) }

 %% The projection function

   get(i: index, s: state): valuetype(i) =
     COND
       a?(i) -> IN_1[polyvalue](a(s)),
       b?(i) -> IN_2[polyvalue](b(s)),
       c?(i) -> IN_3[polyvalue](c(s))
     ENDCOND

   END IndexingExample
```

2.1.4 Properties of Basis Sets

Just as with basis sets in linear algebra, a variety of bases may be capable of accurately modeling a given system. However, unlike basis sets in linear algebra, it is not always possible to translate models expressed in one basis set into an equivalent model expressed in another basis set. Some care is thus needed in the choice of a basis set to ensure that it is useful for expressing interesting properties about the system.

Conversely, the choice of basis set will influence the meaning of the information flow properties expressed in terms of that basis. It is possible, for example, to choose a degenerate basis set that would render nearly any information flow theorem vacuous. To help guard against such deception, it may be useful to consider several properties that different basis sets might exhibit:

- *PolyValued.* It is possible for the value projected by each index to assume at least two unique values. This property ensures that the projection functions are not constant.

- *Divisive.* It is possible for the value projected at two unique index values to vary independently of each other. This property ensures that no projection function simply returns the entire state.
- *Orthogonal.* The portion of state associated with each index value is independent of the portions associated with every other index. A nonorthogonal basis gives the illusion of separation (between two index values) when, in fact, those index values overlap and are therefore dependent.
- *Complete.* If the values projected at every index are equal, then the states are equal. If a basis set is incomplete, then there is some portion of the state that is not observable using that set, and that portion of the state could potentially be used as a covert channel.

Here we formalize these and additional concepts as predicates over projection functions.

```
Ideal[state: TYPE+,index,value: TYPE]  :  THEORY

BEGIN

  projection: TYPE = [[index,state] -> value]

  gettablevalue?(g: projection,
                i: index,
                v:value): bool =
    EXISTS (st: state): v = g(i,st)

  gettablevalue(g: projection, i: index): TYPE =
    { v: value | gettablevalue?(g,i,v) }

  PolyTypeIndex: bool =
    (EXISTS (i,j: index): (j /= i))

  PolyValued(g: projection): bool =
    FORALL (i: index):
      EXISTS (st1,st2: state):
        g(i,st1) /= g(i,st2)

  Divisive(g: projection): bool =
    FORALL (i: index):
      (FORALL (j: index):
        (j /= i) =>
          EXISTS (st1,st2: state):
            g(i,st1) = g(i,st2) AND
              g(j,st1) /= g(j,st2))
```

```
  Reasonable(g: projection): bool =
    PolyValued(g) & Divisive(g)

  Orthogonal(g: projection): bool =
    FORALL (i: index,
            v: gettablevalue(g,i),
            st1: state):
      EXISTS (st2: state):
        FORALL (j: index):
          g(j,st2) = IF (i = j) THEN v
                     ELSE g(j,st1)
                     ENDIF

%% OrthogonalSet should follow from Orthogonal
%% for finite index types.

  OrthogonalSet(g: projection): bool =
    FORALL (s: set[index],
            ps: [i:index -> gettablevalue(g,i)],
            st1: state):
      EXISTS (st2: state):
        FORALL (j: index):
          g(j,st2) = IF member(j,s) THEN ps(j)
                     ELSE g(j,st1)
                     ENDIF

  Complete(g: projection): bool =
    FORALL (st1,st2: state):
      (st1 = st2) =
        (FORALL (i: index): g(i,st1) = g(i,st2))

  Injectable(g: projection): bool =
    FORALL (ps: [i:index -> gettablevalue(g,i)]):
      EXISTS (st2: state):
        FORALL (j: index): g(j,st2) = ps(j)

  OrthogonalSet_implies_Injectable: LEMMA
    FORALL (g: projection):
      OrthogonalSet(g) => Injectable(g)

  Ideal(g: projection): bool =
    Reasonable(g) & OrthogonalSet(g) & Complete(g)

END Ideal
```

A `Reasonable` basis set is both `PolyValued` and `Divisive`. Any alternative would be egregiously degenerate. An `Ideal` basis set will also be `Orthogonal` and `Complete`. We recognize, however, that compelling arguments exist that justify the use of nonideal basis sets under appropriate circumstances. The basis set for the example record state described above is provably `Ideal`.

```
IndexingExampleProperties: THEORY
BEGIN
   IMPORTING IndexingExample
   IMPORTING Ideal[state,index,polyvalue]

   ExampleIsIdeal: LEMMA
     Ideal(get)

END IndexingExampleProperties
```

2.2   A Formal Model of Communication

Communication is a dynamic process involving the movement of information from one location to another. Our model of communication focuses on the location of information, rather than its actual movement. Additionally, a mathematical description of when communication does not take place is much simpler to reason about than a description of when it does. The absence of communication is called noninterference, and noninterference forms the basis upon which all of our analysis is built. Consequently, when we speak of communication taking place between location A and location B, what we are really saying is that we have not proven that communication does not take place. Communication is therefore a weak property in our model while noninterference is a strong property. Fortunately, from a security perspective, the logical separation provided by a noninterference guarantee is often more important than the quality of service provided by a communication guarantee.

Because communication within a computing system is dynamic, it must be associated with some actions in that system. In our model of computation, communication is carried out by the functions used to model the system. We say that a function is partially characterized by the kind of communications it performs. Mathematically we model communication properties of functions as *congruence* relations [5]. A congruence relation is a property of a function that says that the outputs from two different applications instances of the same function will satisfy a particular equivalence if the two input instances to those functions satisfy an equivalence, not necessarily the same equivalence.

The equivalence relations we use in our formalisms are typically parameterized by index values and are defined in terms of the basis' projection function. We say that two objects are equivalent modulo a selected index value if the values projected from the two objects agree at that index. It is straightforward to extend the notion of equivalence modulo an index to equivalence modulo a set of indices.

```
Equiv[index, state, value: TYPE,
          get: [[index, state] -> value]]: THEORY
BEGIN

  IMPORTING IndexSet[index]

  equiv(i: index, st1,st2: state): bool =
    (get(i,st1) = get(i,st2) )

  equivSet(S: set, st1,st2: state): bool =
    (FORALL (i: index):
       member(i,S) => equiv(i,st1,st2))

END Equiv
```

2.2.1 Congruences in PVS

The kind of congruence relation we use to model communication says: given two arbitrary application instances of a specific function (next), the value at a selected index[3] (seg) in the range of the output of the two application instances will be the same if the values of the input domains (st1,st2) within a set (DIA) of index locations are the same.

```
  equivSet(DIA,st1,st2) =>
    equiv(seg,next(st1),next(st2))
```

A set of index values (DIA) satisfying this assertion is called the interferes set of the index (seg), since it contains every index location that might interfere with (or influence or communicate with) the final value of the selected output. This set could also be called the *use set* of the index, since it must contain every input index value used[4] in computing the specified output. Every input index that does not appear in

---

[3] In our original formulation, an index value was referred to as a seg, which is to say, a segment of the state.

[4] In this context, the term "used" is potentially too broad, since not every index value referenced in the course of computation needs to be included if they can be shown to be functionally irrelevant. However, the term "required" is perhaps too narrow, as we allow the use set to be an overapproximation. A more precise description would be those index values which have not been shown to be irrelevant.

this set satisfies a "noninterference" property with respect to the selected output: it is impossible for such input indices to interfere (communicate) with the selected output.

The GWV theorem was our earliest formulation of this congruence [7]. The theorem was motivated by and specifically targeted toward separation kernels. The original theorem formulation broke the equivalence relation in the hypothesis into three components: one similar to the hypothesis shown above, one targeting the currently active partition, and one that required that the value of seg be equivalent in the initial state. While the theorem was sufficient for simple, static separation kernels like the AAMP7G, it suffered from some expressive limitations which spurred development of two more expressive and more general formulations [1, 8].

In the original GWV theorem, DIA (the name was chosen as an acronym for direct interaction allowed) was expressed as a function of the output index. In that formulation, next was the transition function of a state machine model, specifying what the system can do in a single step of its operation. The designation "Direct," therefore, emphasized the single-step nature of the characterization and distinguished it from what may take place transitively over multiple steps of the system.

Origin notwithstanding, this congruence may be used to characterize any function, not just state transition functions. In subsequent revisions of the theorem, the computation of the interferes set has become more dynamic. It is now computed in its full generality as a function of state. For representational convenience, however, this computation has been decomposed into two steps: the computation of a comprehensive information flow model encompassing the behavior of the entire function and the extraction of a specific interferes set based on the output index being considered. The comprehensive information flow model is called the information flow graph. The function that extracts from the graph the interferes set associated with a specific output index is now referred to as the DIA function.

2.2.2 Information Flow as a Graph

When describing the high-level information flow of a system, it is often convenient to use "circle and arrow" diagrams in which circles represent information domains and arrows represent flows between those domains. Such informal representations can be modeled mathematically as graphs. In mathematics, a graph is an object with nodes (circles) connected by edges (arrows). We have adopted graphs as a means of expressing information flows. The nodes in the graph represent index values and the edges (which are directed) represent information flows from one index location to another. Given a graph, the interferes set of a particular output index can be computed by a function that searches the graph and returns the set of nodes with an edge leading to that index.

Graph data structures provide an abstract representation of the information flow behavior of computational systems. Certain information flow properties, properties such as noninterference, can actually be directly expressed as (or deduced from) properties of those flow graphs, without the need to appeal to the computational model.

2.2.3　Graphs in PVS

Our model of graphs in PVS has two types of edges: one that maps an index value to a set of index values (a Compute edge) and one that maps an index value into a single index value (a Copy edge). One model of information flow utilizes Copy edges to model "frame conditions," locations that remain unchanged following the execution of a function. Index values associated with Computed edges are locations in the state that may have in some way been changed during the course of function execution.

```
GraphEdge[index: TYPE]: DATATYPE
BEGIN
  IMPORTING IndexSet[index]
  Compute(UseSet: Set[index])  : Computed?
  Copy(CopyIndex: index)       : Copied?
END GraphEdge
```

Note that a graph models the information flow of a function and that the type of the output of a function may differ from the type of its input. This means that, in general, the type of the index value used to index the graph will differ from the type of the index value found in the set returned by the graph. The strong typing in PVS helps to make this explicit. Our GWV_Graph theory, therefore, is parameterized by both the input and the output index types.

```
GWV_Graph[INindex,OUTindex: TYPE]: THEORY
BEGIN

  IMPORTING IndexSet[INindex]
  IMPORTING IndexSet[OUTindex]

  OSet:  TYPE = Set[OUTindex]
  ISet:  TYPE = Set[INindex]

  IMPORTING GraphEdge[INindex]

  graph: TYPE = [ OUTindex -> GraphEdge[INindex]]
```

The DIA function, defined over graphs, computes the interferes set for a specific output index by mapping the output index to the set of input index values upon which it depends. The DIASet function extends the behavior of DIA to apply to sets of output indices. DIASet is simply the union of the DIA values for each member of the set.

```
DIA(i: OUTindex, g: graph): ISet =
  CASES g(i) OF
    Compute(s): s,
    Copy(j):    singleton(j)
  ENDCASES

DIASet(S: OSet, g: graph): ISet =
  (LAMBDA (i: INindex) :
    (EXISTS (x: OUTindex) :
      member(x,S) AND DIA(x,g)(i)))
```

The overall define set (or defSet) of a function, the set of locations modified by the function, can be computed from the graph. The same is true of the set of locations upon which the define set depends, the use set (or useSet). The inverse DIA function, a function that computes the set of outputs that depend upon a given input, is also provided as it is useful for expressing certain graph properties.

```
defSet(g: graph): Oset =
  (LAMBDA (out: OUTindex): Computed?(g(out)))

useSet(g: graph): Iset =
  (LAMBDA (i: INindex):
    EXISTS (out: OUTindex):
      Computed?(g(out)) &
        member(i,Useset(g(out))))

invDIA(i: INindex, g: graph): Oset =
  (LAMBDA (out: OUTindex): member(i,DIA(out,g)))

END GWV_Graph
```

2.2.4 Graph Functions

As an obligation, a more comprehensive (or larger) interferes set makes the congruence theorem easier to verify. If it is possible to verify that a particular set characterizes a specific system output, any superset will also characterizes that output. Consequently, it is always conservative to add inputs to the use set. However, to say that a system output depends upon every system input is not a particularly useful specification. As a system contract, therefore, a more precise (or smaller) description is generally more useful.

One way to increase the precision of information flow description is to allow them to be conditional. A conditional flow description is one that suggests different interferes sets under different conditions. This is particularly useful when the condition reflects radically different operational modes of the system. For example, the expression (if a b c) depends upon b only when a is true and on c only when a is false. Note, however, that the expression always depends upon a.

Conceptually, conditions can be viewed as labels on the edges of an information flow graph, much like we might find in a labeled transition system.[5] The label in this case identifies a specific condition under which the edge relationship exists. Such conditions are implicitly functions of the state of the system. The information flow characteristics of a function, therefore, are not necessarily static and may, in fact, depend upon the inputs to that function. Precise characteristic graphs for functions must, in general, be computed from the function inputs. Functions that compute information flow graphs are called graph functions, although we often simply refer to such functions as graphs. A graph function can provide the most precise characterization of information flow possible for a given function.

2.2.5 GWVr1 in PVS

Having formalized graphs in PVS, we are now in a position to formalize our extensions of the original GWV information flow theorem, extensions expressed in terms of information flow graphs. Our ability to quantify over functions in PVS allows us to express these extensions in their full generality.

```
GWVr1 [INindex, INState, INvalue: TYPE,
       getIN: [[INindex, INState] -> INvalue],
       OUTindex, OUTState, OUTvalue: TYPE,
       getOUT: [[OUTindex, OUTState] -> OUTvalue]
       ]: THEORY
BEGIN

  IMPORTING GWV_Graph[INindex,OUTindex]
  IMPORTING Equiv[INindex,INState,INvalue,getIN]
    AS Input
  IMPORTING Equiv[OUTindex,OUTState,OUTvalue,getOUT]
    AS Output

  StepFunction: TYPE = [ INState -> OUTState ]
  GraphFunction: TYPE = [ INState -> graph ]
  PreCondition: TYPE  = [ INState -> bool ]
```

---

[5] The representational similarity between models of information flow graphs and models of transition systems is partially responsible for inspiring subsequent work on model checking transitive information flow properties.

We define GWVr1 as predicate over a step function, a precondition, and a graph (more precisely: a graph function) that expresses a specific congruence relation. The congruence relation says that the outputs of two unique applications of the function are equivalent modulo a specific output index if the input instances of those two functions satisfy the preconditions and are equivalent modulo the set of input index values that may, according to the graph, directly influence the output index. For a given function, we say that any graph function satisfying this predicate *characterizes* the information flow of that function under the preconditions.

```
    GWVr1(Next: StepFunction)
         (Hyps: PreCondition,
          Graph: GraphFunction): bool =
   FORALL (x: OUTindex, in1,in2: INState):
     Input.equivSet(DIA(x,Graph(in1)),in1,in2) &
       Hyps(in1) & Hyps(in2) =>
         Output.equiv(x,Next(in1),Next(in2))
```

It is useful to generalize GWVr1 to apply to sets of output index values and it is easy to prove that GWVr1 implies GWVr1Set.

```
    GWVr1Set(Next: StepFunction)
            (Hyps: PreCondition,
             Graph: GraphFunction): bool =
    FORALL (x: set[OUTindex], in1,in2: INState):
      Input.equivSet(DIAset(x,Graph(in1)),in1,in2) &
        Hyps(in1) & Hyps(in2) =>
          Output.equivSet(x,Next(in1),Next(in2))


    GWVr1_implies_GWVr1Set: LEMMA
      FORALL (Next: StepFunction,
              Hyps: PreCondition,
              Graph: GraphFunction):
      GWVr1(Next)(Hyps,Graph) =>
        GWVr1Set(Next)(Hyps,Graph)


  %|- GWVr1_implies_GWVr1Set: PROOF
  %|- (grind)
  %|- QED

END GWVr1
```

It is interesting to note that there is a witnessing graph satisfying GWVr1 for any function (assuming a *Complete* basis set). In particular, it is easy to show that the graph that says that every output depends upon every input characterizes any function, not that such a graph is a particularly useful characterization of any function.

```
    IMPORTING Ideal[INState,INindex,INvalue]

    CompleteGraph(in1: INState): graph =
      (LAMBDA (i: OUTindex):
        Compute((LAMBDA (i: INindex): true)))

    GWVr1Existence: LEMMA
      Complete(getIN) =>
        FORALL (F: StepFunction,H: PreCondition):
          EXISTS (G: GraphFunction):
            GWVr1(F)(H,G)

  %|- GWVr1Existence : PROOF
  %|- (then (ground) (skosimp) (inst 1 "CompleteGraph")
  %|- (expand "GWVr1") (skosimp) (expand "Output.equiv")
  %|- (expand "Complete") (inst -1 "in1!1" "in2!1")
  %|- (expand "Input.equivSet")
  %|- (expand "CompleteGraph") (expand "DIA")
  %|- (expand "member") (expand "equiv") (iff) (ground))
  %|- QED

  END GWVr1Existence
```

2.3 Examples of Applying GWVr1

For further insight on how GWVr1 works, it is illustrative to consider how it could be applied to some simple examples. Consider a state data structure modeled as a mapping from natural numbers to integers. We define a projection function for this state (get), as well as a function that allows us to selectively modify specific portions of the state (set). With these definitions in hand, we import the GWVr1 theory.

```
  GWVr1Test: THEORY
  BEGIN

    state: TYPE = [ nat -> int ]

    get(i:nat,st:state): int = st(i)
```

```
set(i:nat, v: int, st:state)(j:nat): int =
  IF (i = j) THEN v ELSE get(j,st) ENDIF

IMPORTING GWVr1[nat,state,int,get,
               nat,state,int,get]
```

The proofs we will consider require no special preconditions, so we define a precondition function that is always true. For representational convenience, all of our information flow graphs begin as the identity graph, which says that each output location depends only on its initial value. We then define a function that allows us to incrementally add dependencies to the graph for specific output locations.

```
Hyps(st: state): bool = true

ID : graph = (LAMBDA (i: Index): Copy(i))

addDIA(i:nat,d:nat,g:graph): graph =
   (LAMBDA (n:nat):
     IF (n = i) THEN
       Compute(DIA(i,g) + d)
     ELSE
       g(n)
     ENDIF)
```

Our first example is a function that modifies two locations of our state data structure. `twoAssignment` updates location 0 with the value obtained by reading location 1 and it updates location 2 with the sum of the values from locations 3 and 4.

```
twoAssignment(st: state): state =
  LET v0: int = get(1,st) IN
  LET v2: int = get(3,st) + get(4,st) IN
    set(0,v0,set(2,v2,st))

AUTO_REWRITE+ twoAssignment!
```

Most of the locations in our data structure are unmodified by `twoAssignment`. Consequently the identity graph is a reasonable baseline for our information flow model. We also expect location 0 to depend upon location 1 and location 2 to depend upon locations 3 and 4. We can use `addDIA` to express these dependencies.

```
twoAssignmentGraph(st:state): graph =
  addDIA(0,1,addDIA(2,3,addDIA(2,4,ID)))

AUTO_REWRITE+ twoAssignmentGraph!
```

Assuming that we have covered all of the cases, we should now be able to prove
our GWVr1 theorem for twoAssignment. And, in fact, we can.

```
twoAssignmentGWVr1: LEMMA
  GWVr1(twoAssignment)(Hyps,twoAssignmentGraph)

%|- twoAssignmentGWVr1 : PROOF
%|- (then
%|-   (auto-rewrite "GWVr1")
%|-   (auto-rewrite! "DIA" "defSet")
%|-   (auto-rewrite! "addDIA" "defSet" "get" "set")
%|-   (auto-rewrite "ID")
%|-   (auto-rewrite "member")
%|-   (auto-rewrite "equiv")
%|-   (auto-rewrite "copy")
%|-   (auto-rewrite-theory
%|-      "EquivSetRules[nat,state,int,get]")
%|-   (apply (repeat* (then*
%|-      (lift-if) (ground) (skosimp))))))
%|- QED
```

But what if we had not covered all of the cases? Would the theorem catch our
mistakes? Consider what would happen if we failed to account for one of the writes
performed by twoAssignment. The following graph fails to account for the write to
location 2. Consider what happens when we try to prove that this new graph still
characterizes our original function. Note that we employ the same proof tactics.

```
missedUpdateGraph(st: state): graph =
  addDIA(0,1,ID)

AUTO_REWRITE+ missedUpdateGraph!

missedUpdateGWVr1Fails: LEMMA
  GWVr1(twoAssignment)(Hyps,missedUpdateGraph)

%|- missedUpdateGWVr1Fails : PROOF
%|- (then
%|-   (auto-rewrite "GWVr1")
```

```
%|-    (auto-rewrite! "DIA" "defSet")
%|-    (auto-rewrite! "addDIA" "defSet" "get" "set")
%|-    (auto-rewrite "ID")
%|-    (auto-rewrite "member")
%|-    (auto-rewrite "equiv")
%|-    (auto-rewrite "copy")
%|-    (auto-rewrite-theory
%|-       "EquivSetRules[nat,state,int,get]")
%|-    (apply (repeat* (then*
%|-       (lift-if) (ground) (skosimp)))))
%|- QED
```

In this case, the proof fails. The failed subgoal is included below. Note that the
remaining proof obligation is to show that the sum of locations 3 and 4 for the two
different instances is the same. However, the only hypothesis we have is that the
instances agree at location 2 ($x!1 = 2$). This failed proof typifies proof attempts for
graphs that do not account for all state updates.

```
missedUpdateGWVr1Fails :

{-1}    (2 = x!1)
{-2}    (in1!1(x!1) = in2!1(x!1))
{-3}    Hyps(in1!1)
{-4}    Hyps(in2!1)
   |-------
{1}     (in1!1(3) + in1!1(4) = in2!1(3) + in2!1(4))
{2}     (0 = x!1)
{3}     (x!1 = 0)

Rule?
```

Now consider what happens if we fail to appropriately account for all uses of
a location. The following graph fails to account for the use of location 4 in the
computation of location 2. Again we employ the same basic proof strategy.

```
missedUseGraph(st:state) : graph =
  addDIA(0,1,addDIA(2,3,ID))

AUTO_REWRITE+ missedUseGraph!

missedUseGWVr1Fails: LEMMA
  GWVr1(twoAssignment)(Hyps,missedUseGraph)
```

Again the proof fails. In the failed subgoal for this proof, we see that we know that locations 2 and 3 are the same. However, in order to complete the proof we need to know that location 4 is also the same in both instances. This failed proof typifies proof attempts for graphs that do not account for all uses of state locations.

```
missedUseGWVr1Fails :

{-1}    (2 = x!1)
{-2}    (x!1 = 2)
{-3}    (in1!1(3) = in2!1(3))
{-4}    (in1!1(2) = in2!1(2))
{-5}    Hyps(in1!1)
{-6}    Hyps(in2!1)
   |-------
{1}     (in1!1(3) + in1!1(4) = in2!1(3) + in2!1(4))
{2}     (0 = x!1)
{3}     (x!1 = 0)

Rule?
```

As a general rule, the dependencies of conditionals must be unconditionally included in the dependencies of the assignments they guard. The following function updates location 4 with the value from location 7 if location 0 is equal to 3, otherwise it does nothing.

```
conditionalUpdates(st: state): state =
  IF get(0,st) = 3 THEN
    set(4,get(7,st),st)
  ELSE
    st
  ENDIF
AUTO_REWRITE+ conditionalUpdates!
```

The following graph accurately reflects a precise information flow model of the function `conditionalUpdates`. Note that location 4 depends upon location 7 only if the test succeeds. However, it depends upon location 0 regardless of the outcome of the test. The GWVr1 proof for this graph succeeds.

```
conditionalUpdatesGraph(st: state): graph =
  IF get(0,st) = 3 THEN
    addDIA(4,7,add(4,0,ID))
  ELSE
    addDIA(4,0,ID))
  ENDIF
```

```
   AUTO_REWRITE+ conditionalUpdatesGraph!

   conditionalUpdatesGWVr1: LEMMA
     GWVr1(conditionalUpdates)
          (Hyps,conditionalUpdatesGraph)
```

If, however, the conditional dependency is omitted from one (or both) of the branches, the proof fails.

```
   missedConditionalGraph(st: state): graph =
     IF get(0,st) = 3 THEN
       addDIA(4,7,ID)
     ELSE
        ID
     ENDIF

   AUTO_REWRITE+ missedConditionalGraph!

   missedConditionalGWVr1Fails: LEMMA
     GWVr1(conditionalUpdates)
          (Hyps,missedConditionalGraph)
```

In the failed subgoal below, we are trying to prove that location 4 is the same in both instances. Observe, however, that the conditional expression get(0,st) = 3 has produced different outcomes for the two different input instances ($\{-2\}$ and $\{1\}$). This is possible because location 0 is not known to be the same in those instances. This failed proof typifies proof attempts for graphs that do not accurately account for the dependencies of conditional assignments.

```
   missedConditionalGWVr1Fails.1 :

{-1}    (4 = x!1)
{-2}    in1!1(0) = 3
{-3}    (x!1 = 4)
{-4}    (in1!1(7) = in2!1(7))
{-5}    (in1!1(4) = in2!1(4))
{-6}    Hyps(in1!1)
{-7}    Hyps(in2!1)
  |-------
{1}     in2!1(0) = 3
{2}     (in1!1(7) = in2!1(x!1))

Rule?
```

2.4 Graph Composition

We have shown how graphs can be used to model the information flow properties of individual functions. We now explore the transitive information flow relationships that result from function composition. That is to say, if A depends upon B in a first function and B depends upon C in a second function, it should be possible to conclude that A depends upon C when those two functions are combined. The combination of two functions is called function composition. The combination of two graphs is called graph composition. Graph composition, when properly defined, models function composition. That is to say, given two graphs characterizing the information flow of two functions, the composition of the graphs is a model of the information flow of the composition of the two functions.

2.4.1 Graph Composition in PVS

In PVS, we denote graph composition using the infix "o" operator and we define it as follows. We also provide the function `GraphComposition` for use when the infix operator may be ambiguous.

```
GraphComposition[index1,index2,index3: TYPE]:
THEORY BEGIN

  IMPORTING GWV_Graph[index1,index2] AS T12
  IMPORTING GWV_Graph[index2,index3] AS T23
  IMPORTING GWV_Graph[index1,index3] AS T13

  g12: TYPE = T12.graph
  g23: TYPE = T23.graph
  g13: TYPE = T13.graph
 o(g23: g23, g12: g12): g13 =
    (LAMBDA (j3: index3):
      CASES g23(j3) OF
        Compute(S): Compute(DIAset(S,g12)),
        Copy(j):
          CASES g12(j) OF
            Compute(S): Compute(S),
            Copy(k): Copy(k)
          ENDCASES
      ENDCASES)

  GraphComposition(g23: g23, g12: g12): g13 =
    (g23 o g12)

END GraphComposition
```

Note that, in order to compose two graphs, the input index type of the left graph must agree with the output index type of the right graph. Graph composition is not commutative, but it is associative.

```
GraphCompositionProperties[index1,index2,index3,
                              index4: TYPE]:
  THEORY BEGIN

    IMPORTING GraphComposition[index1,index2,index3]
    IMPORTING GraphComposition[index2,index3,index4]
    IMPORTING GraphComposition[index1,index2,index4]
    IMPORTING GraphComposition[index1,index3,index4]

    graph12: TYPE = graph[index1,index2]
    graph23: TYPE = graph[index2,index3]
    graph34: TYPE = graph[index3,index4]

    compose_is_associative: LEMMA
      FORALL (g12: graph12,
              g23: graph23,
              g34: graph34):
        ((g34 o g23) o g12) =
          (g34 o (g23 o g12))

  %|- compose_is_associative : PROOF
  %|- (grind-with-ext)
  %|- QED
  END GraphCompositionProperties
```

When the input and output index types are the same, we can define an identity graph. The identity graph maps each index into its own singleton set. The DIASet of any set, when applied to an identity graph, is the original set. Likewise, any graph composed with the identity graph (either from the left or from the right) remains unchanged.

```
GraphID[Index: TYPE]: THEORY
BEGIN

  IMPORTING GWV_Graph[Index,Index]
  IMPORTING GraphComposition[Index,Index,Index]

  ID : graph = (LAMBDA (i: Index): Copy(i))

  DIAset_ID: LEMMA
```

```
     FORALL (x: set):
       DIAset(x,ID) = x


   compose_ID_1: LEMMA
     FORALL (g: graph):
       ID o g = g


   compose_ID_2: LEMMA
     FORALL (g: graph):
       g o ID = g


 END GraphID
```

Recall that a graph is intended to model the information flow properties of a function. Likewise, graph composition is intended to model the information flow properties of function composition. We demonstrate this fact in the following theory. The composition theorem says that if Graph12 is a characterization of Next12 and Graph23 is a characterization of Next23, then the functional composition of Graph23 and Next12, when graphically composed with Graph12, is a characterization of the functional composition of Next23 and Next12. The awkward function composition between Graph23 and Next12 is a result of the fact that graphs are, in general, functions of the state and the state seen by Graph23 is computed by Next12.

```
 GWVr1_Composition[Index1,Index2,Index3,  ,
                   State1,State2,State3,
                   Value1,Value2,Value3: TYPE,
                   Get1: [[Index1,State1]->Value1],
                   Get2: [[Index2,State2]->Value2],
                   Get3: [[Index3,State3]->Value3]]:

 THEORY BEGIN

   IMPORTING GWVr1[Index1,State1,Value1,Get1,
                   Index2,State2,Value2,Get2] AS P12
   IMPORTING GWVr1[Index2,State2,Value2,Get2,
                   Index3,State3,Value3,Get3] AS P23
   IMPORTING GWVr1[Index1,State1,Value1,Get1,
                   Index3,State3,Value3,Get3] AS P13
   IMPORTING GraphComposition[Index1,Index2,Index3]
     AS G
   IMPORTING Equiv[Index1,State1,Value1,Get1] AS ST1
   IMPORTING Equiv[Index2,State2,Value2,Get2] AS ST2
   IMPORTING Equiv[Index3,State3,Value3,Get3] AS ST3
   IMPORTING function_props[State1,State2,State3] AS F
```

```
GWVr1_Composition_Theorem: LEMMA
  FORALL (Next12  : P12.StepFunction,
          Graph12 : P12.GraphFunction,
          Hyp1    : P12.PreCondition,
          Next23  : P23.StepFunction,
          Graph23 : P23.GraphFunction,
        Hyp2    : P23.PreCondition):
      (P12.GWVr1(Next12)(Hyp1,Graph12) AND
       P23.GWVr1(Next23)(Hyp2,Graph23)) =>
         P13.GWVr1 (Next23 o Next12)
         ((lambda (in1: State1):
            (Hyp2(Next12(in1)) & Hyp1(in1))),
           (lambda (in1: State1):
            (GraphComposition((Graph23 o Next12)(in1),
                              Graph12(in1))))))
END GWVr1_Composition
```

2.5   Graph Abstraction

Given a concrete graph describing the information flow of a system, it may be possible to construct a more abstract graph of the same system containing less detail that preserves the essential information flow properties of the system. A graph abstraction associates groups of concrete index values with abstract index names. Such abstractions can substantially reduce the complexity of a graph, especially when it serves to partition a large basis set into a small number of security domains.

An abstraction has three components: a lifting graph that transforms concrete input indices into abstract input indices, a lifting graph that transforms concrete output indices into abstract output indices, and an abstract graph that models the abstract information flow relation between abstract input and output indices. We say that an abstraction is conservative if the interferes set of each concrete index is preserved (or extended) by the abstraction. With a conservative abstraction, noninterference questions about concrete index values can be translated into noninterference questions about the abstract index values. Because abstract graphs may be more concise than concrete graphs, such questions may be easier to answer in the abstract domain.

```
GraphAbstractionProperty[CiIndex,CoIndex,AiIndex,AoIndex: TYPE]
THEORY BEGIN

   IMPORTING GWV_Graph[CiIndex,CoIndex] AS C
   IMPORTING GWV_Graph[CiIndex,AiIndex] AS Li
   IMPORTING GWV_Graph[CoIndex,AoIndex] AS Lo
   IMPORTING GWV_Graph[AiIndex,AoIndex] AS A
   IMPORTING GraphComposition[CiIndex,CoIndex,AoIndex]
   IMPORTING GraphComposition[CiIndex,AiIndex,AoIndex]
```

```
      IMPORTING IndexSet[CiIndex]
      IMPORTING IndexSet[CoIndex]
      IMPORTING IndexSet[AiIndex]

      ConservativeAbstraction(GCStep: C.graph,
                              GLi   : Li.graph,
                              GLo   : Lo.graph)
                             (AStep : A.graph): bool =
        FORALL (Ao: AoIndex, Ci: CiIndex):
           member(Ci,DIA(Ao,GLo o GCStep)) =>
             member(Ci,DIA(Ao,AStep o GLi))

      AbstractNonInterference: LEMMA
        FORALL (Ci    : CiIndex,
                Co    : CoIndex,
                Ao    : AoIndex,
                GCStep: C.graph,
                GLi   : Li.graph,
                GLo   : Lo.graph,
                AStep : A.graph):
           (ConservativeAbstraction(GCStep,GLi,GLo)(AStep) &
            disjoint(invDIA(Ci,GLi),DIA(Ao,AStep)) &
            member(Co,DIA(Ao,GLo)))
             =>
               not(member(Ci,DIA(Co,GCStep)))

   END GraphAbstractionProperty
```

The composition of two conservative abstract graphs results in a new abstract graph that is a conservative abstraction of the composition of the underlying concrete graphs. We express this property in its full generality, allowing for a third graph to map between the input and output domains of the two abstract graphs. Numbers in the names of graphs in this example represent domains, where each graph maps information flow between two domains. Graphs S12 and S23 represent the two underlying system information flow graphs being composed, ultimately describing information flow between domain 1 and domain 3 via domain 2. A45 represents an abstraction of the information flow of S12. A67 represents an abstraction of the information flow of S23. The graph B56 is a bridging graph that maps index values in abstract domain 5 to index values in abstract domain 6. The final theorem says that, if A45 is a conservative abstraction of S12 modulo the lifting graphs L14 and L25 and A67 is a conservative abstraction of S23 modulo the lifting graphs L26 and L37, and if B56 is a conservative abstraction of the identity graph modulo the lifting graphs L25 and L26, then the composition of A45, B56, and A67 is a conservative abstraction of the composition of S12 and S23.

```
   AbstractGraphComposition[T1,T2,T3,T4,T5,T6,T7: TYPE]: THEORY
BEGIN

 IMPORTING GraphAbstractionProperty[T1,T2,T4,T5]
 IMPORTING GraphAbstractionProperty[T2,T3,T6,T7]
```

```
IMPORTING GraphAbstractionProperty[T1,T3,T4,T7]
IMPORTING GraphAbstractionProperty[T2,T2,T5,T6]
IMPORTING GWV_Graph[T5,T6]
IMPORTING GraphComposition[T1,T2,T3] AS P13
IMPORTING GraphComposition[T4,T5,T6] AS P46
IMPORTING GraphComposition[T4,T6,T7] AS P47

IMPORTING AbstractGraphComposition2[T1,T2,T2,T4,T5,T6] AS P456
IMPORTING AbstractGraphComposition2[T2,T2,T3,T5,T6,T7] AS P567
IMPORTING AbstractGraphComposition2[T1,T2,T3,T4,T6,T7] AS P467
IMPORTING AbstractGraphComposition2[T1,T2,T3,T4,T5,T7] AS P457

IMPORTING GraphID[T2]

I22: graph[T2,T2] = ID[T2]

IMPORTING GraphID2[T1,T2]

  Composition: LEMMA
    FORALL (S12: graph[T1,T2],
            L14: graph[T1,T4],
            L25: graph[T2,T5],
            A45: graph[T4,T5],
            S23: graph[T2,T3],
            L26: graph[T2,T6],
            L37: graph[T3,T7],
            A67: graph[T6,T7],
            B56: graph[T5,T6]):
      ConservativeAbstraction(S12,L14,L25)(A45) &
      ConservativeAbstraction(I22,L25,L26)(B56) &
      ConservativeAbstraction(S23,L26,L37)(A67) =>
        ConservativeAbstraction(P13.o(S23,S12),L14,L37)
                               (P47.o(A67,(P46.o(B56,A45))))
END AbstractGraphComposition
```

2.6 *GWVr1 and Guarded Domains*

While GWVr1 has been shown to be effective at modeling information flow properties of functions, there are example of functions for which GWVr1 requires the use of an information flow graph that seems to overapproximate the information flow of the function.

Recall our illustrative state data structure modeled as a mapping from natural numbers to integers, the associated `get` projection function, and the `set` function that allows us to selectively modify specific portions of the state.

```
state: TYPE = [ nat -> int ]

get(i:nat,st:state): int = st(i)
```

```
set(i:nat, v: int, st:state)(j:nat): int =
  IF (i = j) THEN v ELSE get(j,st) ENDIF
```

Consider a function containing a guarded assignment in which the condition of the guard reflects a domain restriction on the assignment. In other words, the location being updated is in the domain of the system when the condition is true, otherwise it is not. In the `guardedDomain` function, the `nat?` test ensures that the assignment statement is type correct.

```
nat?(x: int): bool = (0 <= x)

guardedDomain(st: state): state =
  LET index:int = get(0,st) IN
    IF nat?(index) THEN
      set(index,1,st)
    ELSE st
    ENDIF


AUTO_REWRITE+ guardedDomain!
```

While this example is similar to the `conditionalUpdates` example considered previously, it has one important distinction. Specifically, it is impossible for the assigned location to unconditionally include the dependency of the condition because doing so would result in a type violation in our graph specification, exactly the type violation that the condition is there to avoid in the original function definition.

We might specify the following type-safe graph and hope that it works.

```
guardedDomainGraph(st: state): graph =
  LET index:int = get(0,st) IN
    IF nat?(index) THEN
      addDIA(index,0,ID)
    ELSE ID
    ENDIF


AUTO_REWRITE+ guardedDomainGraph!

guardedDomainGWVr1Fails: LEMMA
  GWVr1(guardedDomain)(Hyps,guardedDomainGraph)
```

Of course, this fails in the same way our naïve `missedConditionalGraph` failed, except that in this case it is the `nat?` test that differs for the two input instances.

```
guardedDomainGWVr1Fails.2 :

{-1}   (in2!1(0) = x!1)
{-2}   nat?(in2!1(0))
{-3}   (in1!1(x!1) = in2!1(x!1))
{-4}   Hyps(in1!1)
{-5}   Hyps(in2!1)
   |-------
{1}    (in1!1(x!1) = 1)
{2}    nat?(in1!1(0))
{3}    nat?(in1!1(0))

Rule?
```

There is a graph that characterizes this function. However, in this graph *every* index location depends upon location 0.

```
overkillGraph(st: state): graph =
  LET g:graph = (lambda (n: nat): Compute(0+n)) IN
  LET index:int = get(0,st) IN
    IF nat?(index) THEN
      addDIA(index,index,g)
    ELSE g ENDIF;

AUTO_REWRITE+ overkillGraph!

overkillGWVr1: LEMMA
  GWVr1(guardedDomain)(Hyps,overkillGraph)
```

We call this graph `overkillGraph`, because it seems like overkill for every index to have to depend upon this condition. Domain guards turn out to be very common in dynamic systems. For example, null pointer checks on heap resident data structures are one kind of domain guard. If every part of the state of a dynamic system must depend upon every domain guard in the system, pretty soon everything depends upon everything and the information flow graph becomes useless as a specification. Our efforts to model the Green Hills INTEGRITY-178B operating system, with its many heap resident (dynamic) data structures and null pointer checks, brought this issue to a head [10]. It was precisely that modeling experience that motivated our development of GWVr2.

3 GWVr2

The development of GWVr2 was motivated by the difficulty of modeling guarded domains in GWVr1. The two state representation of the congruence relation was assumed to be the culprit, so a single state model of information flow was sought. While the two state congruence theorem is, in fact, the issue, GWVr2 turns out to be the long way around to a satisfactory solution. Because the development of GWVr2 has never been described in its entirety, we document it here. GWVr2, however, has never been used directly in practice. In subsequent sections, we show that GWVr2 can be reformulated in a form much more amenable to automated proof.

In GWVr2, the congruence theorem of GWVr1 is transformed into a simple equality between the original function and another function which interprets the graph data structure relative to the initial state. For each index value in the define set of the graph, the interpreter function constructs a new state object in such a way as to guarantee that the only similarity between the new state object and the original state object is that they satisfy equivSet over the DIA of the index. It then applies the step function to this new object and extracts from the result the value at the index of interest and returns a state object in which the index is bound to that computed value. For each index value not in the define set of the graph, the interpreter simply copies the value at that index from the input to the output.

In addition to the index, state and value types, and the projection function parameters, GWVr2 is also parameterized by a copy function. The theory construction we present is valid only for copy functions that produce values of the right type and basis sets that are orthogonal and complete.

```
GWVr2  [INindex:TYPE, INState: TYPE+, INvalue: TYPE,
        getIN: [[INindex, INState] -> INvalue],
        OUTindex: TYPE, OUTState: TYPE+, OUTvalue: TYPE,
        getOUT: [[OUTindex, OUTState] -> OUTvalue],
        copy:[[INindex,OUTindex] ->
              [INvalue -> OUTvalue]]]:
 THEORY BEGIN

   ASSUMING

     IMPORTING Ideal[OUTState,OUTindex,OUTvalue]

     copy_right: ASSUMPTION
       FORALL (i: INindex, o: OUTindex, ival: INvalue):
         gettablevalue(getOUT,o,copy(i,o)(ival))

     OrthogonalSet_And_Complete: ASSUMPTION
       OrthogonalSet(getOUT) & Complete(getOUT)
```

```
    ENDASSUMING
```

The basic theory types are similar to those found in GWVr1.

```
    IMPORTING GWV_Graph[INindex,OUTindex]
    IMPORTING Equiv[INindex,INState,INvalue,getIN]
      AS Input
    IMPORTING Equiv[OUTindex,OUTState,OUTvalue,getOUT]
      AS Output

    StepFunction:  TYPE = [ INState -> OUTState ]
    GraphFunction: TYPE = [ INState -> graph ]
    PreCondition:  TYPE = [ INState -> bool ]
```

The construction of the new state object is complicated by the fact that we want GWVr2 to be as strong as GWVr1 under appropriate conditions. This leads us to choose (via epsilon, representing an axiom of choice) a value for the new state that is, in a sense, the one most likely to cause our efforts to fail. We call this condition the "bad boy" condition and the resulting state the "bad state" (bad_st). If, however, despite the odds, we succeed in proving equivalence using this malicious state, it ensures that we will be able to prove a similar equivalence with any other state. This claim is reminiscent of GWVr1, which gives us some hope that GWVr2 will at least be similar to GWVr1.

The process begins by defining appropriate types and predicates and articulating exactly how we would recognize a bad boy state: the result of applying Next to that state would differ at the index of interest from an application of Next to our original state. We then use epsilon to choose our malicious state: st_bad.

```
    PreState(Hyp: PreCondition): TYPE =
      {s: INState | Hyp(s)}

    use_equiv(Hyp: PreCondition,
              u: set[INindex],
              st1: PreState(Hyp))
             (stx: INState): bool =
      equivSet(u,st1,stx) & Hyp(stx)

    use_equiv_state(Hyp: PreCondition,
                    u: set[INindex],
                    st1: PreState(Hyp)): TYPE =
      { s: INState | use_equiv(Hyp,u,st1)(s) }

    bad_boy(Hyp: PreCondition, i: OUTindex,
```

```
                Next: StepFunction,u: set[INindex],
                st1: PreState(Hyp))
                (stx: use_equiv_state(Hyp,u,st1)): bool=
          not(equiv(i,Next(st1),Next(stx)))

      st_bad(Hyp: PreCondition, i: OUTindex,
             Next: StepFunction, u: set[INindex],
             st1: PreState(Hyp)): INState =
          epsilon[use_equiv_state(Hyp,u,st1)]
          (bad_boy(Hyp,i,Next,u,st1))
```

st_bad, because of its base type, has several useful properties. It satisfies the
Hyp precondition and it is equivalent to its st1 argument at every index location in
the set u. If possible, it would also satisfy the bad_boy predicate. However, if that is
not possible, we get a very nice property: one that mirrors GWVr1.

```
      st_bad_next_equiv: LEMMA
         FORALL (Hyp: PreCondition, i: OUTindex,
                 Next: StepFunction, u: set[INindex],
                 st1: INState):
            Hyp(st1) &
              equiv(i,Next(st1),
                    Next(st_bad(Hyp,i,Next,u,st1))) =>
            FORALL (st2: INState):
              equivSet(u,st1,st2) & Hyp(st1) & Hyp(st2) =>
                 equiv(i,Next(st1),Next(st2))
```

Our construction of GWVr2 continues with the introduction of a pstate
("projected state") type. Such an object represents a collection of every possible
projection of the state. The ability of pst2st to reconstruct a state object from a
pstate is made possible by the Orthogonality and Completeness of the
basis set.

```
      pstate: TYPE =
         [ i:OUTindex -> gettablevalue(getOUT,i) ]

      pstPredicate(pst: pstate)(st: OUTState): bool =
         FORALL (i: OUTindex):
           getOUT(i,st) = pst(i)

      pst2st(pst: pstate): OUTState =
         epsilon[OUTState](pstPredicate(pst))
```

The function `NextGpst` interprets the graph to construct a `pstate` output. For a computed index, the result is the projection at that index of the Next function applied to a state that is equivalent (via `st_bad`) to the input state `st` at every index location dictated by `DIA(i,G)`. For a copied index, on the other hand, the result is a direct `copy` of that input. The function `NextG` simply converts `NextGpst` back into a state object.

```
NextGpst(Hyp: PreCondition, Next: StepFunction)
        (G: graph, st: PreState(Hyp)): pstate =
   (LAMBDA (i: OUTindex):
     IF Computed?(G(i)) THEN
       getOUT(i,Next(st_bad(Hyp,i,Next,DIA(i,G),st)))
     ELSE
       copy(CopyIndex(G(i)),i)
           (getIN(CopyIndex(G(i)),st))
     ENDIF)

NextG(Hyp: PreCondition, Next: StepFunction)
      (G: graph, st: PreState(Hyp)): OUTState =
   pst2st(NextGpst(Hyp,Next)(G,st))
```

GWVr2 is expressed simply as equality between Next and NextG under the appropriate preconditions.

```
   GWVr2(Next: StepFunction)
       (Hyp: PreCondition,
        Graph: GraphFunction): bool =
     FORALL (st: INState):
       Hyp(st) =>
         Next(st) = NextG(Hyp,Next)(Graph(st),st)
```

3.1 GWVr2 Reduction

In GWVr2, the congruence theorem of GWVr1 is transformed into a simple equality between the original function and another function, which interprets the graph data structure relative to the initial state. But it is still not clear how this helps to address the original issue with GWVr1. To explain this, it is helpful to retrace our steps and to attempt to reestablish a connection between GWVr2 and GWVr1.

GWVr1 and GWVr2 differ in their interpretation of the graph data structure. GWVr2 treats the Copy and Compute edges of the information flow graph differently while GWVr1 sees the graph only through the eyes of DIA, which blurs the distinction between the Copy and Compute edges.

```
FramedGWVr1 [
   INindex: TYPE, INState: TYPE+, INvalue: TYPE,
 getIN: [[INindex, INState] -> INvalue],
 OUTindex, OUTState, OUTvalue: TYPE,
 getOUT: [[OUTindex, OUTState] -> OUTvalue],
 copy:[[INindex,OUTindex] -> [INvalue -> OUTvalue]]]:
THEORY BEGIN

 IMPORTING GWVr1[INindex,INState,INvalue,getIN,
                 OUTindex,OUTState,OUTvalue,getOUT]

 StepFunction: TYPE = [ INState -> OUTState ]
 GraphFunction: TYPE = [ INState -> graph ]
 PreCondition: TYPE  = [ INState -> bool ]
```

For Copy edges, GWVr2 reduces to a single state theorem about how the function being characterized is equivalent to a function that copies an input value to the output. We call this theorem the frame condition (`FrameCondition`). The frame condition provides a strong functional theorem about the behavior of Next at copied index values. Typically copy is defined as the identity function (the first two parameters to copy are included only to satisfy type reasoning). In such case, FrameCondition says that Next leaves the state completely unchanged at copied index locations. We will see in subsequent sections that this strong functional theorem, which is not available with GWVr1, provides a convenient means of expressing noninterference theorems. The defSet of a graph is composed entirely of computed index values. The nonmembership of an index value in the defSet of a graph is equivalent that index being a copy edge in the graph.

```
 FrameCondition(Next: StepFunction)
               (Hyps: PreCondition,
                Graph: GraphFunction): bool =
   FORALL (x: OUTindex, st: INState):
     not(member(x,defSet(Graph(st)))) & Hyps(st) =>
       getOUT(x,Next(st)) =
         copy(CopyIndex(Graph(st)(x)),x)
              (getIN(CopyIndex(Graph(st)(x)),st))
```

For Compute edges, GWVr2 reduces to GWVr1 (due, in part, to the virtues of the carefully chosen `st_bad`). We call a GWVr1 theorem that is conditional on the graph edge associated with the output index a framed congruence (`FramedGWVr1`).

```
FramedGWVr1(Next: StepFunction)
           (Hyps: PreCondition,
            Graph: GraphFunction): bool =
 FORALL (x: OUTindex, in1,in2: INState):
    equivSet(DIA(x,Graph(in1)),in1,in2) &
      Hyps(in1) & Hyps(in2) &
        member(x,defSet(Graph(in1))) =>
          equiv(x,Next(in1),Next(in2))
```

Given these definitions, it is possible to prove that GWVr2 is equal to a framed congruence plus a frame condition. This reformulation is actually more amenable to automated proof and is now the preferred representation of GWVr2.

```
GWVr2_reduction: LEMMA
  FORALL (Next: StepFunction,
          Hyp: PreCondition,
          Graph: GraphFunction):
    GWVr2(Next)(Hyp,Graph) =
      (FramedGWVr1(Next)(Hyp,Graph) &
         FrameCondition(Next)(Hyp,Graph))
```

3.2 Domain Guards and GWVr2

The development of GWVr2 was motivated in part by the need for a better method for modeling systems containing dynamic domain guards. Recall our definition of guardedDomain.

```
nat?(x: int): bool = (0 <= x)

guardedDomain(st: state): state =
  LET index:int = get(0,st) IN
    IF nat?(index) THEN
      set(index,1,st)
    ELSE st
    ENDIF

AUTO_REWRITE+ guardedDomain!
```

And recall the naïve, type-safe graph that failed under GWVr1.

```
guardedDomainGraph(st: state): graph =
  LET index:int = get(0,st) IN
    IF nat?(index) THEN
      addDIA(index,0,ID)
    ELSE ID
    ENDIF


AUTO_REWRITE+ guardedDomainGraph!
```

The proof that failed under GWVr1 succeeds under FramedGWVr1, for an appropriate definition of copy.

```
copy(x,y: nat)(i: int): int = i;

IMPORTING FramedGWVr1[nat,state,int,get,
                      nat,state,int,get,
                      copy]

guardedDomainFramedGWVr1: LEMMA
  FramedGWVr1(guardedDomain)(Hyps,guardedDomainGraph)

%|- guardedDomainFramedGWVr1 : PROOF
%|- (then
%|-   (auto-rewrite "FramedGWVr1")
%|-   (auto-rewrite! "DIA" "defSet")
%|-   (auto-rewrite! "addDIA" "defSet" "get" "set")
%|-   (auto-rewrite "ID")
%|-   (auto-rewrite "member")
%|-   (auto-rewrite "equiv")
%|-   (auto-rewrite "copy")
%|-   (auto-rewrite-theory
%|-     "EquivSetRules[nat,state,int,get]")
%|-   (apply (repeat* (then*
%|-     (lift-if) (ground) (skosimp))))))
%|- QED
```

The proof of the frame condition works as well.

```
guardedDomainFrameCondition: LEMMA
  FrameCondition(guardedDomain)
                (Hyps,guardedDomainGraph)

%|- guardedDomainFrameCondition : PROOF
```

```
%|- (then
%|-    (auto-rewrite "FrameCondition")
%|-    (auto-rewrite! "DIA" "defSet")
%|-    (auto-rewrite! "addDIA" "defSet" "get" "set")
%|-    (auto-rewrite "ID")
%|-    (auto-rewrite "member")
%|-    (auto-rewrite "equiv")
%|-    (auto-rewrite "copy")
%|-    (auto-rewrite-theory
%|-       "EquivSetRules[nat,state,int,get]")
%|-    (apply (repeat* (then*
%|-       (lift-if) (ground) (skosimp))))))
%|- QED
```

Consequently, the GWVr2 property, by selectively partitioning information flow modeling into both a congruence problem and a frame condition, enables the use of more intuitive models of information flow for dynamic systems.

4 Multicycle Information Flow Analysis

As mentioned previously, state machine models are used to represent sequential computing systems. A state machine model suggests a step function that operates over a set of inputs and an initial state to produce a set of outputs and a next state. This step function can be applied iteratively to successive inputs and states to simulate the evolution of the sequential system and its outputs over time. Our theory of multicycle systems is parameterized by everything necessary to specify a basis set as well as an index set representing the inputs to the system, a step function, a precondition, and a graph function.

```
MultiCycle[index:TYPE , state: TYPE+, value: TYPE,
          get: [[index, state] -> value],
          (IMPORTING GWVr1[index,state,value,get,
                           index,state,value,get])
          InputSet: set[index],
          Next:  StepFunction,
          Hyp:   PreCondition,
          Graph: GraphFunction]: THEORY
BEGIN
```

A number of assumptions over the parameters are necessary in the development of our theory. We assume that the graph function is a GWVr1 characterization of the step function under the precondition that the precondition is invariant over the step function, that the precondition is independent of the inputs, that the basis set is Orthogonal and Complete, and that the graph is reactive, which is to say that any input contained in the graph depends upon itself.

```
ASSUMING

  NextCharacterization: ASSUMPTION
    GWVr1(Next)(Hyp,Graph)

  Next_PostCondition: ASSUMPTION
    FORALL (s: state):
      Hyp(s) =>
        Hyp(Next(s))

  HypCongruence: ASSUMPTION
    FORALL (s1,s2: state):
      equivSet(not(InputSet),s1,s2) =>
        Hyp(s1) = Hyp(s2)

  IMPORTING Ideal[state,index,value]

  OrthogonalSet_get: ASSUMPTION
    OrthogonalSet(get)

  Complete_get: ASSUMPTION
    Complete(get)

  reactive_Graph: ASSUMPTION
    FORALL (i: index):
      member(i,InputSet) =>
        FORALL (st: state):
          member(i,DIA(i,Graph(st)))

ENDASSUMING
```

We are developing a model of multicycle operation, and we assume that our model of state includes the system inputs. Consequently, we need a means of describing how those inputs are updated in each step of the system. The function applyInputs is chosen (using an axiom of choice) as a means of applying inputs (its first argument) to a given state (its second argument). Note that for representational convenience we do not define a special input type. Rather, we reuse the state type to model inputs. The crucial property of applyInputs is that, when accessed via get, the value returned will be the associated value from the input when the index is a member of the InputSet and will be the associated value from the original state when the index is not in InputSet.

```
   inputApplication(input,st: state)(s: state):bool =
     FORALL (i: index):
       (IF member(i,InputSet) THEN
           get(i,s) = get(i,input)
         ELSE
           get(i,s) = get(i,st)
         ENDIF)

   applyInputs(inputs,st: state): state =
     epsilon[state](inputApplication(inputs,st))

   get_applyInputs: LEMMA
     FORALL (i: index, input,st: state):
       get(i,applyInputs(input,st)) =
             IF member(i,InputSet) THEN
               get(i,input)
             ELSE
               get(i,st)
             ENDIF
```

We are now in a position to describe the functional behavior of our state machine
over time. Time is modeled as a natural number and a trace is defined as a mapping
from time to states. The first argument to our state machine model is the time up
to which it is to run, beginning at time zero. The second input to our model is a
trace (the oracle) containing the inputs that are to be consumed by the machine at
each step. Additionally, the oracle at time zero is interpreted as the initial state of
the system. In every step, the machine applies the inputs at the current time to the
result of applying a single step (Next) to the state computed in the previous time.

```
   time: TYPE = nat

   trace: TYPE = [ time -> state ]

   Run(t:nat)(oracle: trace): RECURSIVE state =
     IF (t = 0) THEN oracle(0) ELSE
       applyInputs(oracle(t),Next(Run(t-1)(oracle)))
     ENDIF
   MEASURE t
```

Observe that our recursive model of sequential execution can be viewed as a sequence of compositions of the state transition function with the inputs and itself. Not
surprisingly, the information flow of our sequential system can also be modeled as
a sequence of compositions of the graph that characterizes the state transition func-

tion with itself. In other words, the information flow model of a sequential system
is itself a sequential system.

```
IMPORTING GraphID[index]

GraphRun(t: time)(oracle: trace): RECURSIVE graph =
  IF (t = 0) THEN ID ELSE
      Graph(Run(t-1)(oracle)) o
        GraphRun(t-1)(oracle)
  ENDIF
MEASURE t
```

It is relatively easy to prove, given the fact that graph composition is a model of
function composition, that such a state machine model does, in fact, track information flow in a sequential system. However, the development of the proof requires
that we extend our calculus of indices to take into account the fact that our inputs
are applied fresh in every cycle. There are a variety of ways to model this potentially unbounded collection of inputs. We could, for example, extend our basis set
to construct a unique index value for each input at each step in time. The technique
we choose, however, is to extend our interpretation of what an input is. Rather than
an input being a single value, we view each input as a sequence of values that may
vary over time. We call such mappings from time to values signals. Employing this
concept we define a new projection function, sigget, that projects input index values
from the input oracle into signals that may vary over time. For representational convenience, the projection function also projects index values that are not inputs into
signals that assume the value of the index in the initial state (the oracle at time 0) for
all time. Employing this new projection function, we define sigEquiv and sigSetE-
quiv as equivalence relations between two different oracle traces.

```
signal: TYPE = [ time -> value ]

sigget(i: index, oracle: trace): signal =
  IF member(i,InputSet) THEN
    (lambda (t: time): get(i,oracle(t)))
  ELSE
    (lambda (t: time): get(i,oracle(0)))
  ENDIF

sigEquiv(i: index, o1,o2: trace): bool =
  sigget(i,o1) = sigget(i,o2)

sigSetEquiv(set: set[index], o1,o2: trace): bool =
  FORALL (i: index):
    member(i,set) => sigEquiv(i,o1,o2)
```

These new definitions, in conjunction with selected theory assumptions and their consequences, allow us to prove that GraphRun characterizes Run (in a GWVr1 sense) for all time.

```
RunGWVr1: LEMMA
   FORALL (t: time):
     FORALL (i: index,oracle1,oracle2: trace):
       Hyp(oracle1(0)) & Hyp(oracle2(0)) &
         sigSetEquiv(DIA(i,GraphRun(t)(oracle1)),
                     oracle1,oracle2) =>
           equiv(i,Run(t)(oracle1),Run(t)(oracle2))
```

5 Classical Noninterference

Having described our treatment of sequential systems, we are now in a position to show how our model of information flow relates to the classical notion of non-interference found in the literature. Noninterference is a useful and well-studied characterization of information flow for secure systems. Goguen and Meseguer orig-inally defined noninterference as follows:

> One group of users, using a certain set of commands, is noninterfering with another group of users if what the first group does with those commands has no effect on what the second group of users can see [3].

Rushby also provides a formalization of noninterference in which he employs the concept of security domains (in place of users) and associates with those domains some set of actions (in place of commands). Using this terminology:

> A security domain u is noninterfering with domain v if no action performed by u can influ-ence subsequent outputs seen by u [11].

Rushby goes on to claim that the requisite lack of perception exists if the behavior of v remains unchanged, even after purging from the system trace of all the actions performed by u.

Terminology aside, these formalisms are similar in that they both speak of secu-rity domains (or users) that are selectively empowered with a set of computational capabilities (commands or actions). Notions such as actions, commands, and users, while convenient for expressing certain policy statements, are nearly orthogonal to the essential concept of noninterference found in both formulations: that of not per-ceiving (or not seeing) some "effect" and the use of equality between some projected portion of state as a litmus test of that fact.

We view this essential concept of noninterference as an information flow prop-erty. Furthermore, we require that higher level concepts (such as security domains and capabilities) be given formal information flow models before noninterference properties expressed in terms of those concepts can be verified.

5.1 Domains

The concept of a security domain found in the noninterference literature includes some notion of inputs, outputs, and state. Our model of system state is assumed to include all system inputs, outputs, and state variables. We also assume a rigorously defined interpretation of that state relative to a specific basis set of index values. In our model of noninterference, domains are defined by functions that maps a system state into a set of said index values.[6] Because domains are functions of the state, it is possible for the set of index values included in the domain to change over time.

5.2 Capabilities

In the noninterference literature, we also find references to the concept of capabilities in the form of commands or actions that may be performed by (or within) a security domain. Typically a number of capabilities are associated with each security domain.

In our model of noninterference, there is no explicit connection between domains (which are merely collections of index values) and capabilities. However, the evolution of system state is driven by the activation of different capabilities over time. The activation of a specific capability will typically be dependent upon some condition within the state, and such conditions will often depend upon information stored within particular domains. Additionally, capabilities may be encoded as programs that are stored in the system state, again often within specific domains. Thus, from the dependencies of the conditions that drive them and from the dependencies of the capabilities themselves, we may find that specific capabilities are implicitly associated with specific domains.

More important than the association of a specific capability with a particular domain, however, is the information flow characterization of each capability of the system. In our models, we are generally not concerned with the functional behavior of any capability within the system. The kind of computation performed is irrelevant. What is important is the impact that the computation has on information flow in the system. It is the combined information flow contract of all of the various capabilities active within the system that will ultimately determine whether or not the system satisfies a particular noninterference relation.

5.3 Noninterference Example in PVS

Consider a system containing two domains: a Red domain and a Black domain. Informally, the Red domain is noninterfering with the Black domain if no action of the Red domain can ever be perceived by the Black domain. We express this

---

[6] Such functions have also been called crawlers [4,6].

property as a theorem about the sequential behavior of the system after an arbitrary number of steps. In particular, we want to show that the value of each element of the Black domain is the same whether we start from an arbitrary state or from that arbitrary state modified by some capability of the Red domain.

Our noninterference example extends the MultiCycle theory presented previously. Our single-step model function, Next, is employed as a generic model of the application of one or more system capabilities. Run, therefore, models an arbitrary evolution of our system over some amount of time and RunGraph models the information flow of the overall system during that time. One system capability is explicitly identified, however, and we call it RedCapability. Associated with that capability is a graph, RedGraph, that characterizes the information flow behavior of RedCapability. We assume that the system state can be partitioned into some number of security domains, including a Red domain (RedDomain) and a Black domain (BlackDomain).

The type DomainFunction is provided to help identify domain functions, being defined as a function that maps a state into a set of index values. We define a copy function and use it to import the FramedGWVr1 theory. We also define a function that constructs an input oracle trace from a trace and an initial state.

```
DomainFunction: TYPE = [ state -> set[index]]
  copy(i,j: index)(v: value): value = v
  IMPORTING FramedGWVr1[index,state,value,get,
                        index,state,value,get,
                        copy] AS X
  Oracle(input: trace, st: state): trace =
    (LAMBDA (t: time):
       IF (t = 0) THEN st ELSE input(t) ENDIF)
```

Predicates are introduced that become obligations on the various functions appearing in our example. The trueCopies predicate states that the index values stored in the copy edges of a graph function really are copies of the index value used to index the graph. The PostCondition predicate says that the precondition is invariant over the step function. The RedGraphRestriction predicate restricts the defSet of the RedGraph to be a subset of the RedDomain (ensuring no writes outside of the bounds of the Red domain by the Red capability). The final predicate, the NonInterferenceProperty, is a property of the system, RunGraph, that says that no member of the Red domain ever appears in the DIA (interferes set) of any member of the Black domain.

```
trueCopies(G: X.GraphFunction): bool =
  FORALL (s: state):
    FORALL (i: index):
      (Copied?(G(s)(i)) =>
         (CopyIndex(G(s)(i)) = i))
```

```
    PostCondition(Hyp: X.PreCondition,
                  Next: X.StepFunction): bool =
      FORALL (s: state):
        Hyp(s) => Hyp(Next(s))


  RedGraphRestriction
    (RedGraph: X.GraphFunction,
     RedDomain: DomainFunction) : bool =
      FORALL (st: state):
        subset?(defSet(RedGraph(st)),RedDomain(st))


  NonInterferenceProperty
    (RedDomain,BlackDomain: DomainFunction): bool =
      FORALL (blk,red: index,
              in: trace,
              st: state):
        Hyp(st) &
        member(blk,BlackDomain(st)) &
        member(red,RedDomain(st)) =>
        FORALL (t: time):
          not(member(red,
                DIA(blk,GraphRun(t)(Oracle(in,st))))))
```

Employing these definitions, our example noninterference theorem says that, for every RedCapability, RedGraph, RedDomain, and BlackDomain, if the RedGraph contains trueCopies, the RedCapability satisfies the PostCondition, the RedCapability satisfies the FrameCondition (half of GWVr2) suggested by the RedGraph under Hyp, the RedGraph satisfies RedGraphRestriction relative to the RedDomain, and the system RunGraph satisfies the NonInterferenceProperty with respect to the Red-Domain and BlackDomain, then the value extracted by get at every member of the BlackDomain after running t cycles following the execution of the RedCapability will be the same as the value at that index after running t cycles without executing the RedCapability. The proof of this theorem ultimately appeals to the fact that RunGraph characterizes Run.

```
  NonInterferenceTheorem: LEMMA
    FORALL
      (RedCapability  : X.StepFunction,
       RedGraph       : X.GraphFunction,
       RedDomain      : DomainFunction,
       BlackDomain    : DomainFunction
      ):
      trueCopies(RedGraph) &
      PostCondition(Hyp,RedCapability) &
      FrameCondition(RedCapability)(Hyp,RedGraph) &
```

```
RedGraphRestriction(RedGraph,RedDomain) &
NonInterferenceProperty(RedDomain,BlackDomain)
 =>
FORALL (t: time,
        blk: index,
        in: trace,
        st: state):
  Hyp(st) &
  member(blk,BlackDomain(st)) =>

  get(blk,Run(t)(Oracle(in,RedCapability(st)))) =
  get(blk,Run(t)(Oracle(input,st)))
```

It is worth noting that the proof of this particular theorem is made possible by the strong functional property provided by the GWVr2 frame condition. In particular, the frame condition allows us to reduce RedCapability to a copy (no op) when we examine index values outside of its defSet.

The trueCopies, PostCondition, FrameCondition, and RedGraphRestriction properties appearing in this theorem are all obligations that can be dispatched locally. That is, they can be established as properties of RedCapability, RedGraph, RedDomain, and BlackDomain without knowledge of the rest of the system. The NonInterferenceProperty, on the other hand, is a property of the entire system information flow graph (RunGraph). It is a system-wide property and it is the true heart of the noninterference theorem. Goguen and Meseguer claim that noninterference is useful as a system security *policy*. We claim that information flow graphs are useful as a system *specification*. Our example noninterference theorem demonstrates how a noninterference policy can be established from a property (NonInterferenceProperty) of a graphical system specification.

5.4 *Establishing the Noninterference Property*

Our example noninterference theorem follows from a collection of local properties about functions appearing in the theorem and one global noninterference property of the system information flow graph. Note too that, while the local properties are quantified over all states, the noninterference property is quantified over all time. Establishing the noninterference property may be a nontrivial task and the best methodology for doing so is likely to depend upon a variety of factors. For many systems, however, it may be possible to establish the noninterference property using model checking [2]. Model checking of information flow properties is described in detail in [15]. Here we merely establish a connection between our formulation of the noninterference property and a formulation of the property in temporal logic.

Temporal logic is most conveniently expressed in terms of traces. Here we introduce the type GSTrace which is a trace type parameterized by a step relation that constrains the sequential steps in the trace.

```
TraceMU[GState: TYPE]: THEORY
BEGIN

  time: TYPE = nat

  GSPred: TYPE = [ GState -> bool ]

  GSRelation: TYPE = [[GState,GState] -> bool]

  GSStep: TYPE = [ GState -> GState ]

  StepGSRelation(Step: GSStep): TYPE =
    { r: GSRelation |
        FORALL (g1,g2: GState):
          r(g1,g2) = (g2 = Step(g1)) }

  GSTrace(StepRelation: GSRelation) : TYPE =
    { trace: sequence[GState] |
      FORALL (t: time):
        StepRelation(trace(t),trace(t+1)) }
```

We define a simple trace property as one that expresses a precondition in the initial state, Hyp, and a subsequent invariant that must be satisfied at all time, Prop.

```
simple_trace_prop(Hyp,Prop: GSPred,
                  Step?: GSRelation): bool =
  (FORALL (trace: GSTrace(Step?)):
    Hyp(trace(0)) =>
      (FORALL (n: nat): Prop(trace(n))))
```

Using Rushby's formalization of linear temporal logic [12], we can express an equivalent property in LTL and prove that the two are equivalent. There are many model checkers available for checking properties expressed in LTL.

```
    IMPORTING LTL[GState]

  simple_LTL_prop(Hyp,Prop: GSPred,
                  Step?: GSRelation): bool =
    FORALL (s: GSTrace(Step?)):
      (s |= (Holds(Hyp) =>
        G(Holds(Prop))))
```

```
    trace_to_LTL: LEMMA
      FORALL (Hyp,Prop: GSPred,
              Step?: GSRelation):
        simple_LTL_prop(Hyp,Prop,Step?) =>
          simple_trace_prop(Hyp,Prop,Step?)
```

Alternatively, one could express our simple property using MU calculus. This formulation is provably equivalent to our LTL formulation. For properties expressed in the MU calculus, PVS provides built-in model checking capabilities.

```
    simple_MU_prop(Hyp,Prop: GSPred,
                   Step: GSStep): bool =
      (FORALL (gs: GState):
        Hyp(gs) =>
          AG(relation(Step),Prop)(gs))

    LTL_to_MU: LEMMA
      FORALL (Hyp,Prop: GSPred,
              Step: GSStep,
              Step?: StepGSRelation(Step)):
        simple_LTL_prop(Hyp,Prop,Step?) =
          simple_MU_prop(Hyp,Prop,Step)
  END TraceMU
```

Recall our definition of NonInterferenceProperty.

```
  NonInterferenceProperty
    (RedDomain,BlackDomain: DomainFunction): bool =
      FORALL (blk,red: index,
              in: trace,
              st: state):
        Hyp(st) &
        member(blk,BlackDomain(st)) &
        member(red,RedDomain(st)) =>
        FORALL (t: time):
          not(member(red,
                DIA(blk,GraphRun(t)(Oracle(in,st))))))
```

We define a compound type that contains both our system state and our graph state. Using this we define a relation, StepRelation, that constrains our trace to reflect the evolution of our system state machine and system information flow graph. A predicate, GSHyp0, is defined to reflect our preconditions on the initial system state and the initial graph state.

```
   GState: TYPE = [# state: state, graph: graph #]

   gstrace: TYPE = [ nat -> GState ]

   GraphStepRelation(s1: state,g1,g2 :graph): bool =
     (g2 = (Graph(s1) o g1));

   StateStepRelation(s1,s2: state): bool =
     FORALL (i:index):
       not(member(i,InputSet)) =>
         get(i,s2) = get(i,Next(s1))

   StepRelation(gs1,gs2: GState): bool =
     GraphStepRelation
       (state(gs1),graph(gs1),graph(gs2)) &
     StateStepRelation(state(gs1),state(gs2))

    strace(st1: gstrace): trace =
      (LAMBDA (t: time): state(st1(t)))
    StepTrace: TYPE = {gs: gstrace |
      FORALL (t: time): StepRelation(gs(t),gs(t+1))}

   GSHyp0(gs: GState): bool =
      Hyp(state(gs)) & (graph(gs) = ID)
```

Using these definitions, we can express our noninterference property in terms of
a precondition, noninterference–hyp, and an invariant, noninterference–prop. TraceNonInterferenceProperty combines these two properties into a single predicate.
The trace formulation of noninterference is provably equivalent to our original noninterference property.

```
   noninterference_hyp(RedDomain: DomainFunction,
                       BlackDomain:DomainFunction,
                       red,blk: index)
                      (gs: GState): bool =
     GSHyp0(gs) &
       member(blk,BlackDomain(state(gs))) &
         member(red,RedDomain(state(gs)))

   noninterference_prop(red,blk: index)
                       (gs: GState): bool =
     not(member(red,DIA(blk,graph(gs))))
```

```
TraceNonInterferenceProperty
  (RedDomain,BlackDomain: DomainFunction): bool =
    FORALL (blk,red: index, gs: StepTrace):
    noninterference_hyp
      (RedDomain,BlackDomain,red,blk)(gs(0)) =>
        FORALL (t: time):
            noninterference_prop(red,blk)(gs(t))


TraceNonInterference_is_NonInterference: LEMMA
    FORALL (Red: DomainFunction,
            Black: DomainFunction,
        TraceNonInterferenceProperty(Red,Black) =
          NonInterferenceProperty(Red,Black)
```

The TraceNonInterferenceProperty, however, maps easily into the simple_trace_prop that we developed in the TraceMU theory. This proof completes a link that allows our original NonInterferenceProperty to be expressed as TraceNonInterferenceProperty and then as simple_trace_prop and then as either simple_LTL_prop or simple_MU_prop, at which point it may avail itself to model checking.

```
IMPORTING TraceMU[GState]

TraceNonInterference_as_simple_trace_prop: LEMMA
    FORALL (RedD,Black: DomainFunction):
      TraceNonInterferenceProperty(RedD,Black) =
        FORALL (blk,red: index):
          simple_trace_prop(
              noninterference_hyp(RedD,Black,red,blk),
              noninterference_prop(red,blk),
              StepRelation)
```


6 Conclusion

A good mathematical specification is one that can be verified against a formal implementation or validated against an actual system, can be used in the formal verification of properties of larger systems, and accurately expresses and predicts behavior in the domain of interest.

We have demonstrated that models of computing systems can be informally validated and formally verified to satisfy our formal model of communication. Both the Green Hills INTEGRITY-178B and the AAMP7G verification efforts involved informal review of formal models against actual system implementations [10, 16].

The AAMP7G effort also employed commuting proof between the implementation model and an abstract representation as well as a proof that abstract model implemented the desired interpartition communication policy, expressed as an information flow property.

Our models of information flow have been used compositionally to verify interesting security properties of more abstract computing systems. We verified the security properties of a simple firewall implemented on a partitioned operating system by appealing to the information flow properties of the OS and the individual partitions. The hierarchical approach used in our analysis of the Green Hills INTEGRITY-178B operating system derived information flow properties of systems by composing information flow properties of their subsystems. Finally, an analysis of an abstract model of the Turnstile system verified several key information flow properties of the system and confirmed a known information back-channel resulting from "assured delivery" requirements [15].

We have also established that our model of information flow accurately expresses and predicts system behavior in the domain of interest. Recall that our concern is how information may or may not be communicated within computing systems. Our techniques have been shown to address three specific concerns from the secure computing domain [7]:

- *Exfiltration.* A computational principal is able to read information in violation of the system security policy, a policy such as a Bell-LaPadula "read-up" policy.
- *Infiltration.* A computational principal is able to write information in violation of the system security policy, a policy such as a Bell-LaPadula "write-down" policy.
- *Mediation.* A computational principal is able to move information in the system, contrary to a policy which does not allow that principal to perform that action, a policy such as a "Chinese wall" separation of duties policy.

Having demonstrated that our models accurately express and predict behavior in the domain of information assurance, can be used in the verification of larger systems, and can be validated against actual systems, we feel that we have met our objective of developing a good mathematical framework for modeling and reasoning about information security specifications.

References

1. Alves-Foss J, Taylor C (2004) An analysis of the GWV security policy. In: Proceedings of the fifth international workshop on ACL2 and its applications, Austin, TX, Nov. 2004
2. Clarke EM, Grumberg O, Peled DA (1999) Model checking. MIT, Cambridge, MA
3. Goguen JA, Meseguer J (1982) Security policies and security models In: Proceedings of the 1982 IEEE symposium on security and privacy, pp 11–20. IEEE Computer Society Press, Washington, DC
4. Greve D (2004) Address enumeration and reasoning over linear address spaces. In: Proceedings of ACL2'04, Austin, TX, Nov. 2004
5. Greve D (2006) Parameterized congruences in ACL2. In: Proceedings of ACL2'06, Austin, TX, Nov. 2006

6. Greve D (2007) Scalable normalization of heap manipulating functions. In: Proceedings of ACL2'07, Austin, TX, Nov. 2007
7. Greve D, Wilding M, Vanfleet M (2003) A separation kernel formal security policy. In: Proceedings of ACL2'03
8. Greve D, Wilding M, Vanfleet M, Richards R (2005) Formalizing security policies for dynamic and distributed systems. In: Proceedings of SSTC 2005
9. Munoz C (2009) ProofLite product description. http://research.nianet.org/~munoz/ProofLite
10. Richards R (2010) Modeling and security analysis of a commercial real-time operating system kernel. In: Hardin D (ed) Design and verification of microprocessor systems for high-assurance applications. Springer, Berlin, pp 301–322
11. Rushby J (1992) Noninterference, transitivity, and channel-control security policies. Technical report csl-92–2, SRI
12. Rushby J (2001) Formal verification of McMillan's compositional assume-guarantee rule. Technical report, SRI, September 2001
13. Shannon C, Weaver W (1949) The mathematical theory of communication. University of Illinois Press, Champaign, IL
14. SRI, Incorporated (2009) PVS specification and verification system. http://pvs.csl.sri.com
15. Whalen M, Greve D, Wagner L (2010) Model checking information flow. In: Hardin D (ed) Design and verification of microprocessor systems for high-assurance applications. Springer, Berlin, pp 381–428
16. Wilding M, Greve D, Richards R, Hardin D (2010) Formal verification of partition management for the AAMP7G microprocessor. In: Hardin D (ed) Design and verification of microprocessor systems for high-assurance applications. Springer, Berlin, pp 175–191

Modeling and Security Analysis of a Commercial Real-Time Operating System Kernel

Raymond J. Richards

1 Introduction

INTEGRITY-178B is a real-time operating system (RTOS) developed by Green Hills Software [4]. Real-time operation implies that the operating system kernel will schedule tasks as described by a predetermined schedule. System designers depend on the kernel to reliably and faithfully schedule tasks according to the schedule. This is to ensure that the tasks complete their necessary computations before system imposed deadlines.

The initial market for INTEGRITY-178B was safety-critical systems, such as avionics. The FAA accepts the use of DO-178B, Software Considerations in Airborne Systems and Equipment Certification [10], as a means of certifying software in avionics. DO-178B defines five levels of software to describe the impact to aircraft safety should there be a failure. The criticality levels are denoted "A" through "E," "A" is the most critical, and "E" is the least critical.

INTEGRITY-178B is able to host multiple applications of mixed criticality levels. It provides fault containment, preventing faults from cascading to other applications. That is to say, a fault in one application is never noticeable in another application. Recall that this is in a real-time operational environment, meaning that the failure of an application cannot cause any other application to miss a deadline. To achieve this level of fault containment, it is necessary for the kernel to strictly partition not only the time allocated to each application, but also the system memory between the various applications. This stringent time and space partitioning is often referred to as "hard-partitioning."

The high-assurance realms of safety-critical systems and security-critical systems overlap in many interesting ways. In particular, the use of hard partitioning is important in building high-assurance systems in both realms [11]. Mechanisms that provide hard partitioning are often referred to as "separation kernels." In the safety critical realm, separation kernels can integrate functionality of various levels

R.J. Richards (✉)
Rockwell Collins, Inc., Cedar Rapids, IA, USA
e-mail: rjricha1@rockwellcollins.com

D.S. Hardin (ed.), *Design and Verification of Microprocessor Systems for High-Assurance Applications*, DOI 10.1007/978-1-4419-1539-9_10,

of criticality on a single computing platform. Separation provides fault isolation; a fault in a less critical application cannot impact the execution of a more critical function. In the security-critical realm, separation kernels can ensure that there is no unauthorized flow of information between applications. This means that an application cannot inadvertently or maliciously signal another application with which it is not authorized to communicate. Since separation kernels are useful for both safety-critical and security-critical systems, it is reasonable to take a separation kernel that has been certified in one realm and attempt certification in the other realm.

Information assurance products can be certified in accordance with the Common Criteria for Information Technology Security Evaluation [2] or "Common Criteria" for short. In USA, the National Information Assurance Partnership (NIAP) performs Common Criteria evaluations. The Common Criteria defines seven evaluation assurance levels (EALs). The levels are labeled one through seven; EAL 7 is the most stringent level. A study has compared the certification requirements of DO-178B Level A and Common Criteria EAL 7 [1]. This study concluded that a product that is used in a DO-178B certified system could achieve Common Criteria EAL 7 by completing a few missing requirements. The most significant missing requirements are those that pertain to formal analysis.

INTEGRITY-178B was designed to be, and has been used in, systems that have been certified to DO-178B Level A; therefore, it was judged to be a good candidate to be the first separation kernel to obtain a Common Criteria certification. The INTEGRITY-178B analysis effort supported an EAL 6 Augmented (EAL6+) evaluation. EAL6+ means that some of the evaluation requirements were more stringent than prescribed by EAL6.

The Common Criteria defines three levels of rigor in analysis. These three levels are informal, semiformal, and formal. In this context, formal means a precise mathematical treatment with machine-checked proofs. Informal is a natural language-based justification of the security properties. Semiformal is something in between. For the INTEGRITY-178B kernel, this means a mathematical treatment, where some of the proofs are not machine checked.

The INTEGRITY-178B evaluation requirements for EAL 5 and above specify five elements that are either formal or semiformal. These five elements are the Security Policy Model, the Functional Specification, the High-Level Design, the Low-Level Design, and the Representation Correspondence [9]. The level of rigor that was applied to INTEGRITY-178B is as follows:

- Security Policy Model: A formal specification of the relevant security properties of the system.
- Functional Specification: A formal representation of the functional interfaces of the system.
- High-Level Design: A semiformal representation of the system. This representation may be somewhat abstract.
- Low-Level Design: A semiformal, but detailed representation of the system.
- Representation Correspondence: This element demonstrates the correspondence between pairs of the other elements. The Representation Correspondence is

formal when it shows the correspondence between two formal elements; it is semiformal otherwise. The Representation Correspondence shows that:

- The functional specification implements the security policy model.
- The high-level design implements the functional specification.
- The low-level design implements the high-level design.

The Common Criteria explicitly states that one entity may fulfill multiple requirements. For example, a single design specification may fulfill the need for both a high-level and a low-level design; in this case, the correspondence between these two elements is trivial.

INTEGRITY-178B runs on a variety of microprocessors and motherboards. A well-defined hardware abstraction layer with well-defined interfaces facilitates this portability. The formal (and semiformal) analysis was constrained to the hardware-independent portions of the INTEGRIY-178B kernel. A methodical informal analysis was performed on the software in the hardware abstraction layer. This approach allows the formal (and semiformal) analysis to be used as certification evidence on multiple hardware platforms.

This chapter discusses details of the formal analysis approach taken for the INTEGRITY-178B kernel, including:

- The generalization of the GWV theorem to capture the meaning of separation in a dynamic system.
- A discussion on how the system was modeled including:

  - System state
  - Behavior
  - Information flow

- The proof architecture used to demonstrate correspondence.
- The informal analysis of the hardware abstraction layer.

2 Separation Theorem

Existing formal specifications of separation properties were not expressive enough to state anything meaningful about INTEGRITY-178B. The GWV theorem has been shown to hold for the AAMP7G's hardware-based separation kernel [12]. However, the AAMP7G's kernel is very static. Its execution schedule is set a priori; it is impossible for user-level software to have an impact on the state of the kernel's scheduler. The original GWV theorem is only applicable to such strict static schedulers.

INTEGRITY-178B's scheduling model is much more dynamic. A more general GWV theorem was derived that captures the appropriate system level properties. This theorem is known as GWVr2 [5].

For a GWVr2 proof, the system needs to be modeled as a state transition system (Fig. 1). That is, it receives as inputs the current state of the system, as well as any external inputs. It produces a new system state, as well as any external outputs.

Fig. 1 State transition
system

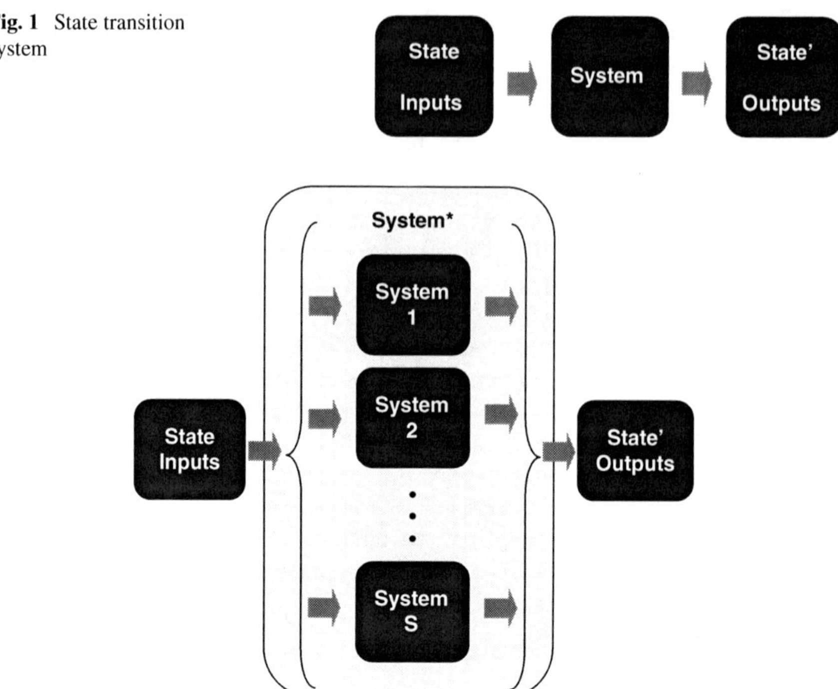

Fig. 2 Modified system model

System execution is a series of these state transitions. As a convenience, we will assume that the external inputs and outputs are contained in the system state structure. This state transition is expressed in the language of the ACL2 theorem prover [7] as:

$$(let \; state'\,(system \; state))$$

The system state, inputs, and outputs can be decomposed into atomic elements. Each of these elements is uniquely identifiable. Let the number of state elements plus the number of output elements be denoted by the symbol S. The system can be represented by S copies of the original system, each producing one element of the next state or output. Each of these S systems can be fed with only the elements of the current state and inputs that are necessary for it to compute its result. A system that takes a current state and input elements, maps the appropriate inputs to S copies of the system, and then maps the S resulting elements into the next state and external outputs is denoted as system* (Fig. 2).

The ACL2 notation for system* is a function that has two inputs. One input is a structure containing current state and external inputs. The other is a graph that specifies how the current state and inputs elements are mapped to the S subsystems. It produces the next system state, expressed in ACL2 as

$$(let \; state'\,(system^* \; graph \; state))$$

If it can be proven that the system and system* produce identical results for all inputs of interest, it implies that the graph used by system* completely captures the information flow of the system. This is the GWVr2 theorem.

$$(\text{equal}$$
$$(system \text{ state})$$
$$(system^* \text{ graph state}))$$

A trivial graph that satisfies this theorem simply gives each subsystem all inputs and state elements. Conversely, there is a minimal graph, for which removing any element from the input of any subsystem causes the theorem to fail. Elements can be added to the minimal graph, without impacting the correctness of the theorem. This means that the input for one of the subsystems defines all of the data necessary for computing one element of the next state or output.

The GWVr2 Theorem is the Common Criteria Security Policy Model for INTGERITY-178B.

3 Modeling System State

To be consistent with the goal that the formal analysis be platform independent, the model of system state is that of nested abstract data structures. Elements within a data structure can either be a scalar or a nested data structure. A data structure that contains other data structures may be a record of heterogeneous data items or an array of homogenous data items. All elements in a data structure have names that uniquely identify them and distinguish them from their peer elements. This is analogous to a Unix file system containing directories and files. The directories represent nested data structures and the files represent scalar data elements.

In such a file system, one can identify any directory or file resident within a particular directory by specifying a path. The path contains the name of every sub-directory that must be traversed in order to reach the item of interest. Similarly, in the model of state, one can reach any piece of state that is resident in a data structure by specifying a path. Arrays are represented in this model by using the array indices as a specifier in the path.

Paths are considered scalar data items; they can be stored as part of state. This is how C language pointers are modeled. Paths can be references to state locations and can be dereferenced. Dereferencing a path produces the value stored at that location in state.

An example data structure is shown in Fig. 3. Four scalar values are stored in nested data structures. The paths to these for values and the data stored in this structure are shown in Table 1.

The ACL2 representation of a path is simply a list of identifiers. The head of a path is the outermost data structure. The tail of a path represents a path that is relative to the head. In this way, paths are analogous to a directory path in a Unix

Fig. 3 Example state
structure

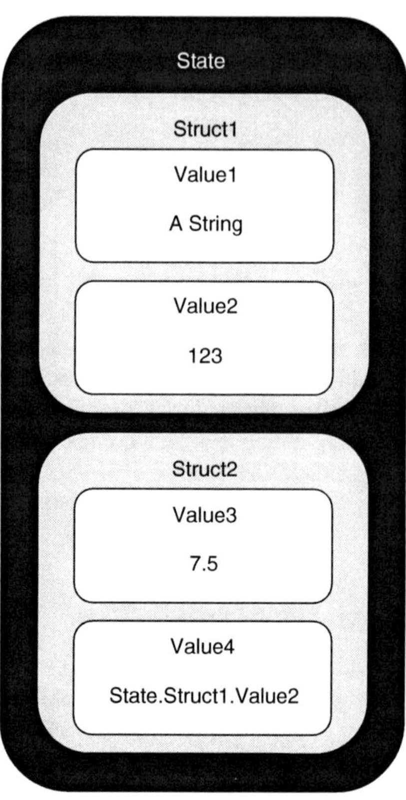

Table 1 Path examples

Path	Value	Data type
State.Struct1.Value1	"A String"	String
State.Struct1.Value2	123	Integer
State.Struct2.Value3	7.5	Float
State.Struct2.Value4	State.Struct1.Value2	Path

file system. Absolute paths are relative to the root of the file system; relative paths
are referenced from the current location.

Operators are defined to update and query the state. Both of these operators use
a path to specify which element of the state they are affecting. The query operator
"GP," or Get from Path, returns the value stored at the specified location. Its implementation is a recursive function that fetches the data specified by the identifier
at the head of the path and recursively calls itself using the tail of the path and the
fetched bit of state as the recursive arguments. The signature of the GP operator is:

```
(GP path st)
```

The update operator "SP," or Set Path, returns a new state, where the element
specified by the given path is replaced with a new value. Its implementation is

also recursive. SP replaces the element specified by the head of the path with the value returned by its recursive call. The arguments to this recursive call are the tail of the given path and the value found in the state at the location specified by the head of the path. The signature of the SP operator is:

```
(SP path value st)
```

4 Modeling Kernel Behavior

The hardware-independent portion of the INTEGRITY-178B kernel is implemented in C code and formally modeled in ACL2. The Common Criteria explicitly forbids the low-level design specification and the implementation representation (source code) to be one and the same. Furthermore, establishing correspondence between the low-level design specification and the implementation representation is typically a manual, labor-intensive endeavor. This process is sometimes referred to as a "code-to-spec" review.

One of the goals in modeling the system is to capture enough of the implementation details so that a clear and compelling argument can be made that the behavior of the system is captured accurately. In an effort to facilitate that argument, it is useful if the low-level design specification has a one-to-one correspondence with the source code. That is to say, for every action in the source code there is a corresponding action in the model, and for every action in the model there is a corresponding action in the source code. In order to make the correspondences clear, it is useful to make the model greatly resemble the source code.

Two areas where it is not possible to make the ACL2 model closely resemble the C code implementation are in modeling loop constructs, as well as certain types of recursion. Since the formal language (ACL2) used to model the system is a functional language, and the INTEGRITTY-178B implementation Language (C) is an imperative language, it is not always possible to directly represent constructs in the system's implementation in the model. For instance, ACL2 does not have looping constructs. Loops are instead modeled by recursive functions.

ACL2 requires a proof of termination before admitting any function. This implies that a certain style of recursion, known as reflexive recursion, cannot be directly modeled [6]. Reflexive recursion occurs when two successive recursive calls are made, the latter one taking as an argument something calculated by the first. The following is an example of a reflexive recursive function:

```
int reflex (int i){
   int j;
   <function body>
   j=reflex(reflex(i-1));
   <function body>
   return j;
}
```

In this example, the outer call to the function `reflex` depends upon the results of the inner call. This results in the proof of termination depending upon the termination of the function. When this occurs in the INTEGRITY-178B kernel, it is modeled by unrolling the recursion. The recursion encountered in INTEGRITY-178B is controlled by a simple counting variable; when that variable reaches a particular value, the recursion is terminated.

4.1 Reader Macro

A set of ACL2 macros is used to allow the functional model to have an imperative look and feel. These macros are known collectively has the *reader macro*. The reader macro expands statements into a functional form. The reader macro is a form that begins with the symbol "%." This allows a syntax that closely resembles C to expand into native ACL2. The native ACL2 uses the state operators "SP" and "GP" to interact with the system state. The following types of statements are handled by the reader macro:

- Global variable access
- Assignment
- Function invocation
- Conditional early exit from a function

4.1.1 Global Variable Access

In the functional language model, all state information, except for local variables, is stored in the state structure that is passed throughout the model. Local, or stack, variables are modeled by local variables in ACL2, as long as there is no address-based accessing of the variable. A local variable that has its address passed to a subordinate function must be modeled as a state element.

The C language's use of variable identifiers does not distinguish between global and local variables. Since global variables are elements in the state structure, syntax was adopted to indicate when an identifier is an access to a global variable. Preceding an identifier with the symbol "@" indicates that the identifier should be treated as a path to a global variable. Preceding any path, including that of a local variable, with the symbol "*" queries the value stored at that location pointed to by the path.

4.1.2 Assignment

Assignment statement syntax depends on the impact of the assignment. That is, assignments to local variables have a different syntax than assignments that change the persistent state.

Assignment statements can be generalized as an lvalue, an assignment operator, and an rvalue.

```
lvalue assign_op rvalue
```

The lvalue denotes where the assigned value is stored. This can be a local variable or a location in state. Local variables are modeled by local ACL2 variables. This means when the scope in which the local variable has been declared is exited, its value is lost. The syntax for local variable assignment uses the variable identifier as the lvalue, followed by an equal sign "=," followed by the rvalue.

```
lvalue = rvalue
```

Assignments to local variables are transformed into *let bindings*. The body of the let binding is the scope where that assignment is valid.

In assignment statements that change state, the lvalue must evaluate to a path. Rvalues are evaluated and the results are stored in the location indicated by the lvalue. Assignments to state are transformed into state updates. The lvalue of an assignment that impacts the state must evaluate to a path into the state. The syntax for such statements is as follows:

```
(path) @= rvalue
global_var @= rvalue
```

4.1.3   Functions

C language functions may or may not return a value. When modeling in ACL2, functions need to at least return the state that is a result of their invocation. The reader macro transforms function invocations that appear to not return a value into a function call that returns the new state, catching it in the appropriate variable. Model functions are declared using a form called "defmodel," which is similar to the ACL2 defun form.

Functions that return a value are modeled using a multivalued return. That is, it returns a list of items. The return list has a length of two; the second item is always the state returned from the function.

4.1.4   Conditional Early Exit

It is a common coding practice for functions to perform checks on the validity of their inputs. If the input checks do not pass satisfactorily, the function is exited, often returning an error code.

```
if (conditional){
  error handling;
  return;
}
```

ACL2 functions only exit at the end of the function body. The reader macro recognizes conditional early exits, translating them into an *if* statement whose *then* clause includes whatever error handling is needed. The *else* clause contains the remainder of the function. The syntax for conditional early exist is:

```
(ifx (conditional)
  error handling)
```

4.2 Model Example

The following example will be used to illustrate the various parts of this analysis. The example is a function that operates on a circular, doubly linked list. This function removes one element from the list, maintaining a well-formed linked list. This function is passed two arguments. The first is a pointer to a structure that contains a pointer to the head of the list. The second is a pointer to the element that is removed from the list. It is assumed that the element pointed to by the second argument is a member of the list pointed to by the first argument. How this assumption is captured in the analysis will be discussed later in this chapter. The example's C language implementation is:

```
void RemoveFromList (LIST *TheList, ELEMENT *Element){
   ELEMENT *NextInList, *PrevInList;

   NextInList = Element -> next;

   if(NULL == NextInList)
     return;

   /* Update list */
   if (Element == NextInList){
     /* only element in the list */
     TheList->First = NULL;
   } else {
     /* not only element in the list */
     if (TheList->First == Element){
        /* Element is first in list */
        TheList->First) = NextInList;
     }

     PrevInList = Element->prev;
     PrevInList->next = NextInList;
     NextInList->prev = PrevInList;
   }
     /* clear this element's links */
     Element->next = NULL;
     Element->prev = NULL;
}
```

The formal model for this function is defined as follows:

```
defmodel RemoveFromList (TheList Element st)
  (%
    (NextInList = (* Element -> next))

      (ifx (NULLP NextInList)
         st)

    (if (equal Element NextInList)
       (%

        ;; only element in the list
        ((TheList -> First) @= (NULL)))

    ;; else
    (%
     ;; not only element in the list
     (if (equal (* TheList -> First) Element)

         ;; Element is first in list
         (%
          ((TheList -> First) @= NextInList))

    ;; else
       st)

    (PrevInList = (* Element -> prev))
    ((PrevInList -> next) @= NextInList)
    ((NextInList -> prev) @= PrevInList)))

    ;; clear this element's links
    ((Element -> next) @= (NULL))
    ((Element -> prev) @= (NULL)))))
```

4.3 *Model Syntax Summary*

The following table describes how various C language constructs are modeled

C	ACL2	Lisp/ACL2 Notes
Variable reference		
x	X	Value of local variable x
	(*@ x)	Value of global variable x
$*x_p$	$(*x_p)$	Value pointed to by local variable x_p
	$(* (*@ x_p))$	Value pointed to by global variable x_p
&x	(@ x)	Address of global y variable

Variable assignment

$x = \ldots;$	$(\% .. (x = \ldots) ..)$	Assign value of local variable x
	$(\% .. (x \, @= \ldots) ..)$	Assign value of global variable x
	$(\% .. ((x_p) \, @= \ldots) ..)$	Assign value pointed to by local variable x_p
$^*x_p = \ldots;$	$(\% .. ((^*@ \, x_p) \, @= \ldots) ..)$	Assign value pointed to by global variable x_p

Simple structure references

x.y	$(^* \, (@ \, x) \, \vert. \vert \, y)$	Value of field y of the structure instance at global variable x
$x_p\!-\!>\!y$	$(^* \, x_p\!-\!>\!y)$	Value of field y of the structure pointed to by local variable x_p
	$(^* \, (^*@ \, x_p)\!-\!>\!y)$	Value of field y of the structure pointed to by global variable x_p
x[y]	$(^* \, (@ \, x) \, [y \,])$	Value of element at index y in the array instance at global variable x
$x_p[y]$	$(^* \, x_p \, [y \,])$	Value of element at index y of the array to which local variable x_p points
	$(^* \, (^*@ \, x_p) \, [y \,])$	Value of element at index y of the array to which global variable x_p points
&x.y	$(\& \, (@ \, x) \, \vert. \vert \, y)$	Address of field y of the structure instance at global variable x
$\&x_p\!-\!>\!y$	$(\& \, x_p\!-\!>\!y)$	Address of field y of the structure pointed to by local variable x_p
	$(\& \, (^*@ \, x_p)\!-\!>\!y)$	Address of field y of the structure pointed to by global variable x_p
&x[y]	$(\& \, (@ \, x) \, [y \,])$	Address of element at index y in the array instance at global variable x
$\&x_p[y]$	$(\& \, x_p \, [y \,])$	Address of element at index y of the array to which local variable x_p points
	$(\& \, (@^*x_p) \, [y \,])$	Address of element at index y of the array to which global variable x_p points

Complex structure references

$^*x.y$	$(^* \, (^* \, (@ \, x) \, \vert. \vert \, y))$	Value pointed to by field y of the structure instance at global variable x
$^*x_p\!-\!>\!y$	$(^* \, (^*x_p\!-\!>\!y))$	Value pointed to by field y of the structure pointed to by local variable x_p
	$(^* \, (^* \, (^*@ \, x_p)\!-\!>\!y))$	Value pointed to by field y of the structure pointed to by global variable x_p
$^*x[y]$	$(^* \, (^* \, (@ \, x) \, [y \,]))$	Value pointed to by element at index y in the array instance at local variable x

| $^*x_p[y]$ | $(* (*x_p [y]))$ | Value pointed to by element at index y of the array to which local variable x_p points |
| | $(* (* (* @ x_p) [y]))$ | Value pointed to by element at index y of the array to which global variable x_p points |

Simple structure assignments

$x.y = \dots;$	$(\% .. (((@ x) \lvert.\rvert y) @= \dots) ..)$	Assign value of field y of the structure instance at global variable x
$x_p -> y = \dots;$	$(\% .. ((x_p -> y) @= \dots) ..)$	Assign value of field y of the structure pointed to by local variable x_p
	$(\% .. (((* @ x_p) -> y) @= \dots) ..)$	Assign value of field y of the structure pointed to by global variable x_p
$x[y] = \dots;$	$(\% .. (((@ x) [y]) @= \dots) ..)$	Assign value of element at index y in the array instance at global variable x
$x_p[y] = \dots;$	$(\% .. ((x_p [y]) @= \dots) ..)$	Assign value of element at index y of the array to which local variable x_p points
	$(\% .. (((* @ x_p) [y]) @= \dots) ..)$	Assign value of element at index y of the array to which global variable x_p points

Complex structure assignments

$^*x.y = \dots;$	$(\% .. (((* (@ x) \lvert.\rvert y)) @= \dots) ..)$	Assign value pointed to by field y of the structure instance at global variable x
$^*x_p -> y = \dots;$	$(\% .. (((* x_p -> y)) @= \dots) ..)$	Assign value pointed to by field y of the structure pointed to by local variable x_p
	$(\% .. (((* (* @ x_p) -> y)) @= \dots) ..)$	Assign value pointed to by field y of the structure pointed to by global variable x_p
$^*x[y] = \dots;$	$(\% .. (((* (@ x) [y])) @= \dots) ..)$	Assign value pointed to by element at index y in the array instance at global variable x
$^*x_p[y]\dots;$	$(\% .. (((* x_p [y])) @= \dots) ..)$	Assign value pointed to by element at index y of the array to which local variable x_p points
	$(\% .. (((* (* @ x_p) [y])) @= \dots) ..)$	Assign value pointed to by element at index y of the array to which global variable x_p points

Arithmetic operators

$X + y;$	$(+ x y)$	Addition
$X - y;$	$(- x y)$	Subtraction
$x^* y;$	$(ACL2::^* x y)$	Multiplication
$x/y;$	$(/ x y)$	Division

Logical operators

$x == y$	$(equal\ x\ y)$	Equal
$x\ != y$	$(not\ (equal\ x\ y))$	Not-equal (Negation)
$x < y$	$(< x y)$	Less-than
$x > y$	$(> x y)$	Greater-than
$x <= y$	$(<= x y)$	Less-than or equal-to
$x_p == NULL$	$(NULLP\ x_p)$	Null pointer test
$!x_p$		

$x_p == 0$		
x_p	(NNULL x_p)	Non-Null pointer test
Conditional control structures		
if (x) y; else z;	(if x y z)	If expression x is *true* do y otherwise do z.
if (x) y;	(if x y st)	If expression x is *true* do y otherwise to nothing. State (st) as the else component is analogous to do-nothing or the absence of as else-component.
{...}	(% ...)	Groups a series of sequential statements into a single global structure or block
If (x) {A$_{stmt}$; B$_{stmt}$; } else { C$_{stmt}$; }	(if x (% A$_{stmt}$B$_{stmt}$...) (% (C$_{stmt}$) ...))	If example with a code block.
x ? y : z	(if x y z)	C alternation
Function Declarations		
type foo ...	(defun foo ...	Return type not specified in function signature
... foo (int x) ...	... foo (x st)	No parameter type declarations State (st) parameter added for access to global state
... foo (int *x) ...	... foo (x_p st)	No parameter type declarations State (st) parameter added for access to global state
return;	(return st)	Function return when function application only changes state
return x;	(return x)	Function returns a single value x when function application does not change state

4.4 Kernel Boundaries

There are several important boundaries to the kernel model that could not be modeled as a straightforward translation of the system source code. These boundaries include asynchronous interactions with the world outside of the kernel, including interrupt processing and execution of application code. Another boundary is the interaction with the portion of the kernel that is specific to the hardware platform.

4.4.1 Breaking the Kernel Loop

The GWVr2 theorem requires that steady-state operation be defined as a step that can be performed repeatedly. Since the steady-state execution of most operating systems is implemented with an infinite loop, it is a natural inclination to have one step in the model represents a single iteration of this loop. We will refer to this as the scheduling loop.

Great care needs to be taken to determine where this loop is broken. The cut point of the loop must represent a well-defined state in the execution. This is the point where the next entity to execute can be identified from the state, and the previous entity has been removed from execution and its state saved. This is the point where all assumptions about the state hold. These assumptions include well formedness and necessary constraints on the data. We refer to the state between two steps as a secure state. It is assumed that the system reaches a *secure initial state* as a result of the system initialization. Evidence justifying this assumption has been generated and is beyond the scope of this chapter. The generalization of the GWV theorem to the GWVr2 theorem does allow for the execution of a previous system step to influence which is the next entity to execute. The successive execution of steps is shown in Fig. 4.

The INTEGRITY-178B scheduling loop picks which thread of the current partition is to execute next and does not necessarily represent a context switch or a change in the logical partition.

4.4.2 Hardware-Dependent Layer

As mentioned earlier, only the hardware-independent portion of the kernel is formally modeled. The hardware-dependent portion of the kernel is present in the formal model only in an abstract form. The functional interface to this portion of the kernel is modeled by ACL2 *encapsulations*. In an encapsulation, the function signatures are defined, as well as key properties, but no implementations are modeled. The properties given in these encapsulations are the properties necessary to prove either termination of a higher level model function or the GWVr2 Theorem.

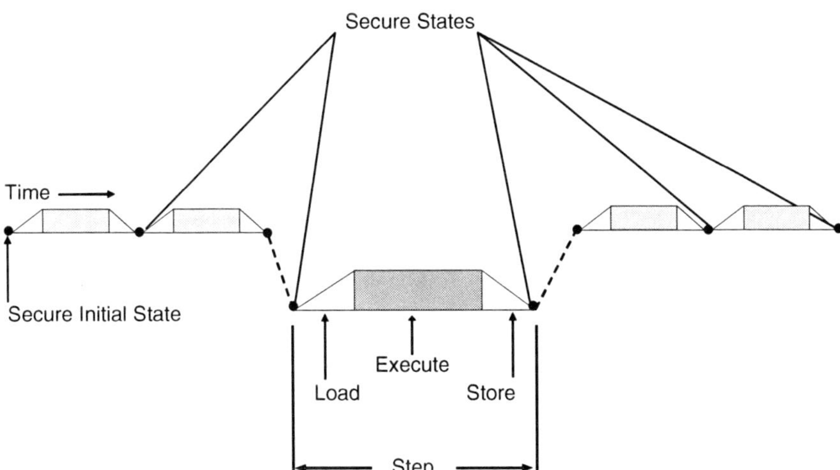

Fig. 4 Dividing Kernel execution into discrete steps

Justification of these properties in the actual implementation occurred as part of the rigorous manual analysis of that code. This analysis is discussed later in this chapter.

4.4.3   Interrupt Processing

Asynchronous events typically interact with INTEGRITY-178B via the interrupt mechanism. Interrupt processing is implemented as functions that are invoked from the platform-specific interrupt handling software. We model all such interrupt processing at the level of the scheduling loop. That is, before the step concludes, it processes all of the interrupts that occurred during execution of the step.

A logical mechanism known as an *oracle* is used to determine the number and type of interrupts that have arrived. An oracle is modeled by an unspecified function that produces the information that is needed. Properties may be stated about the oracle function, which may be as simple as the type of information it produces. The oracle function takes as an input an undefined piece of the system state. This gives us a convenient way to reason about an arbitrary number of arbitrary events.

Interrupts are modeled by a loop whose number of iterations is controlled by one oracle; in the body of the loop, a second oracle determines what type of interrupt has arrived.

4.4.4   Application Software Execution

The purpose of a separation kernel is to orchestrate the execution of multiple partitions. From time to time the kernel relinquishes control of the hardware to application software. However, the actions of the application cannot be modeled, since they are unknown. What is modeled are the application software's interactions with the kernel. These interactions come in the form of system calls. We model an arbitrary number of arbitrary system calls in the same manner in which we model interrupts. One oracle determines how many systems calls are going to be processed, and a second oracle determines the type of each system call.

5   Modeling Information Flow

Information flows are modeled by defining for a given element in system state; what are the state elements that can influence its next value. This can be thought of as a graph with vertices representing state elements and edges representing dependencies. For each function of the model, a new function is defined that calculates its graph, given its set of inputs, including the input state. The naming convention for these graph-computing functions (often referred to as graph functions) is to append "-graph" to the model function's name. The graph function's parameters are identical to that of the model function. The graph function returns a data structure that contains all of the graph edges for each state element that is updated by the model function.

5.1 Crawlers

In order to define a graph, it is necessary to be able to articulate what elements in state belong to which data structures. That is to say, for an operation upon a data structure, we need to be able to create sets of elements that may be updated by the operation and sets of elements whose values are used to create the new values. To do this, we have created a construct known as a *crawler*. Crawlers are used to create collections of state elements. Each member of the collection is identified by its path. Once a collection is created, it can be transformed into a collection of subelements by appending the subelement identifier to each path. We call the action of appending the same identifier to each path *decorating* the paths in the collection.

Let us consider the doubly linked circular list example. Each list element contains a next and previous pointer in order to connect it to the list. A crawler over this data structure might create a collection containing all elements in the list. That is, it creates a set of paths, one representing each element in the list. A decorate operation might refine these paths to refer to the next and previous fields of each element in the list.

5.2 Graphs

A graph describes, for a set of state elements that may have their value changed by an operation, what are the sources of information that are used in calculating the new values. Using our circular linked list example, let us consider a graph for either a sort or remove-element operation. In each case, the state elements that may be changed are the previous and next fields of all of the elements of the list. The new values that may be stored in these locations are the values that are stored in these same locations before the operation. Therefore, the graph states that the new values of the previous and next fields in the list depend upon what is currently stored in the previous and next fields in the list. More precisely, the graph contains an entry for each previous and next pointer as a location that may be updated. Each entry defines a dependency on the set of previous and next pointers as the source of information for the updated values.

Several functions and macros are defined to assist in developing graph functions. Chief among these is "defgraph," which takes four arguments. The first argument is the name of the function whose graph is being defined. The second argument is the list of names of the function's parameters. Any parameter that is a pass-by-value structure has its name in parentheses. The third argument is the list of variable names whose values are returned by this function. Again, pass-by-value structures have their names in parentheses. The last argument is the body of the graph defining function.

The functions "du," "du*," and "merg-u2" are used to create dependencies or use lists. The functions "su" and "su*" are used to associate a dependency set with an

element that has its value defined by the function. The function "mvg" returns the
graph and associates variables with returned values.

The graph for the RemoveFromList example function is defined as follows:

```
(defgraph RemoveFromList (:TheList :Element (:st))
                         ((:st))
  (%
    ;; determine the nodes in the list
    (list-nodes = (crawl-list TheList st))

    ;; define the state elements whose values
    ;; might change
    (list-ass =
        `(,@(decorate-list list-nodes
                          (ElementStruct_Kstr))
          ,@(decorate-list TheList
                          (ListStruct_Kstr))))

    ;; create the dependencies set for the things
    ;; that might change
    (u2 = (merg-u2 :TheList :Element
                          (du* :st list-ass)))

    ;; define the dependencies for things that
    ;; might change
    (g = (su* :st list-ass u2 g))

    (mvg (:st) st)))
```

5.3 *Graph Composition*

In the formal analysis, graphs are created for each function in the kernel. A graph
for a function that calls other functions must be no smaller than the graphs of the
subordinate functions. That is, the dependencies defined by any graph of a called
function must exist in the graph of the calling function.

In the circular linked list example, consider an Add operation. The state elements
that may be updated are not only the previous and next fields of the existing list, but
also the previous and next fields of the element being added to the list. The sources
of new values for these elements are not only the previous and next pointers of the
existing list, but also the parameter to the function pointing to the new element.

The graph of any function calling the Add operation must relate the previous
and next fields of the current list members and the previous and next fields of any
elements that might be added to the list to the sources of possible new values. The

sources of possible new values are, of course, the previous and next fields of the current members of the list and the locations that could supply the new elements to the Add operation.

6 Proof of Separation

In order to prove the GWVr2 theorem, it is useful to first prove two lemmas with respect to the function being analyzed. These lemmas are referred to as the Workhorse Lemma and the ClearP Lemma. We will discuss these lemmas with respect to the circular linked list example. Before we discuss these lemmas, we will define functions needed to support them.

- *RemoveFromList-Hyp.* For every model function "foo" a function "foo-hyp" is defined. This function states the hypothesis that is needed in order to have the model function work appropriately. The hypothesis function takes the same arguments as the model function. Recall that for the RemoveFromList function, it was assumed that the element given to the function is a member of the list; the RemoveFromList-Hyp function is where that assumption is stated.
- *Keys.* The Key function is passed a dependency graph and returns the set of state elements that may be updated, according to the graph.
- *DIA.* The direct interaction allowed (DIA) function takes a state element and a graph. It returns the set of state elements that the passed-in element has dependencies on, as defined by the graph.
- *CP-Set-Equal.* CP-Set-Equal is a predicate that takes a set of state elements and two states. It evaluates to *True* if the two states have the same value for each member of the set. It does not say anything about portions of the state that are not in the set. Therefore, the two states may be different in the parts of the state not defined in the set.
- *CLRP-Set.* CLRP-Set takes a set of state elements and a state. It returns a state that is a copy of the passed-in state, but the elements of the state specified in the set have been cleared. In this case, cleared means that their values have been replaced with *nil*.

6.1 Workhorse Lemma

The Workhorse Lemma states a relationship between the results of two invocations of a function. These two invocations operate on different states, but on the same parameters. For the circular linked list example, the two states satisfy the following constraints:

- Both states satisfy the RemoveFromList-Hyp assumptions. This means we are only considering invocations of this function, where the element to be removed is a member of the list.
- The two states have the same values for all elements that are in one of the dependency sets defined by the RemoveFromList graph.

These constrains are specified in ACL2 as:

```
(AND
    (RemoveFromList-Hyp List Element St1)
    (RemoveFromList-Hyp List Element St2)
    (Member Path (Keys
            (RemoveFromList-Graph List Element St1)))
    (CP-Set-Equal
      (DIA Path
         (RemoveFromList-Graph List Element St1))
         St1 St2))
```

This lemma concludes that having these criteria satisfied implies that the two states resulting from the two RemoveFromList invocations have the same values for all state elements that are defined by the RemoveFromList-Graph function. The ACL2 statement of this lemma is:

```
(DEFTHM RemoveFromList-Workhorse
    (IMPLIES
       (AND
          (RemoveFromList-Hyp List Element St1)
          (RemoveFromList-Hyp List Element St2)
          (Member Path
                  (Keys (RemoveFromList-Graph List
                                               Element
                                               St1)))
          (CP-Set-Equal
            (DIA Path
                 (RemoveFromList-Graph List
                                       Element
                                       St1))
            St1
            St2))
     (IFF (EQUAL
          (GP Path (RemoveFromList List Ele ment St1))
          (GP Path (RemoveFromList List Element St2)))
          T)))
```

The Workhorse lemma demonstrates that the function's graph sufficiently captured the dependencies in the data flows of the function.

6.2 ClearP Lemma

The ClearP Lemma demonstrates that all of the changes to state performed by a function are captured by the function's graph. In the circular linked list example, for a list, element, and a state that satisfy the function's hypothesis function

```
(RemoveFromList-Hyp List Element St1)
```

the ClearP lemma establishes that the operation of the function does not change state in a manner that is not captured by its graph. The lemma considers two states, the input state and the output state the function. If all elements that the graph says might be updated are removed from both states, then satisfying ClearP means that the remaining states are identical. This demonstrates that the footprint of state changed by the function is captured by the graph.

Once these two lemmas are proven, it is straightforward to prove GWVr2 for the function.

7 Hardware-Dependent Code Analysis

In the INTEGRITY-178B evaluation, the small layer of hardware-dependent code was subjected to a rigorous by hand review. NIAP provided a list of source code characteristics that when present tend to promote correctness, stability, and understandability. These characteristics were collected from industry's best practices for real-time high-assurance software. An example of these characteristics is the absence of pointer arithmetic. Importantly, the assumptions stated in the abstract model of the code are validated by this analysis.

All of the source code in this layer (C code and assembly language) was examined to determine if it conformed to the characteristics list. For each function, justification for each characteristic was documented. This documentation was provided as part of the certification evidence.

8 Conclusion

After a thorough review of all of the certification evidence, including the formal, semiformal, and informal analysis described herein, NIAP granted a Common Criteria Certificate for the INTEGRITY-178B kernel at the EAL6+ level on September 1, 2008. The "home page" for the certification documentation can be found online [8]; a summary of the formal verification activities can be found in the Security Target document [3].

References

1. Alves-Foss J, Rinker B, Taylor C (2002) Towards common criteria certification for DO-178B compliant airborne systems, Center for Secure and Dependable Systems, University of Idaho
2. Common Criteria for Information Technology Security Evaluation (CCITSE) (1999). Available at http://www.radium.ncsc.mil/tpep/library/ccitse/ccitse.html

3. Common Criteria Testing Laboratory. Green Hills Software INTEGRITY-178B Security Target, Version 1.0, May 30, 2008. http://www.niap-ccevs.org/cc-scheme/st/st{_}vid10119-st. pdf
4. Green Hills Software, Inc. INTEGRITY real-time operating system. http://www.ghs.com/products/rtos/integrity.html
5. Greve D (2010) Information security modeling and analysis. In: Hardin D (ed) Design and verification of microprocessor systems for high-assurance applications. Springer, Berlin, pp 249–299
6. Greve D, Wilding M (2002) Dynamic data structures in ACL2: a challenge. Available at http://www.hokiepokie.org/docs/festival02.txt
7. Kaufmann M, Manolios P, Moore JS (2000) Computer-aided reasoning: an approach. Kluwer, Dordrecht
8. NIAP CCEVS. Validated Product – Green Hills Software INTEGRITY-178B Separation Kernel. http://www.niap-ccevs.org/cc-scheme/st/vid10119/index.cfm
9. Richards R, Greve D, Wilding M, Vanfleet M (2004) The common criteria, formal methods, and ACL2. In: Proceedings of ACL2'04, Austin, TX, November 2004
10. RTCA, Inc (1992) Software considerations in airborne systems and equipment certification, RTCA/DO-178B, December 1, 1992
11. Rushby J (1981) Design and verification of secure systems. In: Proceedings of the eighth symposium on operating systems principles, vol 15, December 1981
12. Wilding M, Greve D, Richards R, Hardin D (2010) Formal verification of partition management for the AAMP7G microprocessor. In: Hardin D (ed) Design and verification of microprocessor systems for high-assurance applications. Springer, Berlin, pp 175–191

Refinement in the Formal Verification
of the seL4 Microkernel

Gerwin Klein, Thomas Sewell, and Simon Winwood

1 Introduction

seL4, the subject of this verification, is an operating system (OS) microkernel. The OS kernel by definition is the part of the software that runs in the most privileged mode of the hardware. As such, it has full privileges to access and change all parts of the system. Therefore, any defect in the OS kernel is potentially fatal to the operation of the whole system, not just to isolated parts of it. One approach to reduce the risk of such bugs is the microkernel approach: to reduce the privileged kernel code to an absolute minimum. The remaining code base – 8,700 lines of C and 600 lines of assembly in the case of seL4 – is small enough to be amenable to formal verification on the implementation level. The L4.verified project has produced such an implementation proof for the C code of seL4. The overall proof comes to about 200,000 lines of proof script and roughly 10,000 intermediate lemmas.

The proof assumes correctness of compiler, assembly code, and hardware. It also assumes correct use of low-level hardware caches (memory caches and translation-look-aside buffer) and correctness of the boot code (about 1,200 lines of the 8,700). It formally derives everything else. The verified version of the seL4 kernel runs on the ARMv6 architecture and the Freescale i.MX31 platform.

This article gives an overview of the main proof technique and the proof framework that was used in this verification project: refinement.

The proof is not done in a refinement calculus that transforms the program in many small steps, but proceeds in two large refinement steps $\mathcal{R}_A$ and $\mathcal{R}_C$ instead. The three main specification artefacts in the proof are shown in Fig. 1. The top-level specification of kernel behaviour is an abstract, operational model in higher order logic. We call it $\mathcal{A}$ in the following. The intermediate specification $\mathcal{E}$ is an executable, detailed model of kernel behaviour that has been translated from a working prototype written in Haskell into Isabelle/HOL. The bottom layer $\mathcal{C}$ is the C program seL4, automatically parsed into Isabelle/HOL.

G. Klein (✉)
NICTA, Sydney, NSW, Australia
e-mail: gerwin.klein@nicta.com.au

D.S. Hardin (ed.), *Design and Verification of Microprocessor Systems
for High-Assurance Applications*, DOI 10.1007/978-1-4419-1539-9_11,
© Springer Science+Business Media, LLC 2010

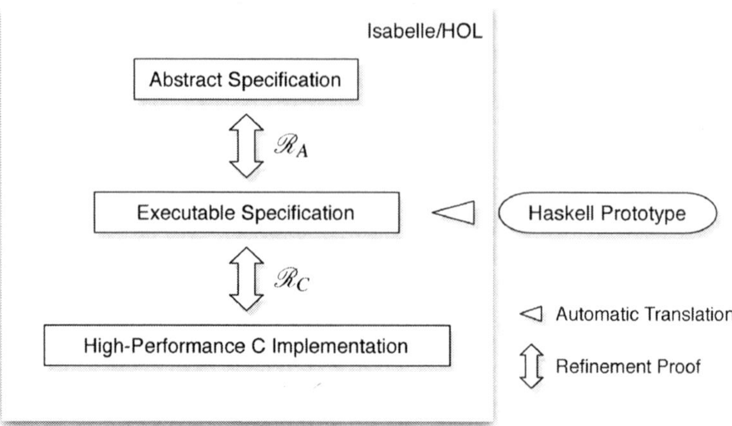

Fig. 1 Refinement steps in L4.verified

On the surface, these two large refinement proofs use different formalisms and connect different kinds of specification artefacts. Technical details on these two proofs have appeared elsewhere [4, 17, 23]. This article recalls some of these details and shows how they are put together into a common, general refinement framework that allows us to connect the results and extract the main overall theorem: the C code of seL4 correctly implements its abstract specification.

Section 2 shows the overall data refinement framework. Section 3 gives some example code on the monadic and C level. Section 4 summarises the refinement proof $\mathcal{R}_A$ and shows how it is mapped into the framework. Section 5 does the same for the C implementation proof $\mathcal{R}_C$.

2 Data Refinement

The ultimate objective of our effort is to prove refinement between an abstract and a concrete process. Following de Roever and Engelhardt [6], we define a process as a triple containing an initialisation function, which creates the process state with reference to some external state, a step function which reacts to an event, transforming the state, and a finalisation function which reconstructs the external state.

```
record process = Init :: 'external ⇒ 'state set
                 Step :: 'event ⇒ ('state × 'state) set
                 Fin :: 'state ⇒ 'external
```

The idea is that the external state is the one observable on the outside, about which one may formulate Hoare logic properties. A process may also contain hidden state to implement its data structures. In the simple case, the full state space of a component is just a pair of external and hidden states and the projection function Fin is just the canonical projection from pairs. With more complex processes, the projection function that extracts the observable state may become more complex as well.

The execution of a process may be non-deterministic, starting from a initial external state, resulting via a sequence of inputs in a *set* of external states:

$$\text{steps } \delta \; s \; events \quad\equiv \text{foldl } (\lambda states \; event. \; (\delta \; event) \; `` \; states) \; s \; events$$
$$\text{execution } A \; s \; events \equiv (\text{Fin } A) \; ` \; (\text{steps } (\text{Step } A) \; (\text{Init } A \; s) \; events)$$

where $R \; `` \; S$ and $f \; ` \; R$ are the images of the set S under the relation R and the function f, respectively.

Process A is refined by C, if with the same initial state and input events, execution of C yields a subset of the external states yielded by executing A:

$$A \sqsubseteq C \equiv \forall \; s \; events. \; \text{execution } C \; s \; events \subseteq \text{execution } A \; s \; events$$

This is the classic notion of refinement as reducing non-determinism. Note that it also includes data refinement: A and C may work on different internal state spaces; they merely both need to project to the same external state space.

A well-known property of refinement is that it is equivalent with the preservation of Hoare logic properties.

Lemma 1. $A \sqsubseteq C \quad iff \quad \forall \; P \; Q. \; A \vdash \{P\} \; events \; \{Q\} \longrightarrow C \vdash \{P\} \; events \; \{Q\},$

where $A \vdash \{P\} \; events \; \{Q\} \equiv \forall \; s \in P. \; \text{execution } A \; s \; events \subseteq Q$. The proof is by unfolding of definitions and basic set reasoning.

This means that once refinement is shown, it is enough to prove a Hoare logic property on the abstract level A for it to hold on the concrete level C. For this to be useful, the external state must be rich enough to represent the properties one is interested in.

2.1 Forward Simulation

Refinement is commonly proven by establishing forward simulation [6], of which it is a consequence. To demonstrate forward simulation we define a relation, SR, between the internal states of the two processes. We must show that the relation is established by Init, is maintained if we advance the systems in parallel, and implies equality of the final external states:

$$\text{fw-sim } SR \; C \; A \equiv (\forall \; s. \; \text{Init } C \; s \subseteq SR \; `` \; \text{Init } A \; s)$$
$$\wedge \; (\forall \; event. \; \text{Step } C \; event \; O \; SR \subseteq SR \; O \; \text{Step } A \; event)$$
$$\wedge \; (\forall \; s \; s'. \; (s, s') \in SR \longrightarrow \text{Fin } C \; s' = \text{Fin } A \; s),$$

where $T \; O \; S$ is the composition of relations S and T.

To prove forward simulation, it is often helpful to use additional facts about the execution of abstract or concrete level. If this information is available in the form of an invariant, it may be established separately and can then be easily integrated into the refinement proof. An invariant is any property which is always established by Init and preserved by Step. In the refinement proof, we may then assume it to be true at the commencement of all steps and before finalisation (Fig. 2).

Fig. 2 Forward simulation

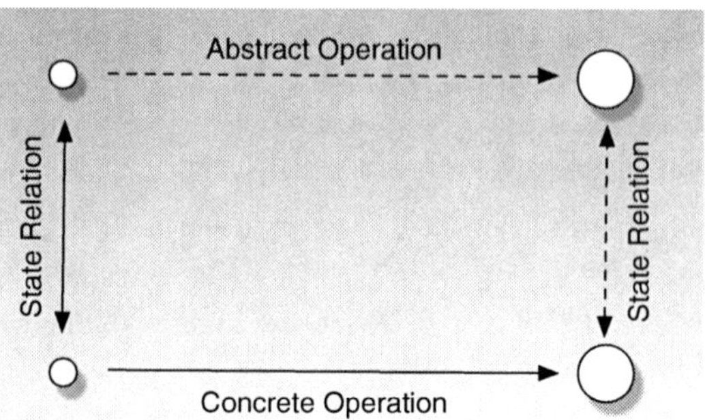

$$
\begin{aligned}
\text{invariant } I\ M \quad &\equiv (\forall s.\ \text{Init } M\ s \subseteq I) \wedge (\forall\ event.\ \text{Step } C\ event\ ``\ I \subseteq I) \\
\text{fw-sim}|SR\ C\ A\ I_c\ I_a \equiv\ &(\forall s.\ \text{Init } C\ s \subseteq SR\ ``\ \text{Init } A\ s) \\
&\wedge\ (\forall\ event.\ \text{Step } C\ event\ O\ (SR \cap (I_a \times I_c)) \subseteq SR\ O\ \text{Step } A\ event) \\
&\wedge\ (\forall\ s\ s'.\ (s, s') \in SR \wedge s \in I_a \wedge s' \in I_c \longrightarrow \text{Fin } C\ s' = \text{Fin } A\ s)
\end{aligned}
$$

The key theorems are, first, that forward simulation implies refinement and, second, that forward simulation assuming invariants implies forward simulation in general.

Lemma 2. fw-sim $SR\ C\ A \longrightarrow A \sqsubseteq C$

The proof is by unfolding definitions and induction on the event sequence in the refinement statement, followed by relation reasoning to apply forward simulation in the induction step.

Lemma 3. *Forward simulation assuming invariants implies forward simulation if the invariants are established separately.*

$$\text{fw-sim}|\ SR\ C\ A\ I_c\ I_a \wedge \text{invariant } I_a\ A \wedge \text{invariant } I_c\ C \longrightarrow \text{fw-sim } SR\ C\ A$$

This lemma is shown by basic relation and set reasoning after unfolding definitions.

2.2 Structure

The three processes we are interested in have a common structure in their Step operations. We model five kinds of events in our processes. The first two are transitions that do not involve the kernel: user thread execution and idle thread execution. We model the execution of user threads with unrestricted non-determinism, allowing all possible behaviours. We distinguish the idle thread as it may run in the kernel's context and thus must be better behaved. The next two kinds of events model the transition from user mode to kernel mode when exceptions occur: user mode exceptions and idle mode exceptions. The final event type is the one we are interested in: kernel execution. This is the only part of the Step operation that differs between our processes.

Formally, we model this in a function global-automaton that takes the kernel behaviour as a parameter and implements the above transitions generically. The kernel transition is:

global-automaton *kernel-call* KernelTransition $\equiv$
{ ((*s*, KernelMode, Some *e*), (*s'*, *m*, None)) |*s s' e m*. (*s,s',m*) $\in$ *kernel-call e*}

The parameter *kernel-call* is a relation between current and final kernel state and the next mode the machine is switched into (kernel mode, user mode, and idle mode). The state space of the process is a triple of the kernel-observed machine state, including memory and devices, a current mode, and a current kernel entry event. The latter is produced by the other transitions in the model. For instance, in idle mode, only an interrupt event can be generated:

global-automaton *kernel-call* IdleEventTransition
{ ((*s*, IdleMode, None), (*s*, KernelMode, Some *Interrupt*)) |*s*. True }

From user mode, any kernel entry event *e* is possible. The transition from user to kernel mode itself does not change the state; the context switch is modelled inside the kernel transition that comes after, because it is modelled differently at each abstraction level. The transition assumes no further conditions and does not depend on the parameter *kernel-call*.

global-automaton *kernel-call* UserEventTransition $\equiv$
{ ((*s*, UserMode, None), (*s*, KernelMode, Some *e*)) |*s e*. True}

The other transitions are analogous.

The definition of kernel execution may vary between our three processes, but they share a common aspect. Each is implemented through a call to the top-level kernel handler function from which a call graph proceeds in a structured language. Exploiting this structure is the key aspect of our approach.

2.3 *Correspondence*

Forward simulation, like most properties that can be expressed in a commuting diagram, composes sequentially. This composition over successive Step actions is important in the proof that forward simulation implies refinement. Sequential composition is also useful in proving refinement within a single kernel execution step.

The kernel execution bodies are, as discussed above, each written in a language which affords substantial internal structure. To exploit similarities in this structure, we define a new notion which we call correspondence. Correspondence is essentially forward simulation defined not on state transformers but on the terms of the languages in which the kernel execution bodies are defined. This leads us to define two different correspondence predicates for $\mathcal{R}_A$ and $\mathcal{R}_C$, which will be discussed in the following sections. It is crucial that these predicates be defined in a manner that allows the correspondence proofs to be composed across the syntactic composition operators of the relevant languages.

3 Example

The seL4 kernel [8] provides the following operating system kernel services: inter-process communication, threads, virtual memory, access control, and interrupt control. In this section, we present a typical function, cteMove, with which we will illustrate the two proof frameworks for refinement. Figure 3 shows the same function in the monadic executable specification and in the C implementation. The first refinement proof relates two monadic specifications; the second refinement proof relates the two layers shown in the figure.

Access control in seL4 is based on *capabilities*. A capability contains an object reference along with access rights. A *capability table entry* (CTE) is a kernel data structure with two fields: a capability and an *mdbNode*. The latter is book-keeping information and contains a pair of pointers which form a doubly linked list.

The cteMove operation, shown in Fig. 3, moves a CTE from *src* to *dest*.

The first six lines in Fig. 3 initialise the destination entry and clear the source entry; the remainder of the function updates the pointers in the doubly linked list. During the move, the capability in the entry may be diminished in access rights. Thus, the argument *cap* is this possibly diminished capability, previously retrieved from the entry at *src*.

In this example, the C source code is structurally similar to the executable specification. This similarity is not accidental: the executable specification describes the low-level design with a high degree of detail. Most of the kernel functions exhibit this property. It is also true, to a lesser degree, for the refinement between two monadic specifications. Even so, the implementation here makes a small optimisation: in the specification, updateMDB always checks that the given pointer is not NULL. In the implementation, this check is done for prev_ptr and next_ptr – which may be NULL – but omitted for srcSlot and destSlot. In verifying cteMove, we will have to prove that these checks are not required.

```
cteMove cap src dest ≡                 void cteMove (cap_t newCap,
do                                                  cte_t *srcSlot, cte_t *destSlot){
  cte ← getCTE src;                      mdb_node_t mdb; uint32_t prev_ptr, next_ptr;
  mdb ← return (cteMDBNode cte);         mdb = srcSlot->cteMDBNode;
  updateCap dest cap;                    destSlot->cap = newCap;
  updateCap src NullCap;                 srcSlot->cap = cap_null_cap_new();
  updateMDB dest (const mdb);            destSlot->cteMDBNode = mdb;
  updateMDB src (const nullMDBNode);     srcSlot->cteMDBNode = nullMDBNode;
                                         prev_ptr = mdb_node_get_mdbPrev(mdb);
  updateMDB                              if(prev_ptr) mdb_node_ptr_set_mdbNext(
    (mdbPrev mdb)                                        &CTE_PTR(prev_ptr)->cteMDBNode,
    (λm. m (| mdbNext := dest |));                       CTE_REF(destSlot));
                                         next_ptr = mdb_node_get_mdbNext(mdb);
  updateMDB                              if(next_ptr) mdb_node_ptr_set_mdbPrev(
    (mdbNext mdb)                                        &CTE_PTR(next_ptr)->cteMDBNode,
    (λm. m (| mdbPrev := dest |))                        CTE_REF(destSlot));
od                                     }
```

Fig. 3 cteMove: executable specification and C implementation

4 Monadic Refinement

4.1 Non-deterministic State Monads

The abstract and executable specifications over which $\mathcal{R}_A$ is proved are written in a monadic style inspired by Haskell. The type constructor $('a, 's)$ nd-monad is a non-deterministic state monad representing computations with a state type $'s$ and a return value type $'a$. Return values can be injected into the monad using the return :: $'a \Rightarrow ('a, 's)$ nd-monad operation. The composition operator bind :: $('a, 's)$ nd-monad $\Rightarrow ('a \Rightarrow ('b, 's)$ nd-monad$) \Rightarrow ('b, 's)$ nd-monad performs the first operation and makes the return value available to the second operation. These canonical operators form a monad over $('a, 's)$ nd-monad and satisfy the usual monadic laws. More details are given elsewhere [4]. The ubiquitous do ... od syntax seen in Sect. 3 is syntactic sugar for a sequence of operations composed using bind.

The type $('a, 's)$ nd-monad is isomorphic to $'s \Rightarrow ('a \times 's)$ set $\times$ bool. This can be thought of as a non-deterministic state transformer (mapping from states to sets of states) extended with a return value (required to form a monad) and a boolean failure flag. The flag is set by the fail :: $('a, 's)$ nd-monad operation to indicate unrecoverable errors in a manner that is always propagated and not confused by non-determinism. The destructors mResults and mFailed access, respectively, the set of outcomes and the failure flag of a monadic operation evaluated at a state.

Exception handling is introduced by using a return value in the sum type. An alternative composition operator op $>>=E$ :: $('e + 'a, 's)$ nd-monad $\Rightarrow ('a \Rightarrow ('e + 'b, 's)$ nd-monad$) \Rightarrow ('e + 'b, 's)$ nd-monad inspects the return value, executing the subsequent operation for normal (right) return values and skipping it for exceptional (left) ones. There is an alternative return operator *returnOk* and these form an alternative monad. Exceptions are thrown with *throwError* and caught with catch.

We define a Hoare triple denoted $\{\!P\!\}\ a\ \{\!R\!\}$ on a monadic operator a, precondition P and postcondition Q. We have a verification condition generator (VCG) for such Hoare triples, which are used extensively both to establish invariants and to make use of them in correspondence proofs.

4.2 Correspondence

The components of our monadic specifications are similar to the non-deterministic state transformers on which forward simulation is defined. To extend to a correspondence framework, we must determine how to handle the return values and failure flags. This is accomplished by the corres predicate. It captures forward simulation between a component monadic computation C, and its abstract counterpart A, with SR instantiated to our standard state relation state-relation. It takes three additional parameters: R is a predicate which will relate abstract and concrete return values, and the preconditions P and P' restrict the input states, allowing use of information such as global invariants:

corres R P P' A C $\equiv$ $\forall$ $(s, s') \in$ state-relation. $P\,s \wedge P'\,s' \longrightarrow$
  $(\forall (r', t') \in$ mResults $(C\,s')$. $\exists (r, t) \in$ mResults $(A\,s)$. $(t, t') \in$ state-relation $\wedge R\,r\,r')$
  $\wedge$ $(\neg$mFailed $(C\,s'))$

Note that the outcome of the monadic computation is a pair of result and failure flag. The last conjunct of the **corres** statement mandates non-failure for C.

The key property of **corres** is that it decomposes over the **bind** constructor through the CORRES-SPLIT rule.

CORRES-SPLIT:

$$\frac{\text{corres } R'\ P\ P'\ A\ C \qquad \forall r\,r'.\, R'\,r\,r' \longrightarrow \text{corres } R\ (S\,r)\ (S'\,r')\ (B\,r)\ (D\,r') \qquad \{Q\}\ A\ \{S\} \qquad \{Q'\}\ C\ \{S'\}}{\text{corres } R\ (P \text{ and } Q)\ (P' \text{ and } Q')\ (A >>= B)\ (C >>= D)}$$

This splitting rule decomposes the problem into four subproblems. The first two are **corres** predicates relating the subcomputations. Two Hoare triples are also required. This is because the input states of the subcomputations appearing in the second subproblem are intermediate states, not input states, of the original problem. Any preconditions assumed in solving the second subproblem must be shown to hold at the intermediate states by proving a Hoare triple over the partial computation. Use of Hoare triples to demonstrate intermediate conditions is both a strength and a weakness of this approach. In some cases, the result is repetition of existing invariant proofs. However, in the majority of cases, this approach makes the flexibility and automation of the VCG available in demonstrating preconditions that are useful as assumptions in proofs of the **corres** predicate.

The decision to mandate non-failure for concrete elements and not abstract ones is pragmatic. Proving non-failure on either system could be done independently; however, the preconditions needed are usually the same as in **corres** proofs and it is convenient to solve two problems simultaneously. Unfortunately we cannot so easily prove abstract non-failure. Because the concrete specification may be more deterministic than the abstract one, there is no guarantee that we will examine all possible failure paths. In particular, if a conjunct mandating abstract non-failure was added to the definition of **corres**, the splitting rule above would not be provable.

Similar splitting rules exist for other common monadic constructs including **bindE**, **catch**, and conditional expressions. There are terminating rules for the elementary monadic functions, for example:

CORRES-RETURN:

$$\frac{R\,a\,b}{\text{corres } R\ \top\ \top\ (\text{return } a)\ (\text{return } b)}$$

The **corres** predicate also has a weakening rule, similar to the Hoare Logic.

CORRES-PRECOND-WEAKEN:

$$\frac{\text{corres } R\ Q\ Q'\ A\ C \qquad \forall s.\, P\,s \longrightarrow Q\,s \qquad \forall s.\, P'\,s \longrightarrow Q'\,s}{\text{corres } R\ P\ P'\ A\ C}$$

Proofs of the corres property take a common form: first the definitions of the terms under analysis are unfolded and the CORRES-PRECOND-WEAKEN rule is applied. As with the VCG, this allows the syntactic construction of a precondition to suit the proof. The various splitting rules are used to decompose the problem, in some cases with carefully chosen return value relations. Existing results are then used to solve the component corres problems. Some of these existing results, such as CORRES-RETURN, require compatibility properties on their parameters. These are typically established using information from previous return value relations. The VCG eliminates the Hoare triples, bringing preconditions assumed in corres properties at later points back to preconditions on the starting states. Finally, as in Dijkstra's postcondition propagation [7], the precondition used must be proved to be a consequence of the one that was originally assumed.

4.3 Mapping to Processes

To prove $\mathcal{R}_A$, we must connect the corres framework described above to the forward simulation property we wish to establish. The Step actions of the processes we are interested in are equal for all events other than kernel executions, and simulation is trivial to prove for equal operations. In the abstract process $\mathcal{A}$, kernel execution is defined in the monadic function call-kernel. The semantics of the whole abstract process $\mathcal{A}$ are then derived by using call-kernel in the call to global-automaton. The context switch is modelled by explicitly changing all user accessible parts, for instance the registers of the current thread, fully non-deterministically. The semantics of the intermediate process for the executable specification $\mathcal{E}$ are derived similarly from a monadic operation callKernel. These two top-level operators satisfy a correspondence theorem KERNEL-CORRES:

$$\forall \textit{event. corres } (\lambda rv\ rv'.\ \text{True})\ \textit{invs invs'}\ (\text{call-kernel } \textit{event})\ (\text{callKernel } \textit{event})$$

The required forward simulation property for kernel execution (assuming the system invariants) is implied by this correspondence rule. Invariant preservation for the system invariants follows similarly from Hoare triples proved over the top-level monadic operations:

$$\forall \textit{event. } \{\!|\textit{invs}|\!\}\ \text{call-kernel } \textit{event}\ \{\!|\lambda\text{-}.\ \textit{invs}|\!\}$$
$$\forall \textit{event. } \{\!|\textit{invs'}|\!\}\ \text{callKernel } \textit{event}\ \{\!|\lambda\text{-}.\ \textit{invs'}|\!\}$$

From these facts, we may thus conclude that $\mathcal{R}_A$ holds:

Theorem 1. *The executable specification refines the abstract one.*

$$\mathcal{A} \sqsubseteq \mathcal{E}$$

5 C Refinement

In this section, we describe our infrastructure for parsing C into Isabelle/HOL and for reasoning about the result.

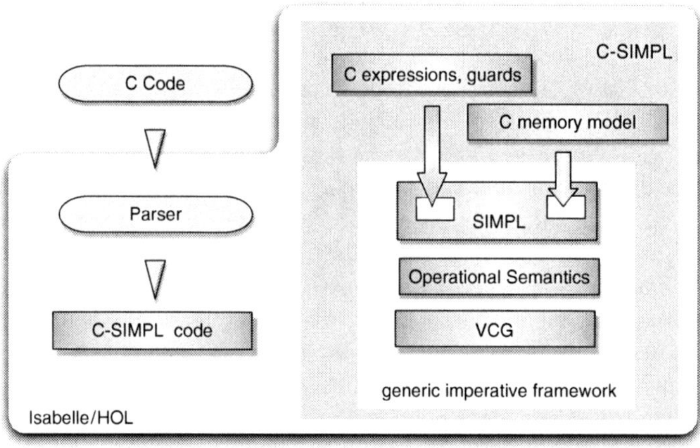

Fig. 4 C language framework

The seL4 kernel is implemented almost entirely in C99 [15]. Direct hardware accesses are encapsulated in machine interface functions, some of which are implemented in ARMv6 assembly. In the verification, we axiomatise the assembly functions using Hoare triples.

Figure 4 gives an overview of the components involved in importing the kernel into Isabelle/HOL. The right-hand side shows our instantiation of SIMPL [18], a generic, imperative language inside Isabelle. The SIMPL framework provides a program representation, a semantics, and a VCG. This language is generic in its expressions and state space. We instantiate both components to form C-SIMPL, with a precise C memory model and C expressions, generated by a parser. The left-hand side of Fig. 4 shows this process: the parser takes a C program and produces a C-SIMPL program.

SIMPL provides a data type and semantics for statement forms; expressions are shallowly embedded. Along with the usual constructors for conditional statements and iteration, SIMPL includes statements of the form Guard F P c which raises the fault F if the condition P is false and executes c otherwise.

Program states in SIMPL are represented by Isabelle records containing a field for each local variable in the program and a field *globals* containing all global variables and the heap. Variables are then simply functions on the state.

SIMPL semantics are represented by judgements of the form $\Gamma \vdash \langle c,x \rangle \Rightarrow x'$, which means that executing statement c in state x terminates and results in state x'; the parameter Γ maps function names to function bodies. These states include both the program state and control flow information, including that for abruptly terminating **THROW** statements used to implement the C statements `return`, `break`, and `continue`.

The SIMPL environment also provides a VCG for partial correctness triples; Hoare-triples are represented by judgements of the form $\Gamma \vdash_{/F} P\ c\ C,A$, where P is the precondition, C is the postcondition for normal termination, A is the postcondition for abrupt termination, and F is the set of ignored faults. If F is *UNIV*, the universal set, then all **Guard** statements are effectively ignored. Both A and F may be omitted if empty.

Our C subset allows type-unsafe operations including casts. To achieve this soundly, the underlying heap model is a function from addresses to bytes. This allows, for example, the C function `memset`, which sets each byte in a region of the heap to a given value. We generally use a more abstract interface to this heap: we use additional typing information to lift the heap into functions from typed pointers to Isabelle terms; see Tuch et al. [20, 21] for more detail.

The C parser takes C source files and generates the corresponding C-SIMPL terms, along with Hoare-triples describing the set of variables mutated by the functions. Although our C subset does not include union types, we have a tool which generates data types and manipulation functions which implement tagged unions via C structures and casts [3]. The tool also generates proofs of Hoare-triples describing the operations.

5.1 Refinement Calculus for C

Refinement phase $\mathcal{R}_C$ involves proving refinement between the executable specification and the C implementation. Specifically, this means showing that the C kernel entry points for interrupts, page faults, exceptions, and system calls refine the executable specification's top-level function **callKernel**.

As with $\mathcal{R}_A$, we introduce a new correspondence notion that implies forward simulation. We again aim to divide the proof along the syntactic structure of both programs as far as possible and then prove the resulting subgoals semantically.

In the following, we first give our definition of correspondence, followed by a discussion of the use of the VCG. We then describe techniques for reusing proofs from $\mathcal{R}_A$ to solve proof obligations from the implementation. Next, we present our approach for handling operations with no corresponding analogue. Finally, we describe our splitting approach and sketch the proof of the example.

5.2 The Correspondence Statement

As with the correspondence statement for $\mathcal{R}_A$, we deal with state preconditions and return values by including guards on the states and a return value relation in the $\mathcal{R}_C$ correspondence statement. In addition, we include an extra parameter used for dealing with early returns and breaks from loops, namely a list of statements called a *handler stack*.

We thus extend the semantics to lists of statements, writing $\Gamma \Vdash \langle c{\cdot}hs, s \rangle \Rightarrow x'$. The statement sequence *hs* is a handler stack; it collects the **CATCH** handlers which

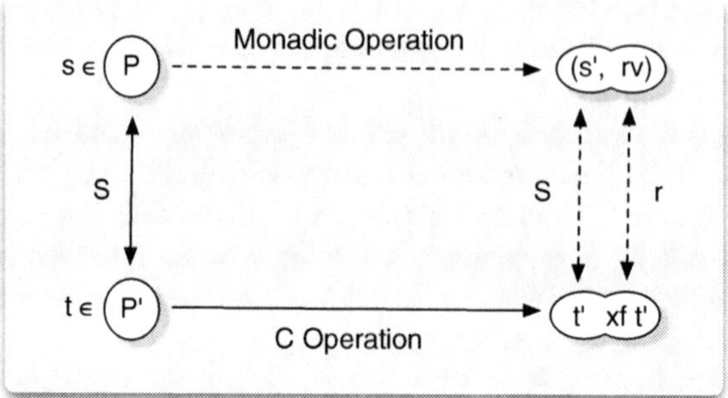

Fig. 5 Correspondence

surround usages of the statements `return`, `continue`, and `break`. If c terminates abruptly, each statement in hs is executed in sequence until one terminates normally.

Relating the return values of functions is dealt with by annotating the correspondence statement with a *return value relation r*. Although evaluating a monadic operation results in both a new state and a return value, functions in C-SIMPL return values by updating a function-specific local variable; because local variables are fields in the state record, this is a function from the state. We thus annotate the correspondence statement with an *extraction function xf*, a function which extracts the return value from a program state.

The correspondence statement is illustrated in Fig. 5 and defined below

$$\text{ccorres } r \, xf \, P \, P' \, hs \, a \, c \equiv$$
$$\forall (s, t) \in \mathcal{S}. \, \forall t'. \, s \in P \wedge t \in P' \wedge \neg \, \text{mFailed} \, (a \, s) \wedge \Gamma \Vdash \langle c \cdot hs, t \rangle \Rightarrow t'$$
$$\longrightarrow \exists (s', rv) \in \text{mResults} \, (a \, s).$$
$$\exists t'_N. \, t' = \text{Normal} \, t'_N \wedge (s', t'_N) \in \mathcal{S} \wedge r \, rv \, (xf \, t'_N)$$

The definition can be read as follows: given related states s and t with the preconditions P and P', respectively, if the abstract specification a does not fail when evaluated at state s, and the concrete statement c evaluates under handler stack hs in extended state t to extended state t', then the following must hold:

1. Evaluating a at state s returns some value rv and new abstract state s'.
2. The result of the evaluation of c is some normal (non-abrupt) state $\text{Normal} \, t'_N$.
3. States s' and t'_N are related by the state relation $\mathcal{S}$.
4. Values rv and $xf \, t'_N$ – the extraction function applied to the final state of c – are related by r, the given return value relation.

Note that a is non-deterministic: we may pick any suitable rv and s'. As mentioned in Sect. 4.2, the proof of $\mathcal{R}_A$ entails that the executable specification does not fail. Thus, in the definition of ccorres, we may assume $\neg \, \text{mFailed} \, (a \, s)$. In practice, this means assertions and other conditions for (non-)failure in the executable specification become known facts in the proof. Of course, these facts are only free because we have already proven them in $\mathcal{R}_A$.

5.3 *Proving Correspondence via the VCG*

Data refinement predicates can, in general [6], be rephrased and solved as Hoare triples. We do this in our framework by using the VCG after applying the following rule:

$$\frac{\forall s. \ \Gamma \vdash \{t \mid s \in P \land t \in P' \land (s, t) \in \mathcal{S}\} \quad c \quad \{t' \mid \exists (rv, s') \in \mathsf{mResults}\ (a\ s).\ (s', t') \in \mathcal{S} \land r\ rv\ (xf\ t')\}}{\mathsf{ccorres}\ r\ xf\ P\ P'\ hs\ a\ c}$$

In essence, this rule states that to show correspondence between a and c, for a given initial specification state s, it is sufficient to show that executing c results in normal termination where the final state is related to the result of evaluating a at s. The VCG precondition can assume that the initial states are related and satisfy the correspondence preconditions.

Use of this rule in verifying correspondence is limited by two factors. First, the verification conditions produced by the VCG may be excessively large or complex. Our experience is that the output of a VCG step usually contains a separate term for every possible path through the target code and that the complexity of these terms tends to increase with the path length. Second, the specification return value and result state are existential and thus outside the range of our extensive automatic support for showing universal properties of specification fragments. Fully expanding the specification is always possible, and in the case of deterministic operations it will yield a single state/return value pair, but the resulting term structure may also be large.

5.4 *Splitting*

As with $\mathcal{R}_A$, we prove correspondence by splitting the proof into corresponding program lines. Splitting allows us to take advantage of structural similarity by considering each match in isolation; formally, given the specification fragment do $rv \leftarrow a;\ b\ rv$ od and the implementation fragment $c;\ d$, splitting entails proving a first correspondence between a and c and a second between b and d.

In the case where we can prove that c terminates abruptly, we discard d. Otherwise, the following rule is used:

$$\frac{\begin{array}{c} \mathsf{ccorres}\ r'\ xf'\ P\ P'\ hs\ a\ c \\ \forall v.\ d'\ v \sim d[v/xf'] \qquad \forall\ rv\ rv'.\ r'\ rv\ rv' \longrightarrow \mathsf{ccorres}\ r\ xf\ (Q\ rv)\ (Q'\ rv\ rv')\ hs\ (b\ rv)\ (d'\ rv') \\ \{\!\!\{R\}\!\!\}\ a\ \{\!\!\{Q\}\!\!\} \qquad \Gamma \vdash_{/\mathcal{U}} R'\ c\ \{s \mid \forall\ rv.\ r'\ rv\ (xf'\ s) \longrightarrow s \in Q'\ rv\ (xf'\ s)\} \end{array}}{\mathsf{ccorres}\ r\ xf\ (P \cap R)\ (P' \cap R')\ hs\ (a >\!\!>= b)\ (c;\ d)}$$

In the second correspondence premise, d' is the result of lifting xf' in d; this enables the proof of the second correspondence to use the result relation from the first correspondence. To calculate the final preconditions, the rule includes VCG premises to move the preconditions from the second correspondence across a and c.

In the C-SIMPL VCG obligation, we may ignore any guard faults as their absence is implied by the first premise. In fact, in most cases the C-SIMPL VCG step can be omitted altogether, because the postcondition collapses to the universal set after simplifications.

We have developed a tactic which assists in splitting: C-SIMPL's encoding of function calls and struct member updates requires multiple specialised rules. The tactic symbolically executes and moves any guards if required, determines the correct splitting rule to use, instantiates the extraction function, and lifts the second correspondence premise.

5.5 Mapping to Processes

We map the C kernel into a process by lifting the operational semantics of the kernel C code into a non-deterministic monad:

$$\text{exec-C } \Gamma\ c \equiv \lambda s.\ (\{()\} \times \{s'\ |\ \Gamma \vdash \langle c, \mathsf{Normal}\ s\rangle \Rightarrow \mathsf{Normal}\ s'\}, \mathsf{False})$$

that is, for a given statement c we construct a function from an initial state s into the set of states resulting from evaluating c at s. We define the return value of this execution as the unit. We set the failure flag to **False** and require a successful **Normal** result from C.

We then construct a function **callKernel-C**, parametrised by the input event, which simulates the hardware exception dispatch mechanism. The function examines the argument and dispatches the event to the corresponding kernel entry point. Finally, we form the process *ADT-C* by instantiating the global automaton with this step function.

We again establish a correspondence result between the kernel entry points, this time between **callKernel** in $\mathcal{E}$ and **callKernel-C** in C. This time, we did not need to prove additional invariants about the concrete level (the C program). The framework presented above enabled us to shift all such reasoning to the level of the executable specification $\mathcal{E}$.

Theorem 2. *The translated C code refines its executable specification.*

$$\mathcal{E} \sqsubseteq \mathcal{C}$$

6 Main Theorem

Putting the two theorems from the previous sections together, we arrive via transitivity of refinement at the main functional correctness theorem.

Theorem 3. $\mathcal{A} \sqsubseteq \mathcal{C}$

7 Related Work

We briefly summarise related work on OS verification; a comprehensive overview is provided by Klein [16].

Early work on OS verification includes PSOS [9] and UCLA Secure Unix [22]. Later, KIT [2] describes verification of process isolation properties down to object code level, but for an idealised kernel they are much simpler than modern microkernels.

The VFiasco project [13] and later the Robin project [19] attempted to verify C++ kernel implementations. They created a precise model of a large, relevant part of C++, but did not verify substantial parts of the kernel.

Heitmeyer et al. [12] report on the verification and Common Criteria certification of a "software-based embedded device" featuring a small (3,000 LOC) separation kernel. Similarly, Green Hills' Integrity kernel [11] recently underwent formal verification during a Common Criteria EAL6+ certification [10]. The Separation Kernel Protection Profile [14] of Common Criteria demands data separation only rather than functional correctness.

A closely related project is Verisoft [1], which is attempting to verify not only the OS, but also a whole software stack from verified hardware up to verified application programs. This includes a formally verified, non-optimising compiler for a Pascal-like implementation language. While Verisoft accepts a simplified (but verified) hardware platform and two orders of magnitude slow-down for the simplified VAMOS kernel, we deal with real C and standard tool chains on ARMv6 and have aimed for a commercially deployable, realistic microkernel. A successor project, Verisoft XT, is aiming to verify the functional correctness of the Microsoft Hypervisor, which contains concurrency and is substantially larger than seL4. While initial progress has been made on this verification [5], it is unclear at this stage if the goal will be reached.

8 Conclusion

We have presented the different refinement techniques used in the verification of the seL4 microkernel. We have given an overview of the overall unifying framework, of the refinement calculus used for stateful, monadic specification, of the refinement calculus for imperative programs, and we have shown how these are put together into the final theorem.

The two frameworks presented here have withstood the test of large-scale application to high-performance C code in the Isabelle/HOL verification of the seL4 microkernel. Proving functional correctness for real-world application down to the implementation level is possible and feasible.

Acknowledgments We thank the other current and former members of the L4.verified and seL4 teams: David Cock, Tim Bourke, June Andronick, Michael Norrish, Jia Meng, Catherine Menon,

Jeremy Dawson, Harvey Tuch, Rafal Kolanski, David Tsai, Andrew Boyton, Kai Engelhardt, Kevin Elphinstone, Philip Derrin, and Dhammika Elkaduwe for their contributions to this verification.

NICTA is funded by the Australian Government as represented by the Department of Broadband, Communications and the Digital Economy and the Australian Research Council through the ICT Centre of Excellence program.

References

1. Alkassar E, Hillebrand M, Leinenbach D, Schirmer N, Starostin A, Tsyban A (2009) Balancing the load – leveraging a semantics stack for systems verification. J Autom Reason 42(2–4): 389–454
2. Bevier WR (1989) Kit: a study in operating system verification. IEEE Trans Softw Eng 15(11):1382–1396
3. Cock D (2008) Bitfields and tagged unions in C: verification through automatic generation. In: Beckert B, Klein G (eds) VERIFY'08, vol 372 of CEUR workshop proceedings, Aug 2008, pp 44–55
4. Cock D, Klein G, Sewell T (2008) Secure microkernels, state monads and scalable refinement. In: Mohamed OA, Muñoz C, Tahar S (eds) 21st TPHOLs, vol 5170 of LNCS. Springer, Berlin, pp 167–182
5. Cohen E, Dahlweid M, Hillebrand M, Leinenbach D, Moskal M, Santen T, Schulte W, Tobies S (2009) VCC: a practical system for verifying concurrent C. In: Theorem proving in higher order logics (TPHOLs 2009), vol 5674 of Lecture notes in computer science, Munich, Germany. Springer, Berlin, pp 23–42
6. de Roever W-P, Engelhardt K (1998) Data refinement: model-oriented proof methods and their comparison. In: Cambridge tracts in theoretical computer science, vol 47. Cambridge University Press, Cambridge
7. Dijkstra EW (1975) Guarded commands, nondeterminacy and formal derivation of programs. Commun ACM 18(8):453–457
8. Elphinstone K, Klein G, Derrin P, Roscoe T, Heiser G (2007) Towards a practical, verified kernel. In: Proceedings of 11th workshop on hot topics in operating systems, San Diego, CA, USA, pp 117–122
9. Feiertag RJ, Neumann PG (1979) The foundations of a provably secure operating system (PSOS). In: AFIPS conference proceedings, 1979 National computer conference, New York, NY, USA, June 1979, pp 329–334
10. Green Hills Software, Inc. (2008) INTEGRITY-178B separation kernel security target version 1.0. http://www.niap-ccevs.org/cc-scheme/st/st_vid10119-st.pdf
11. Green Hills Software, Inc. (2008) Integrity real-time operating system. http://www.ghs.com/products/rtos/integrity.html
12. Heitmeyer CL, Archer M, Leonard EI, McLean J (2006) Formal specification and verification of data separation in a separation kernel for an embedded system. In: CCS '06: proceedings of 13th conference on computer and communications security. ACM, New York, NY, pp 346–355
13. Hohmuth M, Tews H (2005) The VFiasco approach for a verified operating system. In: 2nd PLOS, July 2005
14. Information Assurance Directorate (2007) U.S. government protection profile for separation kernels in environments requiring high robustness, June 2007. Version 1.03. http://www.niap-ccevs.org/cc-scheme/pp/pp.cfm/id/pp_skpp_hr_v1.03/
15. ISO/IEC (2005) Programming languages – C. In: Technical report 9899:TC2, ISO/IEC JTC1/SC22/WG14, May 2005
16. Klein G (2009). Operating system verification – an overview. *Sādhanā* 34(1):27–69
17. Klein G, Elphinstone K, Heiser G, Andronick J, Cock D, Derrin P, Elkaduwe D, Engelhardt K, Kolanski R, Norrish M, Sewell T, Tuch H, Winwood S (2009) seL4: Formal verification of an

OS kernel. In: Proceedings of 22th SOSP, Big Sky, MT, USA, October 2009. ACM, New York, NY, pp 207–220

18. Schirmer N (2006) *Verification of sequential imperative programs in Isabelle/HOL*. PhD thesis, Technische Universität München
19. Tews H, Weber T, Völp M (2008) A formal model of memory peculiarities for the verification of low-level operating-system code. In: Huuck R, Klein G, Schlich B (eds) Proceedings of 3rd international workshop on systems software verification (SSV'08), vol 217 of ENTCS. Elsevier, Amsterdam, pp 79–96
20. Tuch H (2009) Formal verification of C systems code: structured types, separation logic and theorem proving. J Autom Reason (special issue on operating system verification) 42(2–4):125–187
21. Tuch H, Klein G, Norrish M (2007) Types, bytes, and separation logic. In: Hofmann M, Felleisen M (eds) Proceedings of 34th ACM SIGPLAN-SIGACT symposium on principles of programming languages, Nice, France. ACM, New York, NY, pp 97–108
22. Walker B, Kemmerer R, Popek G (1980) Specification and verification of the UCLA unix security kernel. Commun ACM 23(2):118–131
23. Winwood S, Klein G, Sewell T, Andronick J, Cock D, Norrish M (2009) Mind the gap: a verification framework for low-level C. In: Berghofer S, Nipkow T, Urban C, Wenzel M (eds) Proceedings of TPHOls'09, vol 5674 of LNCS, Munich, Germany, August 2009. Springer, Berlin, pp 500–515

Specification and Checking of Software Contracts for Conditional Information Flow

Torben Amtoft, John Hatcliff, Edwin Rodríguez, Robby, Jonathan Hoag, and David Greve

1 Introduction

National and international infrastructures as well as commercial services are increasingly relying on complex distributed systems that share information with multiple levels of security (MLS). These systems often seek to coalesce information with mixed security levels into information streams that are targeted to particular clients. For example, in a national emergency response system, some data will be privileged (e.g., information regarding availability of military assets and deployment orders for those assets) and some data will be public (e.g., weather and mapping information). In such systems, there is a huge tension between providing aggressive information flow in order to gain operational advantage while preventing the flow of information to unauthorized parties. Specifying and verifying security policies and providing end-to-end guarantees in this context are exceedingly difficult tasks.

The multiple independent levels of security (MILS) architecture [32] proposes to make development, accreditation, and deployment of MLS-capable systems more practical, achievable, and affordable by providing a certified infrastructure *foundation for systems that require assured information sharing*. In the MILS architecture, systems are developed on top of (a) a "separation kernel," a concept introduced by Rushby [27] which guarantees isolation and controlled communication between application components deployed in different virtual "partitions" supported by the kernel and (b) MILS middleware services such as "high assurance guards" that allow information to flow between various partitions, and between trusted and untrusted segments of a network, only when certain *conditions* are satisfied.

Researchers at Rockwell Collins Advanced Technology Center (RC ATC) are industry leaders in certifying MILS components according to standards such as the Common Criteria (EAL 6/7) that mandate the use of formal methods. For example,

T. Amtoft (✉)
Kansas State University, Manhattan, KS, USA
e-mail: tamtoft@cis.ksu.edu

D.S. Hardin (ed.), *Design and Verification of Microprocessor Systems for High-Assurance Applications*, DOI 10.1007/978-1-4419-1539-9_12,
© Springer Science+Business Media, LLC 2010

RC ATC engineers carried out the certification of the hardware-based separation kernel in RC's AAMP7 processor (this was the first such certification of a MILS separation kernel and it formed the initial draft of the Common Criteria Protection Profile for Separation Kernels) as well as the software-based kernel in the Green Hills Integrity 178B RTOS.

Seeking to leverage the groundbreaking work on the AAMP7 separation kernel, Rockwell Collins product groups (that include 200+ developers) are building several different information assurance products on top of the AAMP7 following the MILS architecture. These products are programmed using the SPARK subset of Ada [8]. One of the primary motivating factors for the use of SPARK is that it includes annotations (formal contracts for procedure interfaces) for specifying and checking information flow [12]. The use of these annotations plays a key role in the certification cases for the products. The SPARK language and associated tool-set is the only commercial product that we know of which can support checking of code-level information flow contracts, and SPARK provides a number of well-designed and effective capabilities for specifying and verifying properties of implementations of safety-critical and security-critical systems.

However, Rockwell Collins developers are struggling to provide precise arguments for correctness in information assurance certification due to several limitations of the SPARK information flow framework. A key limitation is that SPARK information flow annotations are unconditional (e.g., they capture such statements as "executing procedure P may cause information to flow from input variable X to output variable Y"), but MILS security policies are often conditional (e.g., data from partition A is only allowed to flow to partition B when state variables G_1 and G_2 satisfy certain conditions). Thus, SPARK currently can neither capture nor support verification of critical aspects of MILS policies (treating such conditional flows as unconditional flows in SPARK is an overapproximation that leads to many false alarms).

In previous work, Banerjee and the first author have developed Hoare logics that enable compositional reasoning about information flow [1,4]. Inspired by challenge problems from Rockwell Collins, this logic was extended to support conditional information flow [3]. While the logic as presented in [3] exposed some foundational issues, it only supported intraprocedural analysis, it required developers to specify information flow loop invariants, the verification algorithm was not yet fully implemented (and thus no experience was reported), and the core logic was not mapped to a practical method contract language capable of supporting compositional reasoning in industrial settings.

In this paper, we address these limitations by describing how the logic can provide a foundation for a practical information flow contract language capable of supporting compositional reasoning about conditional information flows. The specific contributions of our work are as follows:

- We propose an extension to SPARK's information flow contract language that supports conditional information flow, and we describe how the logic of [3] can be used to provide a semantics for the resulting framework.
- We extend the algorithm of [3] to support procedure calls and thus modularity.

- We present a strategy for automatically inferring conditional information flow invariants for while loops, thus significantly reducing developers' annotation burden.
- We provide an implementation that can automatically mine conditional information flow contracts (which might then be checked against existing contracts) from source code.
- We report on experiments applying the implementation to a collection of examples.

Recent efforts for certifying MILS separation kernels [18, 19] applied ACL2 [22] or PVS [25] theorem provers to formal models; extensive inspections were then required by certification authorities to establish the correspondence between model and source code. Because our approach is directly integrated with code, it complements these earlier efforts by (a) removing the "trust gaps" associated with inspecting behavioral models (built manually) and (b) allowing many verification obligations to be discharged earlier in the life cycle by developers while leaving only the most complicated obligations to certification teams. In addition, the logic-based approach presented in this paper provides a foundation for producing independently auditable and machine-checkable *evidence* of correctness and MILS policy compliance as recommended [21] by the National Research Council's Committee on Certifiably Dependable Software Systems.

2 Example

Figure 1 illustrates the conceptual information flows in a fragment of an idealized MILS infrastructure component used by Rockwell Collins engineers to demonstrate, for NSA and industry representatives, specification and verification of information flow issues in MILS components running on top of the AAMP7 separation kernel. This demonstration was the first iteration of what is now a much more sophisticated high assurance network guard product line at Rockwell Collins. The "Mailbox" component in the center of the diagram mediates communication between two client

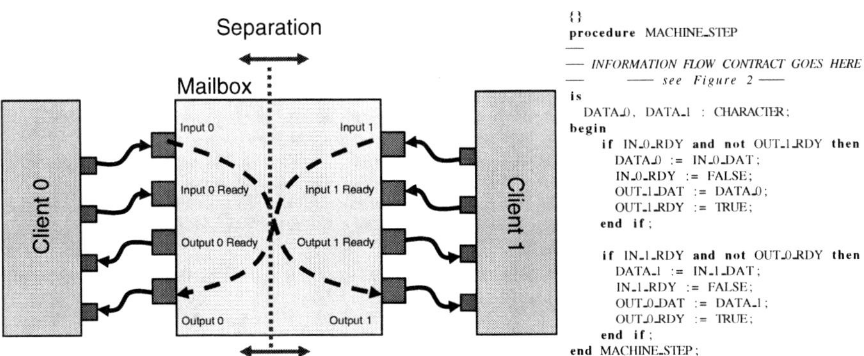


```
{}
procedure MACHINE_STEP

--
-- INFORMATION FLOW CONTRACT GOES HERE
--            -- see  Figure 2 --
is
   DATA_0, DATA_1 : CHARACTER;
begin
    if IN_0_RDY and not OUT_1_RDY then
       DATA_0 := IN_0_DAT;
       IN_0_RDY := FALSE;
       OUT_1_DAT := DATA_0;
       OUT_1_RDY := TRUE;
    end if;

    if IN_1_RDY and not OUT_0_RDY then
       DATA_1 := IN_1_DAT;
       IN_1_RDY := FALSE;
       OUT_0_DAT := DATA_1;
       OUT_0_RDY := TRUE;
    end if;
end MACHINE_STEP;
```

Fig. 1 Simple MILS Guard - mailbox mediates communication between partitions

344
                                                                T. Amtoft et al.

processes – each running on its own partition in the separation kernel. *Client 0* writes data to communicate in the memory segment *Input 0* that is shared between *Client 0* and the mailbox, then it sets the *Input 0 Ready* flag. The mailbox process polls its ready flags; when it finds that, e.g., *Input 0 Ready* is set and *Output 1 Ready* is cleared (indicating that *Client 1* has already consumed data deposited in the *Output 1* slot in a previous communication), then it copies the data from *Input 0* to *Output 1* and clears *Input 0 Ready* and sets *Output 1 Ready*. The communication from *Client 1* to *Client 0* follows a symmetric set of steps. The actions to be taken in each execution frame are encoded in SPARK Ada by the MACHINE_STEP procedure shown in Fig. 1.

Figure 2a shows SPARK Ada annotations for the MACHINE_STEP procedure, whose information flow properties are captured by derives annotations. It requires that each parameter and each global variable referenced by the procedure be classified as in (read only), out (written and initial values [values at the point of procedure call] are unread), or in out (written and initial values read). For a procedure P, variables annotated as in or in out are called *input variables* and denoted as IN_P, while variables annotated as out or in out are *output variables* and denoted as OUT_P. Each output variable x_o must have a derives annotation indicating the input variables whose initial values are used to directly or indirectly calculate the final value of x_o. One can also think of each derives clause as expressing a dependence relation (or program slice) between an output variable and the input variables that it transitively depends on (via both data and control dependence).

For example, the second derives clause specifies that on each MACHINE_STEP execution the output value of OUT_1_DAT is possibly determined by the input values of several variables: from IN_0_DAT when the Mailbox forward data supplied by *Client 0*, from OUT_1_DAT when the conditions on the ready flags are not satisfied (OUT_1_DAT's output value then is its input value), and from OUT_1_RDY and IN_0_RDY because these variables *control* whether or not data flows from *Client 0* on a particular machine step (i.e., they *guard* the flow).

a

```
--# global in out IN_0_RDY,  IN_1_RDY,
--#                OUT_0_RDY, OUT_1_RDY,
--#                OUT_0_DAT, OUT_1_DAT;
--#          in    IN_0_DAT,  IN_1_DAT;
--# derives
--#    OUT_0_DAT from IN_1_DAT, OUT_0_DAT,
--#                   OUT_0_RDY, IN_1_RDY &
--#    OUT_1_DAT from IN_0_DAT, OUT_1_DAT,
--#                   IN_0_RDY, OUT_1_RDY &
--#    IN_0_RDY  from IN_0_RDY, OUT_1_RDY &
--#    IN_1_RDY  from INP_1_RDY, OUT_0_RDY &
--#    OUT_0_RDY from OUT_0_RDY, IN_1_RDY &
--#    OUT_1_RDY from OUT_1_RDY, IN_0_RDY;
```

b

```
--# derives
--# OUT_0_DAT from
--#    IN_1_DAT when
--#       (IN_1_RDY and not OUT_0_RDY),
--#    OUT_0_DAT when
--#       (not IN_1_RDY or OUT_0_RDY),
--#    OUT_0_RDY, IN_1_RDY &
--# OUT_1_DAT from
--#    IN_0_DAT when
--#       (IN_0_RDY and not OUT_1_RDY),
--#    OUT_1_DAT when
--#       (not IN_0_RDY or OUT_1_RDY),
--#    OUT_1_RDY, IN_0_RDY
```

Fig. 2 (**a**) SPARK information flow contract for Mailbox example. (**b**) Fragment of same example with proposed conditional information flow extensions

While upper levels of the MILS architecture require reasoning about lattices of security levels (e.g., *unclassified, secret, top secret*), the policies of infrastructure components such as separation kernels and guard applications usually focus on data separation policies (reasoning about flows between components of program state), and we restrict ourselves to such reasoning in this paper.

No other commercial language framework provides automatically checkable information flow specifications, so the use of the information flow checking framework in SPARK is a significant step forward. As illustrated above, SPARK `derives` clauses can be used to specify flows of information from input variables to output variables, but they do not have enough expressive power to state that information only flows under specific conditions. For example, in the Mailbox code, information from `IN_0_DAT` only flows to `OUT_1_DAT` when the flag `IN_0_RDY` is set and `OUT_1_READY` is cleared, otherwise `OUT_1_DAT` remains unchanged. In other words, the flags `IN_0_RDY` and `OUT_1_RDY` *guard* the flow of information through the mailbox. Unfortunately, the SPARK `derives` cannot distinguish the flag variables as guards nor phrase the conditions under which the guards allow information to pass or be blocked. This means that guarding logic, which is central to many MLS applications including those developed at Rockwell Collins, *is completely absent from the checkable specifications* in SPARK.

In general, the lack of ability to express *conditional* information flow not only inhibits automatic verification of guarding logic specifications, but also results in imprecision which cascades and builds throughout the specifications in the application.

3 Foundations of SPARK Conditional Information Flow

The SPARK subset of Ada is designed for programming and verifying high assurance applications such as avionics applications certified to DO-178B Level A. It deliberately omits constructs that are difficult to reason about such as dynamically created data, pointers, and exceptions. In Fig. 3, we present the syntax of a simple imperative language with assertions that one can consider to be an idealized version of SPARK. We omit some features of SPARK that do not present conceptual challenges, such as records, and the package and inheritance structure.

Referring to Fig. 3, we consider three kind of expressions ($E \in$ **Exp**): arithmetic ($A \in$ **AExp**), boolean ($B \in$ **BExp**), and array expressions ($H \in$ **HExp**). We use x, y to range over scalar variables, h to range over array variables, and w, z to range over both kind of variables; actual variables appearing in programs are depicted using typewriter font. We also use c to range over integer constants, p to range over named (parameterless) procedures, op to range over arithmetic operators in $\{+, \times, \mathsf{mod}, \ldots\}$, and bop to range over comparison operators in $\{=, <, \ldots\}$.

The use of programmer assertions is optional, but often helps to improve the precision of our analysis. For example, a loop **while** B **do** S **od**, which is known to have invariant ϕ, may be transformed into **while** B **do assert**$(\phi \wedge B)$; S **od**;

Expressions	Commands	
arithmetic	$S ::=$ **skip**	
$A ::= x \mid c \mid A \text{ op } A$	$\mid \;\; x := A$	assignment
boolean	$\mid \;\; S \, ; S$	sequential composition
$B ::= A \text{ bop } A$	$\mid \;\;$ **assert**(ϕ)	programmer asssertion
	$\mid \;\;$ **call** p	procedure call
Assertions	$\mid \;\;$ **if** B **then** S **else** S	conditional
$\phi ::= B \mid \phi \wedge \phi$	$\mid \;\;$ **while** B **do** S **od**	iteration
$\mid \;\; \phi \vee \phi \mid \neg \phi$		

Fig. 3 Syntax of a simple imperative language

assert$(\phi \wedge \neg B)$. We refer to the assertions of Fig. 3 as *1-assertions* since they represent predicates on a single program state; they can be contrasted with *2-assertions* that we introduce later for reasoning about information flow in terms of a pair of program states.

Using parameterless procedures simplifies our exposition; our implementation supports procedures with parameters (there are no conceptual challenges in this extended functionality).

The SPARK information flow analysis treats arrays as atomic entities – a conservative approximation that makes analysis significantly easier. In the SPARK analysis, any information flowing to/from a particular array element is treated as flowing to/from *all* elements of the array. Due to the lack of heap-allocated data in SPARK, complex data structures are often implemented in arrays. For information assurance applications, the SPARK treatment of information flow for arrays can significantly impede the ability to verify interesting end-to-end information flow properties. Elsewhere [6], we describe how our logic provides a basis for reasoning about individual elements of arrays, thus giving more precision than SPARK.

For an expression E, we write fv(E) for the variables occurring free in E, which is the union of the free scalar variables fsv(E) and the free array variables fav(E), and write $E[A/x]$ for the result of substituting in E all occurrences of x by A. We use similar notations for assertions ϕ, where as usual we define $\phi_1 \rightarrow \phi_2$ as $\neg\phi_1 \vee \phi_2$, and also define *true* as $0 = 0$ and *false* as $0 = 1$.

3.1 Semantics

The semantics of an arithmetic expression $[\![A]\!]$ is a function from stores into values, where a value ($v \in Val$) is an integer n and where a *store* $s \in Store$ maps variables to values; we write $dom(s)$ for the domain of s and write $[s|x \mapsto v]$ for the store that is like s except that it maps x into v. Similarly, $[\![B]\!]_s$ denotes a boolean.

Figure 4 summarizes the semantics of commands. A command transforms the store into another store; hence its semantics is given in relational style, in the form $s\,[\![S]\!]\,s'$. For some S and s, there may not exist any s' such that $s\,[\![S]\!]\,s'$; this can

$$s \; [\![\textbf{skip}]\!] \; s' \text{ iff } s' = s$$

$$s \; [\![x := A]\!] \; s' \text{ iff } \exists v : v = [\![A]\!]_s \text{ and } s' = [s \mid x \mapsto v]$$

$$s \; [\![S_1 \, ; \, S_2]\!] \; s' \text{ iff } \exists s'' : s \; [\![S_1]\!] \; s'' \text{ and } s'' \; [\![S_2]\!] \; s'$$

$$s \; [\![\textbf{assert}(\phi)]\!] \; s' \text{ iff } s \models \phi \text{ and } s' = s$$

$$s \; [\![\textbf{call } p]\!] \; s' \text{ iff } s \; P(p) \; s'$$

$$s \; [\![\textbf{if } B \textbf{ then } S_1 \textbf{ else } S_2]\!] \; s' \text{ iff } ([\![B]\!]_s = \mathsf{True} \text{ and } s \; [\![S_1]\!] \; s')$$
$$\text{or } ([\![B]\!]_s = \mathsf{False} \text{ and } s \; [\![S_2]\!] \; s')$$

$$s \; [\![\textbf{while } B \textbf{ do } S \textbf{ od}]\!] \; s' \text{ iff } \exists i \geq 0 : s \; f_i \; s' \text{ where } f_i \text{ is inductively defined by:}$$
$$s \; f_0 \; s' \text{ iff } [\![B]\!]_s = \mathsf{False} \text{ and } s' = s$$
$$s \; f_{i+1} \; s' \text{ iff } \exists s'' : ([\![B]\!]_s = \mathsf{True} \text{ and }$$
$$s \; [\![S]\!] \; s'' \text{ and } s'' \; f_i \; s')$$

Fig. 4   Command semantics

happen if a **while** loop does not terminate, or an **assert** fails. We assume an implicit global procedure environment P that for each p returns a relation between input and output stores.

Assertions ϕ are also called 1-assertions since they represent predicates on a single program state; we write $s \models \phi$ to denote that ϕ holds in s following the standard semantics. We write $\phi \rhd_1 \phi'$ if whenever $s \models \phi$ also $s \models \phi'$. As usual we define $\phi_1 \rightarrow \phi_2$ as $\neg\phi_1 \vee \phi_2$; we also define *true* as $0 = 0$ and *false* as $0 = 1$.

3.2   Reasoning About Information Flow in Terms of Noninterference

MILS seeks to prevent security breaches that can occur via unauthorized/unintended information flow from one partition to another; thus previous certification efforts for MILS components have among the core requirements included the classical property of *non-interference* [17] which (in this setting) states: for every pair of runs of a program, if the runs agree on the initial values of one partition's data (but may disagree on the data of other partitions) then the runs also agree on the final values of that partition's data.

3.3   Capturing Noninterference and Secure Information Flow in a Compositional Logic

The logic developed in [1] was designed to verify specifications of the following form: *given two runs of P that initially agree on variables $x_1, \ldots, x_n$, the runs agree on variables $y_1, \ldots, y_m$ at the end of the runs.* This includes noninterference as a special case, as can be seen by letting $x_1, \ldots, x_n$, and $y_1, \ldots, y_m$, be the variables of one partition. We may express such a specification, which makes the "end-to-end" (input to output) aspect of verifying confidentiality explicit, in Hoare-logic style as

$$\{x_1\bowtie, \ldots, x_n\bowtie\}\ P\ \{y_1\bowtie, \ldots, y_m\bowtie\},$$

where the *agreement assertion* $x\bowtie$ is satisfied by a *pair* of states, s_1 and s_2, if $s_1(x) = s_2(x)$. With P the example program from Sect. 2, we would have, e.g.,

$$\{\texttt{INP\_1\_DAT}\bowtie, \texttt{OUT\_0\_DAT}\bowtie, \texttt{INP\_1\_RDY}\bowtie, \texttt{OUT\_0\_RDY}\bowtie\}\ P\ \{\texttt{OUT\_0\_DAT}\bowtie\}.$$

To capture conditional information flow, recent work [3] by Banerjee and the first author introduced *conditional* agreement assertions, also called *2-assertions*. They are of the form $\phi \Rightarrow E\bowtie$ which is satisfied by a pair of stores if either at least one of them does not satisfy ϕ, or they agree on the value of E:

$$s\ \&\ s_1 \models \phi \Rightarrow E\bowtie \text{ iff whenever } s \models \phi \text{ and } s_1 \models \phi \text{ then } [\![E]\!]_s = [\![E]\!]_{s_1}.$$

We use $\theta \in \mathbf{2Assert}$ to range over 2-assertions. For $\theta = (\phi \Rightarrow E\bowtie)$, we call ϕ the antecedent of θ and write $\phi = ant(\theta)$, and we call E the consequent of θ and write $E = con(\theta)$. We often write $E\bowtie$ for $true \Rightarrow E\bowtie$. We use $\Theta \in \mathcal{P}(\mathbf{2Assert})$ to range over sets of 2-assertions (where we often write θ for the singleton set $\{\theta\}$), with conjunction implicit. Thus, $s\&s_1 \models \Theta$ iff $\forall \theta \in \Theta : s\&s_1 \models \theta$.

Figure 5 depicts a simple derivation using conditional information flow assertions that answers the question: what is the source of information flowing into variable $\texttt{OUT\_0\_DAT}$? The natural way to read the derivation is from the bottom up (since our algorithm works "backward"). Thus, for $\texttt{OUT\_0\_DAT}\bowtie$ to hold after execution of P, we must have $\texttt{DATA\_1}\bowtie$ before line 3 (since data flows from $\texttt{DATA\_1}$ to $\texttt{OUT\_0\_DAT}$), $\texttt{INP\_1\_DAT}\bowtie$ before line 2 (since data flows from $\texttt{INP\_1\_DAT}$ to $\texttt{DATA\_1}$), and finally $\texttt{INP\_1\_RDY}\bowtie$ and $\texttt{OUT\_0\_RDY}\bowtie$ (since they *control* which branch of the condition is taken), along with conditional assertions. The precondition shows that the value of $\texttt{OUT\_0\_DAT}$ depends *unconditionally* on $\texttt{INP\_1\_RDY}$ and $\texttt{OUT\_0\_RDY}$, and *conditionally* on $\texttt{INP\_1\_DAT}$ and $\texttt{OUT\_0\_DAT}$, just as we would expect.

```
     { INP_1_RDY ∧ ¬OUT_0_RDY ⟹ INP_1_DAT⋈,
       ¬INP_1_RDY ∨ OUT_0_RDY ⟹ OUT_0_DAT⋈,
       INP_1_RDY⋈, OUT_0_RDY⋈ }
1. if INP_1_RDY and not OUT_0_RDY then
       { INP_1_DAT⋈ }
2.     DATA_1 := INP_1_DAT; INP_1_RDY := false;
       { DATA_1⋈ }
3.     OUT_0_DAT := DATA_1; OUT_0_RDY := true;
       { OUT_0_DAT⋈ }
4. fi
       { OUT_0_DAT⋈ }
```

Fig. 5 A derivation for the mailbox example, illustrating the handling of conditionals

3.4 Relations Between Agreement Assertions

We define $\Theta \rhd_2 \Theta'$, pronounced "Θ 2-implies Θ'," to hold iff for all s, s_1: whenever $s\&s_1 \models \Theta$ then also $s\&s_1 \models \Theta'$. In development terms, when $\Theta \rhd_2 \Theta'$ holds we can think of Θ as a *refinement* of Θ' and Θ' an *abstraction* of Θ. For example, $\{x\bowtie, y\bowtie\}$ refines $x\bowtie$ by adding an (unconditional) agreement assertion, and $y < 10 \Rightarrow x\bowtie$ refines $y < 7 \Rightarrow x\bowtie$ by weakening the antecedent of a 2-assertion. In general, logical implication on **1Assert** conditions in agreement assertions is related in a contravariant manner to logical implication in agreement assertions; this is a special case (with $E = E_0$) of the following result:

Lemma 1. *Assume that* (a) $\phi \rhd_1 \phi_0$, *and* (b) *whenever* $s \models \phi$ *and* $s_1 \models \phi$ *and* $[\![E_0]\!]_s = [\![E_0]\!]_{s_1}$ *then also* $[\![E]\!]_s = [\![E]\!]_{s_1}$. *Then* $(\phi_0 \Rightarrow E_0\bowtie) \rhd_2 (\phi \Rightarrow E\bowtie)$.

Proof. Assuming (i) $s\&s_1 \models \phi_0 \Rightarrow E_0\bowtie$ and (ii) $s \models \phi$ and $s_1 \models \phi$, we must prove $[\![E]\!]_s = [\![E]\!]_{s_1}$. By (a), we infer from (ii) that $s \models \phi_0$ and $s_1 \models \phi_0$, which by (i) implies $[\![E_0]\!]_s = [\![E_0]\!]_{s_1}$. By (b), we get the desired $[\![E]\!]_s = [\![E]\!]_{s_1}$. $\square$

We define a function *decomp* that converts arbitrary 2-assertions into assertions with only variables as consequents: $decomp(\Theta)=\{\phi \Rightarrow x\bowtie \,|\, \phi \Rightarrow E\bowtie \in \Theta, x \in \mathrm{fv}(E)\}$. For example, $decomp(\phi \Rightarrow (x+y)\bowtie)=\{\phi \Rightarrow x\bowtie, \phi \Rightarrow y\bowtie\}$.

Fact 2 *For all* Θ, *decomp*(Θ) *is a refinement of* Θ.

The converse does not hold, with a counterexample being $s\&s_1 \models (x+y)\bowtie$ but not $s\&s_1 \models x\bowtie$ or $s\&s_1 \models y\bowtie$, as when $s(x) = s_1(y) = 3, s(y) = s_1(x) = 7$.

4 Conditional Information Flow Contracts

4.1 Foundations of Flow Contracts

The syntax of a SPARK `derives` annotation for a procedure P (as illustrated in Fig. 2a) can be represented formally as a relation $\mathcal{D}_P$ between OUT_P and $\mathcal{P}(\mathrm{IN}_P)$. A particular clause $\mathsf{derives}(z, \bar{w}) \in \mathcal{D}_P$ declares that the final value of output variable z depends on the input values of variables $\bar{w} = w_1, \ldots, w_k$. The correctness of such a clause as a contract for P can be expressed in terms of the logic of the preceding section, as requiring the triple $\{\bar{w}\bowtie\}\, S\, \{z\bowtie\}$ where S is the body of procedure P and where $\bar{w}\bowtie$ is a shorthand for $\{w_1\bowtie, \ldots, w_k\bowtie\}$.

Because $\mathcal{D}_P$ contains multiple clauses (one for each output variable of P), it captures multiple "channels" of information flow through P. Therefore, we cannot simply describe the semantics of a multiclause derives contract $\{\mathsf{derives}(z, \bar{w}), \mathsf{derives}(x, \bar{y})\}$ as $\{(\bar{w}\bar{y})\bowtie\}\, S\, \{z\bowtie, x\bowtie\}$ because this would confuse the dependencies associated with z and x, i.e., it would allow z to depend on $\bar{y}$. Accordingly, the full semantics of SPARK derives contracts is supported by what we term a *multichannel version* of the logic which is extended to include *indexed agreement assertions* $z\bowtie_c$ indexed by a channel identifier c – which one can usually associate with a particular output variable. In the multichannel logic, the confused triple above can now be correctly stated as $\{\bar{w}\bowtie_z, \bar{y}\bowtie_x\}\ S\ \{z\bowtie_z, x\bowtie_x\}$. (Alternatively, we could have *two* single-channel triples: $\{\bar{y}\bowtie\}\ S\ \{x\bowtie\}$ and $\{\bar{w}\bowtie\}\ S\ \{z\bowtie\}$.) The algorithm to be given in Sect. 5 extends to the multichannel version of the logic in a straightforward manner; hence the implementation described subsequently supports the multichannel version of the logic. For notational simplicity, we continue the discussion of the semantics of contracts using the single-channel version of the logic.

We now give a more convenient notation for triples of the form $\{\Theta\}P\{\Theta'\}$. This will provide a formal interpretation for method contracts that capture conditions of flows from beginning to end of a method P. A flow judgment κ is of the form $\Theta \rightsquigarrow \Theta'$, with Θ the precondition and with Θ' the postcondition. We say that $\Theta \rightsquigarrow \Theta'$ is valid for command S, written $S \models \Theta \rightsquigarrow \Theta'$, if whenever $s_1 \& s_2 \models \Theta$ and $s_1[\![S]\!]s_1'$ and $s_2[\![S]\!]s_2'$ then also $s_1' \& s_2' \models \Theta'$ (if the 2-assertions in the precondition hold for input states s_1 and s_2, the postcondition must also hold for associated output states s_1' and s_2').

4.2 Language Design for Conditional SPARK Contracts

The logic of the preceding section is potentially much more powerful than what we actually want to expose to developers – instead, we view it as a "core calculus" in which information flow reasoning is expressed. Our design goals that determine how much of the power of the logic we wish to expose to developers in enhanced SPARK conditional information flow contracts are (1) the effort required to write the contracts should be as simple as possible, (2) the contracts should be able to capture common idioms of MILS information guarding, (3) the contract checking framework should be compositional so as to support MILS goals, and (4) there should be a natural progression (e.g., via formal refinements) from unconditional `derives` statements to conditional statements.

4.2.1 Simplifying Assertions

The agreement assertions from the logic of Sect. 3 have the form $\phi \Rightarrow E\bowtie$. Here E is an arbitrary expression (not necessarily a variable), whereas SPARK `derives` statements are phrased in terms of IN/OUT variables only. We believe that including arbitrary expressions in SPARK conditional `derives` statements would add significant complexity for developers, and our experimental studies have shown that little increase in precision would be gained by such an approach. Instead, we retain the use of expression-based assertions $\phi \Rightarrow E\bowtie$ only during intermediate(automated) steps of the analysis. Appealing to Fact 2, we have a canonical way of strengthening, at procedure boundaries, $\phi \Rightarrow E\bowtie$ to $\phi \Rightarrow w_1\bowtie, \ldots, \phi \Rightarrow w_k\bowtie$ where $\text{fv}(E) = \{w_1, \ldots, w_k\}$. A second simplification relates to the fact that the core logic allows both pre- and postconditions to be conditional (e.g., $\{\phi_1 \Rightarrow E_1\bowtie\}$ P $\{\phi_2 \Rightarrow E_2\bowtie\}$ where ϕ_1 and ϕ_2 may differ.) Based on discussions with developers at Rockwell Collins and initial experiments, we believe that this would expose too much power/complexity to developers leading to unwieldy contracts and confusion about the underlying semantics. Accordingly, we are currently pursuing an approach in which only preconditions can be conditional. Combining these two simplifications, SPARK `derives` clauses are extended to allow conditions on input variables as follows:

```
derives x from y₁ when φ₁,
        ...,
        yₖ when φₖ
```

Here $\phi_1, \ldots, \phi_k$ are boolean expressions on the prestate of the associated procedure P. Thus, the above specification can be read as "The value of variable x at the conclusion of executing P (for *any* final state s') is derived from those y_j where ϕ_j holds in the prestate s from which s' is computed." Additional syntactic sugar can be introduced to simplify the contract notation, e.g., when input variables are conditioned (guarded) using the same expression. Figure 2b shows how this can be used to specify conditional flows for procedure MACHINE_STEP in Fig. 1.

4.2.2 Design Methodology Separating Guard Logic from Flow Logic

The lack of conditional assertions in postconditions has the potential to introduce imprecision. Yet, we believe that the above approach to conditional expressions can be effective for the following reason: we have observed that information assurance application design tends to factor out the *guarding logic* (i.e., the pieces of state and associated state changes that determine *when* information can flow) from the code which propagates information. This follows a common pattern in embedded systems in which the control logic is often factored out from data computation logic.

This informal design strategy can be firmed up and presented as an effective design methodology: some procedures act to modify the conditions under which information flows (the guard logic) while other procedures actually realize the flows. This could be enhanced by an explicit declaration of the *guard state*, i.e., the program variables that can be observed by guards. Guard state variables would be modified in guard logic procedures, but not be modified in any procedure that declares conditions based on those guards. SPARK's existing IN/OUT variable annotations can capture this requirement (i.e., we could require that no variable appearing in a condition can be a member of OUT).

4.2.3 Contract Abstraction and Refinement

For a practical design and development methodology, it is important to consider notions of contract abstraction (generalization) and refinement – ideally, conditional contracts should be a refinement of unconditional contracts. For example, we believe it will be easier to introduce conditional contracts into workflows if developers can (1) make a rough cut at specifying information flows without conditions and (2) systematically refine to produce conditional contracts. In addition, in situations where developers have trouble capturing flow policies, they can state flows without conditions and expert verification engineers can later refine those into conditional contracts. Conversely, it is important for managers to understand that they are not locked into our emerging technology; if they decide not to pursue a verification approach based on conditional SPARK contracts, they can safely abstract all conditional contracts back to unconditional contracts.

We now establish the desired notion of contract refinement (in terms of the general underlying calculus instead of its limited exposure in SPARK), by defining a relation between flow judgments: $\kappa_1 \rhd_\kappa \kappa_2$, pronounced "$\kappa_1$ refines κ_2," to hold iff for all commands S, whenever $S \models \kappa_1$ then also $S \models \kappa_2$.

To gain the proper intuition about contract refinement, it is important to note that the refinement relation is contravariant in the precondition and covariant in the postcondition: given $\kappa_1 \equiv \Theta_1 \rightsquigarrow \Theta_1'$ and $\kappa_2 \equiv \Theta_2 \rightsquigarrow \Theta_2'$, if $\Theta_2 \rhd_2 \Theta_1$ and $\Theta_1' \rhd_2 \Theta_2'$ then $\kappa_1 \rhd_\kappa \kappa_2$. For example, $x\bowtie \rightsquigarrow y\bowtie \rhd_\kappa x\bowtie, y\bowtie \rightsquigarrow y\bowtie$ holds because $x\bowtie, y\bowtie \rhd_2 x\bowtie$ (Sect. 3). Intuitively, this captures the fact that a contract can always be *abstracted* to a weaker one by stating that the output variables may depend on additional input variables. This illustrates that our contracts capture "may" dependence modalities: output y *may* depend on both inputs x and y, but a refinement $x\bowtie \rightsquigarrow y\bowtie$ shows that output y need not depend on input y (the contract before refinement is an *over-approximation* of dependence information). Also, we have $(y < 7 \Rightarrow x\bowtie \rightsquigarrow z\bowtie) \rhd_\kappa (x\bowtie \rightsquigarrow z\bowtie)$, which realizes our design goals of achieving (a) a formal refinement by adding conditions to a contract and (b) a formal (safe) abstraction by removing conditions.

5 A Precondition Generation Algorithm

We define in Fig. 6 an algorithm Pre for inferring preconditions from postconditions. We write $\{\Theta\}\,(R) \Longleftarrow S\,\{\Theta'\}$ when, given command S and postcondition Θ', Pre returns a precondition Θ for S that is designed so as to be sufficient to establish Θ' and a relation R that associates each 2-assertion $\theta \in \Theta'$ with the 2-assertions in Θ needed to establish θ. R captures dependences between variables before and after the execution of S, and it also supports reasoning about multiple channels of information flow as discussed in Sect. 4.1, e.g., if $\{y_1 y_2 \bowtie_x, y_1 y_3 \bowtie_z\}\,S\,\{x\bowtie_x, z\bowtie_z\}$, then R will relate y_1 to x and to z, y_2 to x, and y_3 to z. More precisely, we have $R \subseteq \Theta \times \{m, u\} \times \Theta'$, where tags m,u are mnemonics for "modified"

$\{\Theta\}\,(R) \Longleftarrow \mathbf{skip}\,\{\Theta'\}$ iff $R = \{(\theta, u, \theta) \mid \theta \in \Theta'\}$ and $\Theta = \Theta'$

$\{\Theta\}\,(R) \Longleftarrow \mathbf{assert}(\phi_0)\,\{\Theta'\}$ iff $R = \{((\phi \wedge \phi_0) \Rightarrow E\bowtie, u, \phi \Rightarrow E\bowtie) \mid \phi \Rightarrow E\bowtie \in \Theta'\}$ and $\Theta = dom(R)$

$\{\Theta\}\,(R) \Longleftarrow x := A\,\{\Theta'\}$ iff $R = \{(\phi[A/x] \Rightarrow E[A/x]\bowtie, \gamma, \phi \Rightarrow E\bowtie) \mid \phi \Rightarrow E\bowtie \in \Theta'\}$,
where $\gamma = m$ iff $x \in \mathrm{fv}(E)$ and $\Theta = dom(R)$

$\{\Theta\}\,(R) \Longleftarrow S_1\,;\,S_2\,\{\Theta'\}$ iff $\{\Theta''\}\,(R_2) \Longleftarrow S_2\,\{\Theta'\}$ and $\{\Theta\}\,(R_1) \Longleftarrow S_1\,\{\Theta''\}$
and $R = \{(\theta, \gamma, \theta') \mid \exists \theta'', \gamma_1, \gamma_2 : (\theta, \gamma_1, \theta'') \in R_1,\ (\theta'', \gamma_2, \theta') \in R_2\}$, where $\gamma = m$ iff $\gamma_1 = m$ or $\gamma_2 = m$

$\{\Theta\}\,(R) \Longleftarrow \mathbf{if}\ B\ \mathbf{then}\ S_1\ \mathbf{else}\ S_2\,\{\Theta'\}$
iff $\{\Theta_1\}\,(R_1) \Longleftarrow S_1\,\{\Theta'\},\ \{\Theta_2\}\,(R_2) \Longleftarrow S_2\,\{\Theta'\},\ R = R_1' \cup R_2' \cup R_0' \cup R_0$, and $\Theta = dom(R)$,
where $R_1' = \{((\phi_1 \wedge B) \Rightarrow E_1\bowtie, m, \theta') \mid \theta' \in \Theta_m',\ (\phi_1 \Rightarrow E_1\bowtie, \_, \theta') \in R_1\}$
and $R_2' = \{((\phi_2 \wedge \neg B) \Rightarrow E_2\bowtie, m, \theta') \mid \theta' \in \Theta_m',\ (\phi_2 \Rightarrow E_2\bowtie, \_, \theta') \in R_2\}$
and $R_0' = \{(((\phi_1 \wedge B) \vee (\phi_2 \wedge \neg B)) \Rightarrow B\bowtie, m, \theta')$
$$\mid \theta' \in \Theta_m',\ (\phi_1 \Rightarrow E_1\bowtie, \_, \theta') \in R_1, (\phi_2 \Rightarrow E_2\bowtie, \_, \theta') \in R_2\}$$
and $R_0 = \{(((\phi_1 \wedge B) \vee (\phi_2 \wedge \neg B)) \Rightarrow E\bowtie, u, \theta')$
$$\mid \theta' \in \Theta_u',\ (\phi_1 \Rightarrow E\bowtie, u, \theta') \in R_1, (\phi_2 \Rightarrow E\bowtie, u, \theta') \in R_2\}$$
and $\Theta_m' = \{\theta' \in \Theta' \mid \exists(\_, m, \theta') \in R_1 \cup R_2\}$ and $\Theta_u' = \Theta' \setminus \Theta_m'$

$\{\Theta\}\,(R) \Longleftarrow \mathbf{call}\ p\,\{\Theta'\}$
iff $R = R_u \cup R_0 \cup R_m$ and $\Theta = dom(R)$,
where $R_u = \{(rm^+_{\mathrm{OUT}_p}(\phi) \Rightarrow E\bowtie, u, \phi \Rightarrow E\bowtie) \mid (\phi \Rightarrow E\bowtie) \in \Theta' \wedge \mathrm{fv}(E) \cap \mathrm{OUT}_p = \emptyset\}$
and $R_0 = \{(rm^+_{\mathrm{OUT}_p}(\phi) \Rightarrow w\bowtie, m, \phi \Rightarrow E\bowtie)$
$$\mid (\phi \Rightarrow E\bowtie) \in \Theta' \wedge \mathrm{fv}(E) \cap \mathrm{OUT}_p \neq \emptyset \wedge w \in \mathrm{fv}(E) \wedge w \notin \mathrm{OUT}_p\}$$
and $R_m = \{(rm^+_{\mathrm{OUT}_p}(\phi) \wedge \phi_w^z \Rightarrow z\bowtie, m, \phi \Rightarrow E\bowtie)$
$$\mid (\phi \Rightarrow E\bowtie) \in \Theta' \wedge w \in \mathrm{fv}(E) \cap \mathrm{OUT}_p \wedge \phi_w^z \Rightarrow z\bowtie \text{ among preconditions for } w\bowtie \text{ in } p\text{'s summary}\}$$

$\{\Theta\}\,(R) \Longleftarrow \mathbf{while}\ B\ \mathbf{do}\ S_0\ \mathbf{od}\,\{\Theta'\}$
iff $R = R_u \cup R_m$ and $\Theta = dom(R)$,
where for each $w \in X$ (with X the variables "involved") we inductively in i define $\phi_w^i, \Theta^i, R^i, \psi_w^i$ by
$\phi_w^0 = \bigvee\{\phi \mid \exists E : (\phi \Rightarrow E\bowtie) \in \Theta' \wedge w \in \mathrm{fv}(E)\},\ \Theta^i = \{\phi_w^i \Rightarrow w\bowtie \mid w \in X\},\ \{\_\}\,(R^i) \Longleftarrow S_0\,\{\Theta^i\}$
$\psi_w^i = \bigvee\{\phi \mid \exists(\phi \Rightarrow E\bowtie, \_, \_) \in R^i \text{ with } w \in \mathrm{fv}(E)$
$$\text{or } w \in \mathrm{fv}(B) \text{ and } \exists(\theta, m, \theta') \in R^i \text{ with } \phi = ant(\theta) \text{ or } \phi = ant(\theta')\}$$
$\phi_w^{i+1} = \mathbf{if}\ \psi_w^i \rhd_1 \phi_w^i\ \mathbf{then}\ \phi_w^i\ \mathbf{else}\ \phi_w^i\ \nabla\ \psi_w^i$
and j is the least i such that $\Theta^i = \Theta^{i+1}$
and $R_u = \{(\phi \Rightarrow E\bowtie, u, \theta') \mid \theta' \in \Theta_u', E = con(\theta'), (\mathrm{fv}(E) = \emptyset, \phi = true) \vee (\mathrm{fv}(E) \neq \emptyset, \phi = \bigvee_{w \in \mathrm{fv}(E)}(\phi_w^j))\}$
and $R_m = \{(\theta, m, \theta') \mid \theta' \in \Theta_m' \wedge \theta \in \Theta^j \cup \{true \Rightarrow 0\bowtie\}\}$
and $\Theta_m' = \{\theta' \in \Theta' \mid \exists w \in \mathrm{fv}(con(\theta')) : \exists(\_, m, \_ \Rightarrow w\bowtie) \in R^j\}$ and $\Theta_u' = \Theta' \setminus \Theta_m'$

Fig. 6 The precondition generator

and "unmodified"; if $(\theta, u, \theta') \in R$ then additionally it holds that S modifies no "relevant" variable, where a "relevant" variable is one occurring in the consequent of θ'. We use γ to range over $\{m, u\}$ and write $dom(R) = \{\theta \mid \exists(\theta, \_, \_) \in R\}$ and $ran(R) = \{\theta' \mid \exists(\_, \_, \theta') \in R\}$.

5.1 Correctness Results

If $\{\Theta\}\,(\_) \Longleftarrow S\,\{\Theta'\}$, then Θ is indeed a precondition (but not necessarily the *weakest* such) that is strong enough to establish Θ', as stated by:

Theorem 1 (Correctness). *Assume* $\{\Theta\}\,(\_) \Longleftarrow S\,\{\Theta'\}$. *Then* $S \models \Theta \rightsquigarrow \Theta'$. *That is, if* $s \& s_1 \models \Theta$, *and* s', s_1' *are such that* $s\ [\![S]\!]\ s'$ *and* $s_1\ [\![S]\!]\ s_1'$, *then* $s' \& s_1' \models \Theta'$.

Note that Theorem 1 is termination *in*sensitive; this is not surprising given our choice of a relational semantics (but see [2] for a logic-based approach that is termination sensitive). Also note that correctness is phrased directly wrt the underlying semantics, unlike [1, 4] which first establish the semantic soundness of a logic and next provide a sound implementation of that logic. Theorem 1 is proved in the technical report accompanying this paper [5], much as the corresponding result [3] (that handled a language with heap manipulation but without procedure calls and without automatic computation of loop invariants), by establishing some auxiliary properties that have largely determined the design of Pre. The first such property is a variant of the "*-property" by Bell and La Padula [11], also called "write confinement" [7], which is used to preclude, e.g., "low writes under high guards." In our setting, it captures the role of the u tag and reads as follows:

Lemma 3. *Assume* $\{\Theta\}\ (R) \Longleftarrow S\ \{\Theta'\}$. *Then* $dom(R) = \Theta$ *and* $ran(R) = \Theta'$. *Given* $\theta' \in \Theta'$, *there exists at most one* θ *such that* $(\theta, u, \theta') \in R$. *If there exists such* θ, *then* $con(\theta) = con(\theta')$, *and with* $E = con(\theta)$ *we have that if* $s\ [\![S]\!]\ s'$ *then* s *agrees with* s' *on* $\mathrm{fv}(E)$.

Lemma 3, proved in [5], is needed in the proof of Theorem 1 to handle the case where the two runs in question follow *different branches* in a conditional, as we must then ensure that neither run modifies a variable on which we want the two runs to agree afterward. We shall also use a lemma, proved in [5], which expresses that there will always be one applicable condition in the precondition:

Lemma 4. *Assume* $\{\Theta\}\ (R) \Longleftarrow S\ \{\Theta'\}$. *Given* $\theta' \in \Theta'$, *there exists* $(\theta, \_, \theta') \in R$ *such that whenever* $s\ [\![S]\!]\ s'$ *and* $s' \models ant(\theta')$ *then* $s \models ant(\theta)$.

5.2 Intraprocedural Analysis

We now explain the various clauses of Pre in Fig. 6, where the clause for **skip** is trivial. For an assignment $x := A$, each 2-assertion $\phi \Rightarrow E\kappa$ in Θ' produces exactly one 2-assertion in Θ, given by substituting A for x (as in standard Hoare logic) in ϕ as well as in E; the connection is tagged m when x occurs in E. For example, if S is $x := w$ then R might contain the triplets $(y > 4 \Rightarrow w\kappa, m, y > 4 \Rightarrow x\kappa)$ and $(w > 3 \Rightarrow z\kappa, u, x > 3 \Rightarrow z\kappa)$.

The rule for $S_1 ; S_2$ works backward, first computing S_2's precondition which is then used to compute S_1's; the tags express that a consequent is modified iff it has been modified in either S_1 or S_2. The rule for **assert** allows us to weaken 2-assertions, by strengthening their antecedents; this is sound since execution will abort from stores not satisfying the new antecedents.

To illustrate and motivate the rule for conditionals, we shall use Fig. 5 where, given postcondition $\mathsf{OUT\_0\_DAT}\kappa$, the **then** branch generates (as the domain of R_1) precondition $\mathsf{INP\_1\_DAT}\kappa$, which by R_1' contributes the first conditional assertion of the overall precondition. The **skip** command in the implicit **else** branch generates (as the domain of R_2) precondition $\mathsf{OUT\_0\_DAT}\kappa$ which by R_2' contributes the second conditional assertion of the overall precondition. We must also capture that two runs, in order to agree on OUT_O_DAT after the conditional, must agree on the value of the test B; this is done by R'_0 which generates the precondition $(true \wedge B) \vee (true \wedge \neg B) \Rightarrow B\bowtie$; optimizations (not shown) in our algorithm simplify this to $B\bowtie$ and then use Fact 2 to split out the variables in the conjuncts of B into the two unconditional assertions of the overall precondition. Finally, assume the postcondition contained an assertion $\phi \Rightarrow E\bowtie$, where E is not modified by either branch: if also ϕ is not modified then $\phi \Rightarrow E\bowtie$ belongs to both R_1 and R_2, and hence by R_0 also to the overall precondition; if ϕ is modified by one or both branches, R_0 generates a more complex antecedent for $E\bowtie$.

5.3 Interprocedural Analysis

A procedure summary for p must satisfy:

1. If $s \ P(p) \ s'$ then $s(w) = s'(w)$ for all $w \notin \mathrm{OUT}_p$.
2. A postcondition is of the form $w\bowtie$ with $w \in \mathrm{OUT}_p$, and for each $w \in \mathrm{OUT}_p$, there is exactly one postcondition $w\bowtie$.
3. A precondition is of the form $\phi \Rightarrow z\bowtie$.
4. For each postcondition $w\bowtie$, there is a precondition of the form $true \Rightarrow z\bowtie$.
5. For each postcondition $w\bowtie$, with Θ_w its preconditions: if $s\&s_1 \models \Theta_w$ and $s \ P(p) \ s'$ and $s_1 \ P(p) \ s'_1$, then $s'(w) = s'_1(w)$.

Here Requirement 1 expresses that OUT_p does indeed contain all output variables. Requirements 2 and 3 were motivated in Sect. 4.2. Requirement 4 is needed for Lemma 4 to hold; it might seem restrictive but can always be established without losing precision, as by adding $true \Rightarrow q\bowtie$, where q is a variable not occurring in the program. Requirement 5 expresses that the summary computes correct information flow.

At a call site **call** p, antecedents in the call's postcondition will carry over to the precondition, *provided* that they do not involve variables in OUT_P. Otherwise, since our summaries express variable dependencies but not functional relationships, we cannot state an exact formula for modifying antecedents (unlike what is the case for assignments). Instead, we must conservatively strengthen the preconditions, by weakening their antecedents; this is done by an operator rm^+ such that if $\phi' = rm_X^+(\phi)$ (where $X = \mathrm{OUT}_p$) then ϕ logically implies ϕ' where ϕ' does not contain any variables from X. A trivial definition of rm^+ is to let it always return $true$ (which drops all conditions associated with X), but we can often get something more precise; for instance, we can choose $rm_{\{x\}}^+(x > 7 \wedge z > 5) = (z > 5)$ as is done by the following definition of rm^+.

We define rm^+, having the property that if $\phi' = rm_X^+(\phi)$ then ϕ' does not contain any variables from X, and is logically implied by ϕ, simultaneously with its dual rm^- which has the property that if $\phi' = rm_X^-(\phi)$ then ϕ' does not contain any variables from X and logically implies ϕ.

Summary information for p:

$\text{OUT}_p = \{x\}$

```
derives x from y,   z when y> 0,   w when y≤ 0
```

Procedure call

$\{z > 7 \Rightarrow v\ltimes, \; z > 5 \Rightarrow u\ltimes, \; z > 5 \Rightarrow y\ltimes, \; z > 5 \wedge y > 0 \Rightarrow z\ltimes, \; z > 5 \wedge y \leq 0 \Rightarrow w\ltimes\}$
  call p
$\{x > 5 \wedge z > 7 \Rightarrow v\ltimes, \; x > 7 \wedge z > 5 \Rightarrow (x + u)\ltimes\}$

Fig. 7 An example illustrating the handling of procedure calls

$$rm_X^+(B) = true \;\; \text{if fv}(B) \cap X \neq \emptyset \qquad\qquad rm_X^-(B) = false \;\; \text{if fv}(B) \cap X \neq \emptyset$$

$$rm_X^+(B) = B \;\; \text{if fv}(B) \cap X = \emptyset \qquad\qquad rm_X^-(B) = B \;\; \text{if fv}(B) \cap X = \emptyset$$

$$rm_X^+(\phi_1 \wedge \phi_2) = rm_X^+(\phi_1) \wedge rm_X^+(\phi_2) \qquad rm_X^-(\phi_1 \wedge \phi_2) = rm_X^-(\phi_1) \wedge rm_X^-(\phi_2)$$

$$rm_X^+(\phi_1 \vee \phi_2) = rm_X^+(\phi_1) \vee rm_X^+(\phi_2) \qquad rm_X^-(\phi_1 \vee \phi_2) = rm_X^-(\phi_1) \vee rm_X^-(\phi_2)$$

$$rm_X^+(\neg\phi_0) = \neg rm_X^-(\phi_0) \qquad\qquad rm_X^-(\neg\phi_0) = \neg rm_X^+(\phi_0)$$

Equipped with rm^+, we can now define the analysis of procedure call, as done in Fig. 6 and illustrated in Fig. 7. Here R_u deals with assertions (such as $x > 5 \wedge z > 7 \Rightarrow v\ltimes$ in the example) whose consequent has not been modified by the procedure call (its "frame conditions" determined by the OUT declaration). For an assertion whose consequent E has been modified (such as $x > 7 \wedge z > 5 \Rightarrow (x + u)\ltimes$), we must ensure that the variables of E agree after the procedure call (when the antecedent holds). For those not in OUT_p (such as u), this is done by R_0 (which expresses some "semiframe conditions"); for those in OUT_p (such as x), this is done by R_m which utilizes the procedure summary (contract) of the called procedure.

5.4 Synthesizing Loop Invariants

For while loops (the only iterative construct), the idea is to consider assertions of the form $\phi_x \Rightarrow x\ltimes$ and then repeatedly analyze the loop body so as to iteratively weaken the antecedents until a fixed point is reached. To illustrate the overall behavior, consider the example in Fig. 8 where we are given $r\ltimes$ as postcondition; hence the initial value of r's antecedent is *true*, whereas all other antecedents are initialized to *false*. The first iteration updates v's antecedent to $\mathsf{odd}(i)$, since v is used to compute r when i is odd, and also updates i's antecedent to *true*, since (the parity of) i is used to decide whether r is updated or not. The second iteration updates x's antecedent to $\neg\mathsf{odd}(i)$, since in order for two runs to agree on v when i is odd, they must have agreed on x in the previous iteration when i was even. The third iteration

```
while i < 7 do
   if odd(i)
     then r := r + v;
          v := v + h
     else v := x;
   i := i + 1
{r⋈}
```

Iteration	0	1	2	3	
	false	*false*	*false*	*false*	⇒ h⋈
	false	*true*	*true*	*true*	⇒ i⋈
	true	*true*	*true*	*true*	⇒ r⋈
	false	odd(i)	odd(i)	odd(i)	⇒ v⋈
	false	*false*	¬odd(i)	*true*	⇒ x⋈

Fig. 8 Iterative analysis of while loop. (We use odd(i) as a shorthand for $i \bmod 2 = 1$)

updates x's antecedent to *true*, since in order for two runs to agree on x when i is even, they must agree on x always (as x does not change). We have now reached a fixed point. It is noteworthy that even though the postcondition mentions r⋈, and r is updated using v which in turn is updated using h, the generated precondition does not mention h, since the parity of i was exploited. This shows [3] that even if we should only aim at producing contracts where all assertions are unconditional, precision may still be improved if the analysis engine makes internal use of *conditional* assertions.

In the general case, however, fixed point iteration may not terminate. To ensure termination, we need a "widening operator" $\triangledown$ on 1-assertions, with the following properties:

(a) For all ϕ and ψ, ψ logically implies $\psi \triangledown \phi$ and also ϕ logically implies $\psi \triangledown \phi$
(b) If for all i we have that ϕ^{i+1} is of the form $\psi \triangledown \phi^i$, then the chain $\{\phi^i \mid i \geq 0\}$ eventually stabilizes.

A trivial widening operator is the one that always returns *true*, in effect converting conditional agreement assertions into unconditional. A less trivial option will utilize a number of assertions, say $\psi_1, \ldots, \psi_k$, and allow $\psi \triangledown \phi = \psi_j$ if ψ_j is logically implied by ψ as well as by ϕ; such assertions may be given by the user if he has a hint that a suitable invariant may have one of $\psi_1, \ldots, \psi_k$ as antecedent.

We can now explain the various lines in the clause for while loops in Fig. 6. The iteration starts with antecedents ϕ_w^0 that are computed such that the corresponding 2-assertion, Θ^i, implies the postcondition Θ'. The ith iteration updates the antecedents ϕ_w^i into antecedents ϕ_w^{i+1} that are potentially weaker in that for each $w \in X$, each disjunct of ψ_w^i must imply ϕ_w^{i+1}; here ψ_w^i captures the "business logic" of the while loop:

1. If the precondition computed for the iteration contains an assertion $\phi \Rightarrow E⋈$ with $w \in \mathrm{fv}(E)$, then ϕ is an element of ψ_w^i.
2. If a consequent has been modified by the loop body, then the antecedent must belong to ψ_w^i for all $w \in \mathrm{fv}(B)$.

Here (2) ensures that if one run stays in the loop and updates a variable on which the two runs must agree, then also the other run stays in the loop (similar to the role of R_0' in the clause for conditionals), whereas (1) caters for the soundness when both runs stay in the loop, cf. the role of R_1' and R_2' in the case for conditionals.

Alternatively, to more closely follow the rule for conditionals, for (1) we could instead demand that $\phi \wedge B$ belongs to ψ_w^i; our current choice reflects that we expect the bodies of **while** loops to be prefixed by **assert** statements (which will automatically add B to the antecedents), but do not expect such transformations for branches of a conditional.

With the iteration stabilizing after j steps (thanks to the widening operator), the while loop's precondition Θ and its R component can now be computed; the former is given as the domain of the latter which is made up from two parts:

- First, R_u deals with those assertions in Θ' whose consequents have not been modified (a kind of "frame condition" for the while loop); each such assertion is connected to an assertion with the same consequent (so as to establish Lemma 3) but with an antecedent that is designed to be so weak that we can establish Lemma 4.
- Next, R_m deals with those assertions in Θ' whose consequents have been modified; each such assertion is connected to *all* other assertions in Θ^j so as to express that the subsequent iterations of the while loop may give rise to chains of variable dependences. (It would be possible to give a definition that in most cases it produces only a subset of those connections, but this would increase the conceptual complexity of Pre, without – we conjecture – any improvement in the overall precision of the algorithm.) In addition, again to establish Lemma 4, we introduce a trivial assertion $true \Rightarrow 0\ltimes$.

6 Evaluation

6.1 *Summary of Performance*

The algorithm of Sect. 5 provides a foundation for automatically *inferring* contracts from implementations, but can also be used for *checking* (conditional and unconditional) `derives` contracts supplied by a developer: first run the inference algorithm per method and then check (currently by hand but we are looking into automatization) if the contract precondition implies the precondition generated by the algorithm. In principle, this approach may reject a sound contract since the inference algorithm (due to approximations) does not always generate the weakest precondition (i.e., the strongest antecedents), but we have verified by manual inspection that the algorithm *never* produces a contract that is *less precise* than the original SPARK contract.

There is much merit in a methodology that encourages writing of the contract *before* writing/checking the implementation. However, one of our strategies for injecting our techniques into industrial development groups is to pitch the tools as being able to discover more precise conditional specifications to supplement conventional SPARK `derives` contracts already in the code; thus we focus the experimental studies of this section on the more challenging problem of automatically inferring contracts starting from code with no existing derives annotations.

For each procedure P, with $OUT_P = \{w_1, \ldots, w_k\}$, the algorithm analyzes the body wrt a postcondition $w_1 \bowtie_1, \ldots, w_k \bowtie_k$. Since SPARK disallows recursion, we simply move in a bottom-up fashion through the call-graph – guaranteeing that a contract exists for each called procedure. When deployed in actual development, one would probably allow developers to tweak the generated contracts (e.g., by removing unnecessary conditions for establishing end-to-end policies) before proceeding with contract inference for methods in the next level of the call hierarchy. However, in our experiments, we used autogenerated contracts for called methods without modification. All experiments were run under JDK 1.6 on a 2.2-GHz Intel Core2 Duo.

6.1.1 Code Bases

Embedded security devices are the initial target domain for our work, and the security-critical sections to be certified from these code bases are often relatively small, e.g., roughly 1,000 LOC for the guard partition of the Rockwell Collins high assurance guard mentioned earlier and 3,000 LOC for the (undisclosed) device recently certified by Naval Research Labs researchers [19]. For our evaluation, we consider a collection of five small to moderate size applications from the SPARK distribution in addition to an expanded version of the mailbox example of Sect. 2. Of these, the *Autopilot* and *Missile Control* applications are the most realistic. There are well over 250 procedures in the code bases, but due to space constraints, in Table 1 we list metrics for only the most complex procedures from each application (see [29] for the source code of all the examples). Columns **LOC**, **C**, **L**, and **P** report the number of noncomment lines of code, conditional expressions, loops, and procedure calls in each method. Our tool can run in two modes. The first mode (identified as version **1** in Table 1) implements the rules of Fig. 6 directly, with just one small optimization: a collection of boolean simplifications are introduced, e.g., simplifying assertions of the form $true \wedge \phi \Rightarrow E \bowtie$ to $\phi \Rightarrow E \bowtie$. The second mode (version **2** in Table 1) enables a collection of simplifications aimed at compacting and eliminating redundant flows from the generated set of assertions. One simplification performed is elimination of assertions with *false* in the antecedent (these are trivially true) and elimination of duplicate assertions. Also, it eliminates simple entailed assertions, such as $\phi \Rightarrow E \bowtie$ when $true \Rightarrow E \bowtie$ also appears in the assertion set.

6.1.2 Typical Refinement Power of the Algorithm

Column **O** gives the number of OUT variables of a procedure (this is equal to the number of `derives` clauses in the original SPARK contract), and Column **SF** gives the number of *flows* (total number of IN/OUT pairs) appearing in the original

Table 1 Experiment data (excerpts)

Package.Procedure Name	LoC	C	L	P	O	SF	Flows 1	Flows 2	Cond. Flows 1	Cond. Flows 2	Gens. 1	Gens. 2	Time (seconds) 1	Time (seconds) 2
Autopilot.AP.Altitude.Pitch.Rate.History_Average	10	0	1	0	1	2	5	3	0	0	0	0	0.047	0.063
Autopilot.AP.Altitude.Pitch.Rate.History_Update	8	1	1	0	1	2	3	3	0	0	0	0	0.000	0.157
Autopilot.AP.Altitude.Pitch.Rate.Calc_Pitchrate	13	2	0	2	2	7	17	8	0	0	15	15	0.000	0.015
Autopilot.AP.Altitude.Pitch.Target_ROC	9	2	0	0	1	2	7	3	6	2	0	0	0.000	0.000
Autopilot.AP.Altitude.Pitch.Target_Rate	17	4	0	1	1	3	53	4	42	0	142	46	0.015	0.015
Autopilot.AP.Altitude.Pitch.Calc_Elevator_Move	7	0	0	1	1	3	4	4	0	0	0	0	0.000	0.000
Autopilot.AP.Altitude.Pitch.Pitch_AP	7	0	0	4	2	11	54	11	42	0	0	0	0.015	0.000
Autopilot.AP.Altitude.Maintain	9	2	0	1	4	19	36	23	23	19	0	0	0.000	0.015
Autopilot.AP.Heading.Roll.Target_ROR	15	3	0	1	1	2	4	3	0	0	26	26	0.000	0.000
Autopilot.AP.Heading.Roll.Target_Rate	11	2	0	1	1	3	9	4	0	0	14	14	0.000	0.000
Autopilot.AP.Heading.Roll.Calc_Aileron_Move	7	0	0	1	1	3	4	4	0	0	0	0	0.000	0.000
Autopilot.AP.Heading.Roll.Roll_AP	7	0	0	4	2	7	9	7	0	0	0	0	0.000	0.000
Autopilot.AP.Control	19	1	0	13	8	46	58	54	0	0	63	51	0.016	0.032
Autopilot.AP.Heading.Yaw.Calc_Rudder_Move	7	0	0	1	1	2	4	3	0	0	0	0	0.000	0.000
Autopilot.AP.Heading.Yaw.Yaw_AP	5	0	0	3	2	5	5	5	0	0	0	0	0.000	0.000
Autopilot.Scale.Inverse	4	0	0	0	1	1	1	1	0	0	0	0	0.000	0.016
Autopilot.Scale.Scale_Movement	22	4	0	2	1	4	47	10	46	9	0	0	0.016	0.000
Autopilot.Scale.Heading_Offset	7	1	0	0	1	1	3	1	2	0	0	0	0.000	0.000
Autopilot.Heading.Maintain	6	1	0	2	4	15	28	20	16	16	0	0	0.000	0.000
Autopilot.Main	5	0	1	1	8	47	176	54	0	0	0	0	0.031	0.031
Minepump.Logbuffer.ProtectedWrite	8	1	0	0	5	9	9	9	4	4	0	0	0.031	0.047
Minepump.Logbuffer.ProtectedRead	6	0	0	0	5	6	7	7	0	0	0	0	0.000	0.000
Minepump.Logbuffer.Write	2	0	0	1	5	9	11	9	3	1	0	0	0.000	0.000
Mailbox.MACHINE_STEP	17	2	0	0	6	16	18	18	12	12	0	0	0.047	0.062
Mailbox.Main	6	0	1	1	6	16	54	22	0	0	2	2	0.031	0.016
BoilerWater-Monitor.FaultIntegrator.Test	11	3	0	0	4	11	46	22	42	18	0	0	0.047	0.047
BoilerWater-Monitor.FaultIntegrator.ControlHigh	8	1	0	2	2	4	6	5	0	0	2	2	0.000	0.000
BoilerWater-Monitor.FaultIntegrator.ControlLow	8	1	0	2	2	4	6	5	0	0	2	2	0.000	0.000
BoilerWater-Monitor.FaultIntegrator.Main	11	0	1	6	2	2	14	4	0	0	0	0	0.016	0.016
Lift-Controller.Next_Floor	9	2	0	0	1	2	7	4	6	3	0	0	0.047	0.047
Lift-Controller.Poll	22	2	1	3	2	9	77	12	43	0	0	0	0.031	0.031
Lift-Controller.Traverse	18	0	1	11	3	10	210	13	66	0	0	0	0.281	0.063
Missile_Guidance.Clock_Read	12	2	0	0	3	5	13	11	10	8	0	0	0.047	0.047
Missile_Guidance.Clock_Utils_Delta_Time	7	1	0	0	1	2	4	2	2	0	0	0	0.000	0.000
Missile_Guidance.Extrapolate_Speed	13	2	0	2	2	7	14	10	6	4	36	16	0.000	0.000
Missile_Guidance.Code_To_State	12	3	0	0	1	7	15	9	14	8	0	0	0.000	0.000
Missile_Guidance.Transition	20	4	0	2	1	9	3,527	63	3,524	62	4	4	0.156	0.125
Missile_Guidance..Relative_Drag_At_Altitude	8	2	0	0	1	1	7	3	6	2	0	0	0.000	0.000
Missile_Guidance.Drag_cfg.Calc_Drag	21	4	0	1	1	3	37	3	34	0	0	0	0.000	0.000
Missile_Guidance.If_Airspeed_Get_Speed	6	1	0	0	2	3	4	4	2	2	0	0	0.000	0.000
Missile_Guidance.Nav.Handle_Airspeed	18	4	0	4	3	13	117	28	110	25	18	18	0.000	0.000
Missile_Guidance.Nav.Estimate_Height	21	5	0	2	2	11	60	18	57	16	4	4	0.000	0.000

contract. Column **Flows** gives the number of flows generated by different versions of our algorithm. This number increases over **SF** as SPARK flows are refined into conditional flows (often creating two or more conditioned flows for a particular IN/OUT variable pair). The data shows that the compacting optimizations often substantially reduce the number of flows; the practical impact of this is to substantially increase the readability/tractability of the contracts. Column **Cond. Flows** indicates the number of flows from **Flows** that are conditional. Not only we expect to see the refining power of our approach in procedures with conditionals (column **C**) primarily, but we also see increases in precision that is due to conditional contracts of called procedures (column **P**). In few cases, we see a blow-up in the number of conditional flows. The worse case is `MissileGuidance.Transition`, which contains a case statement with each branch containing nested conditionals and procedure calls with conditional contracts – leading to an exponential explosion in path conditions. Only a few variables in these conditions lie in what we consider to be the "control logic" of the system. The tractability of this example would improve significantly with the methodology suggested earlier in which developers declare explicitly the guarding variables (such as the `xx_RDY` variables of Fig. 1) and the algorithm then omits tracking of conditional flows not associated with declared guard variables. Overall, a manual inspection of each inferred contract showed that the algorithm usually produces conditions that an expert would expect.

6.1.3 Efficiency of Inference Algorithm

As can be see in the **Time** columns, the algorithm is quite fast for all the examples, usually taking a little longer in version **2** (all optimizations on). However, for some examples, version **2** is actually faster; these are the cases of procedures with calls to other procedures. Due to the optimizations, the callees now have simpler contracts, simplifying the processing of the caller procedures.

6.1.4 Sources of Loss of Precision

We would like to determine situations where our treatment of loops or procedure calls leads to abstraction steps that discard conditional information. While this is difficult to determine for loops (one would have to compare to the most precise loop invariant – which would need to be written by hand), Column **Gens.** indicates the number of conditions dropped across processing of procedure calls. The data shows, and our experience confirms, that the loss of precision is not drastic (in some cases, one wants conditions to be discarded), but more experience is needed to determine the practical impact on verification of end-to-end properties.

6.1.5 Threats to Validity of Experiments

While the applications we consider are representative of small embedded controller systems, only the mailbox example is an information assurance application. While these initial results are encouraging, we are still in the process of negotiating access to the source code of actual products being developed at Rockwell Collins; that will allow us to answer the important question: does our approach provide the precision needed to better verify local and end-to-end MILS policies, without generating large contracts that become unwieldy for developers and certifiers?

6.2 Detailed Discussion of Selected Examples

In this section, we give a detailed discussion of two case studies: the Mailbox example (briefly discussed in Sect. 2) and part of the control code for the Autopilot code base (for which number figures were given in Table 1). For the Mailbox, we will discuss in detail the MACHINE_STEP procedure, previously introduced, comparing the results of running our tool with the original SPARK specification. In the Autopilot case study, we will discuss four procedures and two functions, spanning an entire call chain in the package, starting at the Main procedure, and going through the code that controls the altitude in this simplified example of an aircraft autopilot.

6.2.1 Mailbox Example

This example was discussed in detail in Sect. 2, so we will focus in comparing the resulting information flow specifications obtained from running our tool on the code with the original SPARK specification. Figure 9 shows the procedure MACHINE_STEP with the original SPARK information flow specification. Figure 10 shows the information flow specification obtained by running our tool on the same procedure (using the slightly modified version of the SPARK language described in Sect. 4.2). For simplicity, in Fig. 10 we have omitted the body of the procedure, as well as the global annotations. In addition to using unabbreviated variable names, the code of Fig. 9 differs from that of Fig. 1 in its use of procedures to manipulate both the control variables (e.g., Mailbox.CHARACTER_INPUT_0_READY) as well as the data variables of the system. For example, the procedure NOTIFY_INPUT_0_CONSUMED clears the Mailbox.CHARACTER_INPUT_0_READY flag where as NOTIFY_OUTPUT_1_READY sets the Mailbox.CHARACTER_OUTPUT_1_READY flag.

Upon close examination of Fig. 10, we can see the usage of the symbol {}. These *empty braces* are used to represent flow from a constant value. For example, in the following information flow declaration from Fig. 10:

```
derives ... Mailbox.CHARACTER_INPUT_0_READY from
    ... {} when ( Mailbox.CHARACTER_INPUT_0_READY
          and not Mailbox.CHARACTER_OUTPUT_1_READY)
```

indicates that the variable Mailbox.CHARACTER_INPUT_0_READY, in the case when the condition specified holds, has its postcondition value derived from a constant instead of another variable. By examining the code in Fig. 9, we can see that this is the case when Mailbox.CHARACTER_INPUT_0_READY is assigned the literal false.

The results displayed in Fig. 10 show that the information flow specifications for every variable in this example have been refined with at least one conditional flow.

```
procedure MACHINE_STEP
--# global in out Mailbox.CHARACTER_INPUT_0_READY,
--#                Mailbox.CHARACTER_INPUT_1_READY,
--#                Mailbox.CHARACTER_OUTPUT_0_READY,
--#                Mailbox.CHARACTER_OUTPUT_1_READY,
--#                Mailbox.CHARACTER_OUTPUT_0_DATA,
--#                Mailbox.CHARACTER_OUTPUT_1_DATA :
--#         in     Mailbox.CHARACTER_INPUT_0_DATA,
--#                Mailbox.CHARACTER_INPUT_1_DATA :
--# derives Mailbox.CHARACTER_OUTPUT_0_DATA from Mailbox.CHARACTER_INPUT_1_DATA,
--#                                              Mailbox.CHARACTER_OUTPUT_0_READY,
--#                                              Mailbox.CHARACTER_OUTPUT_0_DATA,
--#                                              Mailbox.CHARACTER_INPUT_1_READY &
--#         Mailbox.CHARACTER_OUTPUT_1_DATA from Mailbox.CHARACTER_INPUT_0_DATA,
--#                                              Mailbox.CHARACTER_INPUT_0_READY,
--#                                              Mailbox.CHARACTER_OUTPUT_1_DATA,
--#                                              Mailbox.CHARACTER_OUTPUT_1_READY &
--#         Mailbox.CHARACTER_INPUT_0_READY from Mailbox.CHARACTER_INPUT_0_READY,
--#                                              Mailbox.CHARACTER_OUTPUT_1_READY &
--#         Mailbox.CHARACTER_INPUT_1_READY from Mailbox.CHARACTER_INPUT_1_READY,
--#                                              Mailbox.CHARACTER_OUTPUT_0_READY &
--#         Mailbox.CHARACTER_OUTPUT_0_READY from Mailbox.CHARACTER_OUTPUT_0_READY,
--#                                               Mailbox.CHARACTER_INPUT_1_READY &
--#         Mailbox.CHARACTER_OUTPUT_1_READY from Mailbox.CHARACTER_OUTPUT_1_READY,
--#                                               Mailbox.CHARACTER_INPUT_0_READY :
is
  DATA_0 : CHARACTER;
  DATA_1 : CHARACTER;
begin
  if Mailbox.INPUT_0_READY and Mailbox.OUTPUT_1_CONSUMED then
    DATA_0 := Mailbox.READ_INPUT_0;
    Mailbox.NOTIFY_INPUT_0_CONSUMED;
    Mailbox.WRITE_OUTPUT_1(DATA_0);
    Mailbox.NOTIFY_OUTPUT_1_READY;
  end if;

  if Mailbox.INPUT_1_READY and Mailbox.OUTPUT_0_CONSUMED then
    DATA_1 := Mailbox.READ_INPUT_1;
    Mailbox.NOTIFY_INPUT_1_CONSUMED;
    Mailbox.WRITE_OUTPUT_0(DATA_1);
    Mailbox.NOTIFY_OUTPUT_0_READY;
  end if;
end MACHINE_STEP;
```

Fig. 9 Original SPARK specification for Mailbox example

```
procedure MACHINE_STEP;
--# derives Mailbox.CHARACTER_OUTPUT_0_DATA
--#             from Mailbox.CHARACTER_INPUT_1_DATA
--#                     when (Mailbox.CHARACTER_INPUT_1_READY and not Mailbox.CHARACTER_OUTPUT_0_READY),
--#                  Mailbox.CHARACTER_OUTPUT_0_READY,
--#                  Mailbox.CHARACTER_OUTPUT_0_DATA
--#                     when (not (Mailbox.CHARACTER_INPUT_1_READY and not Mailbox.CHARACTER_OUTPUT_0_READY)),
--#                  Mailbox.CHARACTER_INPUT_1_READY &
--#         Mailbox.CHARACTER_OUTPUT_1_DATA
--#             from Mailbox.CHARACTER_INPUT_0_DATA
--#                     when (Mailbox.CHARACTER_INPUT_0_READY and not Mailbox.CHARACTER_OUTPUT_1_READY),
--#                  Mailbox.CHARACTER_INPUT_0_READY,
--#                  Mailbox.CHARACTER_OUTPUT_1_DATA
--#                     when (not (Mailbox.CHARACTER_INPUT_0_READY and not Mailbox.CHARACTER_OUTPUT_1_READY)),
--#                  Mailbox.CHARACTER_OUTPUT_1_READY &
--#         Mailbox.CHARACTER_INPUT_0_READY
--#            from Mailbox.CHARACTER_INPUT_0_READY,
--#                  {} when (Mailbox.CHARACTER_INPUT_0_READY and not Mailbox.CHARACTER_OUTPUT_1_READY),
--#                  Mailbox.CHARACTER_OUTPUT_1_READY &
--#         Mailbox.CHARACTER_INPUT_1_READY
--#            from Mailbox.CHARACTER_INPUT_1_READY,
--#                  {} when (Mailbox.CHARACTER_INPUT_1_READY and not Mailbox.CHARACTER_OUTPUT_0_READY),
--#                  Mailbox.CHARACTER_OUTPUT_0_READY &
--#         Mailbox.CHARACTER_OUTPUT_0_READY
--#            from Mailbox.CHARACTER_OUTPUT_0_READY,
--#                  {} when (Mailbox.CHARACTER_OUTPUT_0_READY and not Mailbox.CHARACTER_INPUT_1_READY),
--#                  Mailbox.CHARACTER_INPUT_1_READY &
--#         Mailbox.CHARACTER_OUTPUT_1_READY
--#            from Mailbox.CHARACTER_OUTPUT_1_READY,
--#                  {} when (Mailbox.CHARACTER_OUTPUT_1_READY and not Mailbox.CHARACTER_INPUT_0_READY),
--#                  Mailbox.CHARACTER_INPUT_0_READY;
```

Fig. 10 Results of running tool on Mailbox example

Now, we wish to determine what benefits are gained from having such a refined information flow specification; that is, what do we gain from having information flow specifications split into cases denoted by particular conditions? We must keep in our mind the objective of our research and engineering effort: we want to build a foundation for an information assurance specification and verification framework.

From our point of view, an adequate information flow assurance framework must capture and describe the following information about an information-critical system:

- *Admissible channels of information flow.* The framework must provide mechanisms to appropriately specify when a flow of information from one part of the system to another (or from one variable to another) is acceptable. The original `derives` annotations from SPARK, and its corresponding checking mechanism, can already be used for this purpose (although they were not originally intended to fulfill this functionality).
- *Enabling conditions for information flow channels.* The framework must provide mechanisms to specify under what conditions a particular information flow channel is active. In information flow assurance applications, information flow channels are often controlled by system conditions. However, as it is, SPARK does not posses any mechanism that allows specifying under what conditions a particular information flow channel is active.

In the case of the mailbox example, we have a device intended to serve as a communication channel between two entities. If we were to try to describe the information flow policy requirements for the mailbox, we could write something like:

The mailbox will guarantee that information produced at the by Client 0 will be forwarded to Client 1, and the information produced by Client 1 will be forwarded to Client 0.

However, when we look at the information flow specification for the Client 0's output on Fig. 9, we have:

```
derives Mailbox.CHARACTER_OUTPUT_0_DATA from Mailbox.CHARACTER_INPUT_1_DATA,
                                             Mailbox.CHARACTER_OUTPUT_0_READY,
                                             Mailbox.CHARACTER_OUTPUT_0_DATA,
                                             Mailbox.CHARACTER_INPUT_1_READY
```

The output of Client 0 is not derived only from Client 1's input, but from other three variables. It is not necessarily obvious where these other dependences are coming from, and they certainly do not match with our first attempt at describing the mailbox's behavior. As it turns out, what happens here is that this specification describes *more than one* information flow channel, and the conditions on which they are active, but all this information has been merged into a single annotation. Let us look at the equivalent annotation from Fig. 10 to see what is going on:

```
derives Mailbox.CHARACTER_OUTPUT_0_DATA
        from Mailbox.CHARACTER_INPUT_1_DATA
             when (Mailbox.CHARACTER_INPUT_1_READY
                  and not Mailbox.CHARACTER_OUTPUT_0_READY),
             Mailbox.CHARACTER_OUTPUT_0_READY,
             Mailbox.CHARACTER_OUTPUT_0_DATA
             when (not (Mailbox.CHARACTER_INPUT_1_READY
                  and not Mailbox.CHARACTER_OUTPUT_0_READY)),
             Mailbox.CHARACTER_INPUT_1_READY
```

In the original SPARK specification, we cannot tell whether there are several information channels, or that the target variable is derived from a combination of the source variables, because there are no conditions. However, by looking at the specification produced by our tool, we can see that there are actually **two information flow channels** acting on this variable, controlled by two different conditions. We can also see that the dependence on the extra two variables is produced from *control dependence* on the variables that are used to compute the conditions.

It is clear now that there are two information flow channels working on this variable (1) when information available from Client 1 and Client 0 is ready to receive this information, then the output read by Client 0 is derived from the input produced by Client 1, and (2) if either there is no input from Client 1 or Client 0 is not ready to receive, then the output read by Client 0 keeps its old value. And clearly, which of these two channels is active depends on the aforementioned conditions, which in turn produce a control dependence on the variables that keep track of whether Client 1 has produced any information and whether Client 0 is ready to receive.

After the previous discussion, the benefits of having conditional information flow specifications are immediately clear. We have a more precise description of the behavior of the system and are able to check both aspects of the information assurance behavior of a system that we described before: the channels of information flow and the conditions under which those channels are active.

Another improvement that could be made is to differentiate the parts of the specification that deal with the control logic from those that deal exclusively with information flow. For example, in the case of the mailbox annotation for output 0, we get dependences on a couple of extra variables that arise from control dependence. Perhaps one could mark these flows with a special annotation to explicitly state that they arise from the control logic. Similarly, we can see in Figs. 9 and 10 that, besides those for the output variables, we have flow annotations for each of the control variables. These annotations are needed because these variables may be reset by the procedure. However, these modifications of control variables are also part of the control logic, and perhaps these flows could also be annotated in a special way. Furthermore, one could imagine a tool that would use these annotations to filter views and show all annotations or hide flows corresponding to the control logic, etc.

6.2.2 Autopilot Example

The Autopilot system is one of the examples included in the SPARK distribution (discussed in detail in [8, Chapter 14]). This is a control system controlling both the altitude and heading of an aircraft. The altitude is controlled by manipulating the elevators and the heading is controlled by manipulating the ailerons and rudder. The autopilot has a control panel with three switches each of which has two positions – on and off.

- The master switch – the autopilot is completely inactive if this is off.
- The altitude switch – the autopilot controls the altitude if this is on.
- The heading switch – the autopilot controls the heading if this is on.

Desired autopilot heading values are entered in a console by the pilot, whereas desired altitude values are determined by the current altitude (similar to how an automobile cruise control takes its target speed from the current speed when the cruise control is activated). For this example, we will take a look at a total of four procedures and two functions.

The procedure in Fig. 15 is interesting for conditional information flow analysis for multiple reasons:

- It contains nested case statements with a call at the lowest level of nesting to procedure `Pitch.Pitch_AP` that updates global variables.
- The actual updates to global variables occur several levels down the call chain from `Pitch.Pitch_AP`.
- The call chain includes several procedures with conditional flows – some of the conditions propagate up through the call chain, whereas others do not.

We discuss in detail the conditional information flow along the following call path:

- `Main` (`main.adb`): It contains an infinite loop that does nothing but call `AP.Control` on each iteration.
- `AP.Control` (`ap.adb`): It reads values for the three switches above from the environment. If `Master_Switch` is on, then it uses the values read for `Altitude_Switch` and `Heading_Switch` to set switch variables `Altitude_Selected` and `Heading_Selected`, otherwise `Altitude_Selected` and `Heading_Selected` are set to "off." Instruments needed to calculate altitude and heading are read, then `Altitude.Maintain` (with `Altitude_Selected` as the actual parameter for `Switch_Pressed`) and `Heading.Maintain` are called to update the autopilot state.
- `AP.Altitude.Maintain` (`ap-altitude.adb`): If `Altitude_Switch` has transitioned from off to on, the `Present_Altitude` is used as value for `Target_Altitude`. Otherwise, the previous value of `Target_Altitude` is used for value of `Target_Altitude`. `Pitch.Pitch_AP` is called to calculate the value of `Surfaces.Elevators` based on the parameter values of `Pitch.Pitch_AP` and the pitch history.
- `Pitch.Pitch_AP` (`ap-altitude-pitch.ads`): It calls a series of helper functions which update the local variables `Present_Pitchrate`, `Target_Pitchrate`, and `Elevator_Movement` and these are used in `Surfaces.Move_Elevators` to calculate the value of the global output variable `Surfaces.Elevators`. The behavior of `Surfaces.Move_Elevators` lies outside the SPARK boundary and thus the interface to `Surfaces.Move_Elevators` represents the leaf of the call tree path under consideration.
- We will also consider two functions called from `Pitch.Pitch_AP`: `Altitude.Target_Rate` and `Altitude.Target_ROC`. This will allow us to illustrate some interesting aspects of computing information flow specifications for SPARK functions.

The first procedure we look at is the main procedure. This is, like in most languages, the top most procedure and the point of access for the whole system.

```
procedure Main
  --# global in out AP.State;
  --#            out Surfaces.Elevators,
  --#                Surfaces.Ailerons,
  --#                Surfaces.Rudder;
  --#         in     Instruments.Altitude,
  --#                Instruments.Bank,
  --#                Instruments.Heading,
  --#                Instruments.Heading_Bug,
  --#                Instruments.Mach,
  --#                Instruments.Pitch,
  --#                Instruments.Rate_Of_Climb,
  --#                Instruments.Slip;
  --# derives AP.State
  --#           from *,
  --#                Instruments.Altitude,
  --#                Instruments.Bank,
  --#                Instruments.Pitch,
  --#                Instruments.Slip &
  --#   Surfaces.Elevators
  --#           from
  --#                AP.State,
  --#                Instruments.Altitude,
  --#                Instruments.Bank,
  --#                Instruments.Mach,
  --#                Instruments.Pitch,
  --#                Instruments.Rate_Of_Climb,
  --#                Instruments.Slip
  --#           &
  --#   Surfaces.Ailerons
  --#           from
  --#                AP.State,
  --#                Instruments.Altitude,
  --#                Instruments.Bank,
  --#                Instruments.Heading,
  --#                Instruments.Heading_Bug,
  --#                Instruments.Mach,
  --#                Instruments.Pitch,
  --#                Instruments.Slip   &
  --#   Surfaces.Rudder
  --#           from AP.State,
  --#                Instruments.Altitude,
  --#                Instruments.Bank,
  --#                Instruments.Mach,
  --#                Instruments.Pitch,
  --#                Instruments.Slip
  --#   ;
is
begin

  loop
      AP.Control;
  end loop;
end Main;
```

Fig. 11 Original SPARK specification for procedure main from the autopilot code base

Figure 11 shows the original SPARK specifications and the code, and Fig. 12 shows
the corresponding information flow specifications computed by our tool. The first
thing we note is that there are more *derived* variables in the annotations derived by
our tool than in the original annotations. This is not a mistake. The reason for this is
that we still have not incorporated SPARK's "own refinement" abstraction mechanism in our tool. All the variables in the `derives` annotations from Fig. 12 that start
with `AP.` are abstracted into the variable `AP.State` in Fig. 11. As a consequence,
we get more flow specifications because they are refined from those in the original
annotations.

An interesting effect of not having abstraction in our annotations is that
some of the false flows introduced by the abstraction process are not present
in our annotations. For instance, in Fig. 11 one of the annotations suggests that
`Surfaces.Ailerons` may be derived from `Instruments.Altitude`. However, as
we can see in Fig. 12, this is not the case; such flow is absent from the specification.
The reason this false flow appears in the abstracted version is that, when all the `AP.`

```
procedure Main;
  --# derives AP.Heading.Yaw.Rate.Yaw_History
  --#          from *,
  --#               {},
  --#                    AP.Controls.Master_Switch,
  --#                    AP.Controls.Heading_Switch,
  --#                    Instruments.Slip &
  --#  AP.Altitude.Pitch.Rate.Pitch_History
  --#          from *,
  --#               {},
  --#                    AP.Controls.Master_Switch,
  --#                    AP.Controls.Altitude_Switch,
  --#                    Instruments.Pitch &
  --#  AP.Heading.Roll.Rate.Roll_History
  --#          from *,
  --#               {},
  --#                    AP.Controls.Master_Switch,
  --#                    AP.Controls.Heading_Switch,
  --#                    Instruments.Bank &
  --#  AP.Altitude.Target_Altitude
  --#          from *,
  --#               {},
  --#                    AP.Controls.Master_Switch,
  --#                    AP.Controls.Altitude_Switch,
  --#                    AP.Altitude.Switch_Pressed_Before,
  --#                    Instruments.Altitude &
  --#  AP.Altitude.Switch_Pressed_Before
  --#          from *,
  --#               {},
  --#                    AP.Controls.Master_Switch,
  --#                    AP.Controls.Altitude_Switch &
  --#  Surfaces.Elevators
  --#          from {},
  --#                    AP.Controls.Master_Switch,
  --#                    AP.Controls.Altitude_Switch
  --#                    AP.Altitude.Pitch.Rate.Pitch_History,
  --#                    AP.Altitude.Switch_Pressed_Before,
  --#                    AP.Altitude.Target_Altitude,
  --#                    Instruments.Altitude,
  --#                    Instruments.Mach,
  --#                    Instruments.Pitch,
  --#                    Instruments.Rate_Of_Climb &
  --#  Surfaces.Ailerons
  --#          from {},
  --#                    AP.Controls.Master_Switch,
  --#                    AP.Controls.Heading_Switch,
  --#                    AP.Heading.Roll.Rate.Roll_History,
  --#                    Instruments.Bank,
  --#                    Instruments.Heading,
  --#                    Instruments.Heading_Bug,
  --#                    Instruments.Mach &
  --#  Surfaces.Rudder
  --#          from {},
  --#                    AP.Controls.Master_Switch,
  --#                    AP.Controls.Heading_Switch,
  --#                    AP.Heading.Yaw.Rate.Yaw_History,
  --#                    Instruments.Mach,
  --#                    Instruments.Slip
  --#  ;
```

Fig. 12 Results of running tool on procedure main from the autopilot code base

variables are abstracted into AP.State, then Surfaces.Ailerons depends on this new abstract variable (because it depends on some AP. variables), and since some AP. variables depend on Instruments.Altitude, then Surfaces.Ailerons gets a possible dependence on Instruments.Altitude, even though this dependence is really nonexistent.

Other than the differences described above, the specifications obtained by our tool are basically the same as those in the original annotations. Also, we see {} annotations (derivations from constant values). In the original SPARK, these *constant derivations* are simply ignored; however, we leave them explicit for the sake of completeness. In fact, these annotations become more interesting when they are associated by themselves with a condition, as they might actually represent "reset" conditions (as in the case of the mailbox example).

```
procedure Control
--# global in        Controls.Master_Switch,
--#                  Controls.Altitude_Switch,
--#                  Controls.Heading_Switch;
--#         in out   Altitude.State,
--#                  Heading.State:
--#            out   Surfaces.Elevators,
--#                  Surfaces.Ailerons,
--#                  Surfaces.Rudder;
--#         in       Instruments.Altitude,
--#                  Instruments.Bank,
--#                  Instruments.Heading,
--#                  Instruments.Heading_Bug,
--#                  Instruments.Mach,
--#                  Instruments.Pitch,
--#                  Instruments.Rate_Of_Climb,
--#                  Instruments.Slip;
--# derives Altitude.State
--#            from *,
--#                  Controls.Master_Switch,
--#                  Controls.Altitude_Switch,
--#                  Instruments.Altitude,
--#                  Instruments.Pitch &
--#         Heading.State
--#            from *,
--#                  Controls.Master_Switch,
--#                  Controls.Heading_Switch,
--#                  Instruments.Bank,
--#                  Instruments.Slip &
--#   Surfaces.Elevators
--#            from  Controls.Master_Switch,
--#                  Controls.Altitude_Switch,
--#                  Altitude.State,
--#                  Instruments.Altitude,
--#                  Instruments.Mach,
--#                  Instruments.Pitch,
--#                  Instruments.Rate_Of_Climb &
--#   Surfaces.Ailerons
--#            from  Controls.Master_Switch,
--#                  Controls.Heading_Switch,
--#                  Heading.State,
--#                  Instruments.Bank,
--#                  Instruments.Heading,
--#                  Instruments.Heading_Bug,
--#                  Instruments.Mach &
--#   Surfaces.Rudder
--#            from  Controls.Master_Switch,
--#                  Controls.Heading_Switch,
--#                  Heading.State,
--#                  Instruments.Mach,
--#                  Instruments.Slip
--#   ;
is
   Master_Switch, Altitude_Switch, Heading_Switch,
      Altitude_Selected, Heading_Selected : Controls.Switch;
   Present_Altitude : Instruments.Feet;
   Bank             : Instruments.Bankangle;
   Present_Heading  : Instruments.Headdegree;
   Target_Heading   : Instruments.Headdegree;
   Mach             : Instruments.Machnumber;
   Pitch            : Instruments.Pitchangle;
   Rate_Of_Climb    : Instruments.Feetpermin;
   Slip             : Instruments.Slipangle;
begin
   Controls.Read_Master_Switch(Master_Switch);
   Controls.Read_Altitude_Switch(Altitude_Switch);
   Controls.Read_Heading_Switch(Heading_Switch);
   case Master_Switch is
      when Controls.On =>
         Altitude_Selected := Altitude_Switch;
         Heading_Selected := Heading_Switch;
      when Controls.Off =>
         Altitude_Selected := Controls.Off;
         Heading_Selected := Controls.Off;
   end case;
   Instruments.Read_Altimeter(Present_Altitude);
   Instruments.Read_Bank_Indicator(Bank);
   Instruments.Read_Compass(Present_Heading);
   Instruments.Read_Heading_Bug(Target_Heading);
   Instruments.Read_Mach_Indicator(Mach);
   Instruments.Read_Pitch_Indicator(Pitch);
   Instruments.Read_VSI(Rate_Of_Climb);
   Instruments.Read_Slip_Indicator(Slip);
   Altitude.Maintain(Altitude_Selected, Present_Altitude, Mach, Rate_Of_Climb, Pitch);
   Heading.Maintain(Heading_Selected, Mach, Present_Heading, Target_Heading, Bank, Slip);
end Control;
```

Fig. 13 Original specification for procedure AP.Control from the autopilot code base

```
procedure Control;
 --# derives Altitude.Switch_Pressed_Before
 --#          from *,
 --#               {},
 --#               Controls.Master_Switch,
 --#               Controls.Altitude_Switch &
 --# Altitude.Pitch.Rate.Pitch_History
 --#          from *,
 --#               {},
 --#               Controls.Master_Switch,
 --#               Controls.Altitude_Switch,
 --#               Instruments.Pitch &
 --# Altitude.Target_Altitude
 --#          from *,
 --#               {},
 --#               Controls.Master_Switch,
 --#               Controls.Altitude_Switch,
 --#               Altitude.Switch_Pressed_Before,
 --#               Instruments.Altitude &
 --# Heading.Yaw.Rate.Yaw_History
 --#          from *,
 --#               {},
 --#               Controls.Master_Switch,
 --#               Controls.Heading_Switch,
 --#               Instruments.Slip &
 --# Heading.Roll.Rate.Roll_History
 --#          from *,
 --#               {},
 --#               Controls.Master_Switch,
 --#               Controls.Heading_Switch,
 --#               Instruments.Bank &
 --# Surfaces.Elevators
 --#          from {},
 --#               Controls.Master_Switch,
 --#               Controls.Altitude_Switch,
 --#               Altitude.Target_Altitude,
 --#               Altitude.Switch_Pressed_Before,
 --#               Altitude.Pitch.Rate.Pitch_History,
 --#               Instruments.Altitude,
 --#               Instruments.Mach,
 --#               Instruments.Pitch,
 --#               Instruments.Rate_Of_Climb &
 --# Surfaces.Ailerons
 --#          from {},
 --#               Controls.Master_Switch,
 --#               Controls.Heading_Switch,
 --#               Heading.Roll.Rate.Roll_History,
 --#               Instruments.Bank,
 --#               Instruments.Heading,
 --#               Instruments.Heading_Bug,
 --#               Instruments.Mach &
 --# Surfaces.Rudder
 --#          from {},
 --#               Controls.Master_Switch,
 --#               Controls.Heading_Switch,
 --#               Heading.Yaw.Rate.Yaw_History,
 --#               Instruments.Mach,
 --#               Instruments.Slip
 --# ;
```

Fig. 14 Results of running tool on procedure AP.Control from the autopilot code base

Next we look at the procedure AP.Control. The original SPARK annotations, as well as the code for the procedure, are given in Fig. 13, and the result from our tool is presented in Fig. 14. Just as with main procedure explained before, we get basically the same annotations, except for getting extra variables due to differences in abstraction and the presence of {} annotations in our tool's results. However, it is rather disappointing that we do not get conditional information flow specifications in this example. Upon a closer look at the code, we see that there is a case statement in the body that should give rise to conditions. Furthermore, as we will see later, most of the procedures called from this procedure generate conditional specifications. So, why do not we get any conditional specifications here?

The bottom line is that we are getting hurt by the generalization rules triggered by the procedure call rule discussed in Sect. 5. The procedure does generate conditions;

however, all these conditions are dropped once the top three procedure calls are analyzed: the procedures that read the value of the switches (the guard variables). What happens is that the conditions generated are basically predicates in terms of the guard variables (the value of the switches), and since the top three procedures set these switch variables, we have to drop the conditions and turn the annotation into an unconditional one.

To see this with more detail, let us take a look at what happens at the return point of the procedure call to `Controls.Read_Heading_Switch`. Recall that our algorithm is a weakest precondition algorithm and, as such, it works bottom-up. So, when we reach the point right before the call to this procedure, the algorithm has, among all of the derivations generated, the following flow specification:

```
derives Surfaces.Ailerons from Instruments.Heading
     when Heading_Switch = Controls.On and Master_Switch = Controls.On
```

This specification basically tells us that if the heading and master switches are in the ON position, then `Surfaces.Ailerons` derives its value from the current heading reading (`Instruments.Heading`). Furthermore, another similar specification tells us that if the heading switch is OFF, then `Surfaces.Ailerons` maintains its current value. However, when we process the call to `Controls.Read_Heading_Switch`, all this information is lost, because this procedure sets the value of `Heading_Switch` with the actual state of the switch. In doing this, since our analysis is modular (i.e., we do not look at the body of `Controls.Read_Heading_Switch`), we do not know in general what this modification to `Heading_Switch` did, and so we have to drop the condition. At the very least we would have to drop the part of the condition that refers to `Heading_Switch`; however, in this case it does not matter, because as soon as `Controls.Read_Master_Switch` was analyzed, the rest of the condition would be dropped.

As it turns out, in this particular case, it would be safe not to drop the condition and simply substitute `Heading_Switch` with `Controls.Heading_Switch`, but this cannot be determined without looking at the procedure's body and breaking modularity. One way by which this situation could be ameliorated would be to refactor the procedure into two different procedures: one that reads in the value of the switches and another that implements the rest of the logic. For example,

```
procedure Control
is
   ...
begin
   Read_Controls_Swintches(Master_Switch, Altitude_Switch, Heading_Switch);
   Execute_Control_Logic(Master_Switch, Altitude_Switch, Heading_Switch);
end Control;
```

where `Read_Controls_Switches` simply performs the top three procedure calls in `Control`, and the rest of the functionality is implemented in `Execute_Control_Logic`. Then the conditional information flow specifications of `Control` would be exposed in the procedure `Execute_Control_Logic`.

Another option would be to exploit other annotations in the code (like postconditions and/or assertions) to avoid unnecessary generalizations. For example, if the procedure `Controls.Read_Heading_Switch` had the following annotation:

```
procedure  Read_Heading_Switch  (Heading_Switch)
—#  post:  Heading_Switch  =  Controls.Heading_Switch;
```

then from this annotation we could determine exactly what the value of `Heading_Switch` is in the postcondition (`Controls.Heading_Switch`) and perform a direct substitution in the condition expressions instead of having to drop them. These are all options that we are considering for future versions of the tool.

The next procedure to discuss is `Altitude.Maintain`, which is called from `AP.Control`. This is the first procedure in our study of the Autopilot that has generated conditional specifications. The original SPARK annotations as well as the code are displayed in Fig. 15, and the annotations generated by our tool are presented in Fig. 16. The purpose of this procedure is to maintain the altitude of the airplane depending on the current configuration of the autopilot, so there are quite a few cases this procedure has to handle, which is why we get several conditional information flow specifications.

```
procedure  Maintain ( Switch_Pressed      :  in  Controls.Switch;
               Present_Altitude  :  in  Instruments.Feet;
               Mach              :  in  Instruments.Machnumber;
               Climb_Rate        :  in  Instruments.Feetpermin ;
               The_Pitch         :  in  Instruments.Pitchangle )
—#  global  in  out  Target_Altitude ,
—#                   Switch_Pressed_Before ,
—#                   Pitch.State;
—#             out  Surfaces.Elevators:
—#  derives    Target_Altitude
—#               from *,
—#                   Switch_Pressed ,
—#                   Switch_Pressed_Before ,
—#                   Present_Altitude &
—#             Pitch.State
—#               from *,
—#                   Switch_Pressed ,
—#                   The_Pitch &
—#             Switch_Pressed_Before
—#               from
—#                   Switch_Pressed &
—#             Surfaces.Elevators
—#               from  Switch_Pressed ,
—#                   Switch_Pressed_Before ,
—#                   Target_Altitude ,
—#                   Present_Altitude ,
—#                   Mach,
—#                   Climb_Rate ,
—#                   The_Pitch ,
—#                   Pitch.State
—#  ;
is
begin
   case  Switch_Pressed  is
      when  Controls.On =>
         case  Switch_Pressed_Before  is
            when  Controls.Off =>
               Target_Altitude  :=  Present_Altitude ;
            when  Controls.On =>
               null ;
         end  case ;
               Pitch.Pitch_AP ( Present_Altitude , Target_Altitude , Mach , Climb_Rate , The_Pitch );
      when  Controls.Off =>
         null ;
   end  case ;
   Switch_Pressed_Before  :=  Switch_Pressed ;
end  Maintain ;
```

Fig. 15 Original SPARK specification for procedure Altitude.Maintain from the autopilot code base

```
procedure Maintain ( Switch_Pressed    : in Controls . Switch ;
                     Present_Altitude  : in Instruments . Feet ;
                     Mach              : in Instruments . Machnumber ;
                     Climb_Rate        : in Instruments . Feetpermin ;
                     The_Pitch         : in Instruments . Pitchangle );
--# derives   Target_Altitude
--#               from * when ( not Switch_Pressed = Controls.On ),
--#                        Switch_Pressed ,
--#                        Switch_Pressed_Before when ( Switch_Pressed = Controls.On ),
--#                        Present_Altitude
--#                           when ( Switch_Pressed = Controls.On and Switch_Pressed_Before = Controls.Off ),
--#                        * when ( Switch_Pressed = Controls.On and ( not Switch_Pressed_Before = Controls.Off )) &
--#           Pitch . Rate . Pitch_History
--#               from {*, The_Pitch} when ( Switch_Pressed = Controls.On ),
--#                        Switch_Pressed ,
--#                        * when ( not Switch_Pressed = Controls.On ) &
--#           Switch_Pressed_Before
--#               from
--#                        Switch_Pressed &
--#           Surfaces . Elevators
--#               from Switch_Pressed ,
--#                    { Switch_Pressed_Before ,
--#                      Present_Altitude ,
--#                      Mach ,
--#                      Climb_Rate ,
--#                      The_Pitch ,
--#                      Pitch . Rate . Pitch_History} when ( Switch_Pressed = Controls.On ),
--#                    Target_Altitude
--#                      when ( Switch_Pressed = Controls.On and ( not Switch_Pressed_Before = Controls.Off )),
--#                    Present_Altitude
--#                      when ( Switch_Pressed = Controls.On and Switch_Pressed_Before = Controls.Off ),
--#                    Surfaces . Elevators when ( not Switch_Pressed = Controls.On )
--# ;
```

Fig. 16 Results of running tool on procedure Altitude.Maintain from the autopilot code base

Here again we have a case in which it is extremely beneficial to have conditional
information flow contracts. The information flow contracts in this example describe
not only the information flow channel present, but also the conditions under which
they are active, revealing details of the control logic of the system. Not only that, in
this particular example, the details of the control logic revealed actually give a good
insight of the actual functionality of the procedure.

Let us start by looking at the flow specifications for Target_Altitude. We can
see that Target_Altitude either derives its new value from Present_Altitude
or it keeps from its own previous value. We can see that the dependences on
Switch_Pressed and Switch_Pressed_Before are simply control dependences, as
these are the variables that appear in the conditions. Switch_Pressed contains the
state of the altitude switch, passed as an argument, and Switch_Pressed_Before is
a global used to store the state of Switch_Pressed in the previous state.

So, under what conditions is the value of Target_Altitude modified using
Present_Altitude? Looking at the first specification, we can see that whenever Switch_Pressed is OFF, Target_Altitude's new value is derived from
itself. Now we have two interesting cases. If Switch_Pressed is ON, but
Switch_Pressed_Before is OFF (i.e., the value of Switch_Pressed in the previous state was OFF) then Target_Altitude derives gets its new value derived
from Present_Altitude. On the other hand, if Switch_Pressed is ON and
Switch_Pressed_Before is OFF, then Target_Altitude gets its new value derived from itself. Thus, we have exposed the logic of the altitude control system:
when the altitude switch transitions from OFF to ON, the target altitude (the altitude at which the plane will be automatically maintained) is set to the present
altitude, after that initial transition the target altitude does not change (transition

from ON to ON), unless the system transitions again from OFF to ON. So Switch_Pressed_Before is basically used to detect the transitions from OFF to ON and set the target altitude. A similar analysis applied to the information flow specifications for Surfaces.Elevators.

Now let us examine procedure Altitude.Pitch.Pitch_AP which is called from Altitude.Maintain. The original SPARK annotations as well as the code can be seen in Fig. 17, and the results of our tool are presented in Fig. 18. This example is actually relatively simple, and as seen by looking at the figures the results of our tool are exactly the same as the original SPARK annotations. As Pitch_AP's purpose is just to update a set of variables, depending on its input, there is really no conditional information flow behavior in this procedure. This procedure sets the

```
procedure  Pitch_AP(Present_Altitude  :  in  Instruments.Feet;
                    Target_Altitude    :  in  Instruments.Feet;
                    Mach               :  in  Instruments.Machnumber;
                    Climb_Rate         :  in  Instruments.Feetpermin;
                    The_Pitch          :  in  Instruments.Pitchangle)
--# global in out Rate.Pitch_History;
--#            out Surfaces.Elevators;
--# derives Rate.Pitch_History
--#         from *,
--#              The_Pitch &
--#         Surfaces.Elevators
--#         from Rate.Pitch_History,
--#              Present_Altitude,
--#              Target_Altitude,
--#              Mach,
--#              Climb_Rate,
--#              The_Pitch
--#  ;
is
    Present_Pitchrate   :  Degreespersec;
    Target_Pitchrate    :  Degreespersec;
    Elevator_Movement   :  Surfaces.Controlangle;
begin
    Calc_Pitchrate(The_Pitch, Present_Pitchrate);
    Target_Pitchrate := Target_Rate(Present_Altitude, Target_Altitude, Climb_Rate);
    Elevator_Movement := Calc_Elevator_Move(Present_Pitchrate, Target_Pitchrate, Mach);
    Surfaces.Move_Elevators(Elevator_Movement);
end Pitch_AP;
```

Fig. 17 Original SPARK specification for procedure AP.Altitude.Pitch_AP from the autopilot code base

```
procedure  Pitch_AP(Present_Altitude  :  in  Instruments.Feet;
                    Target_Altitude    :  in  Instruments.Feet;
                    Mach               :  in  Instruments.Machnumber;
                    Climb_Rate         :  in  Instruments.Feetpermin;
                    The_Pitch          :  in  Instruments.Pitchangle)
--# derives Rate.Pitch_History
--#         from *,
--#              The_Pitch &
--#    Surfaces.Elevators
--#         from Rate.Pitch_History,
--#              Present_Altitude,
--#              Target_Altitude,
--#              Mach,
--#              Climb_Rate,
--#              The_Pitch
--#  ;
is
    Present_Pitchrate   :  Degreespersec;
    Target_Pitchrate    :  Degreespersec;
    Elevator_Movement   :  Surfaces.Controlangle;
begin
    Calc_Pitchrate(The_Pitch, Present_Pitchrate);
    Target_Pitchrate := Target_Rate(Present_Altitude, Target_Altitude, Climb_Rate);
    Elevator_Movement := Calc_Elevator_Move(Present_Pitchrate, Target_Pitchrate, Mach);
    Surfaces.Move_Elevators(Elevator_Movement);
end Pitch_AP;
```

Fig. 18 Results of running tool on procedure AP.Altitude.Pitch_AP from the autopilot code base

pitch, depending on the values of the present and target altitude. What is interesting about this procedure is that it calls a SPARK function, which is the one we look at next.

To conclude we look at a couple of SPARK functions, which are at the bottom of this call chain in Autopilot. The reason we look at this functions is to discuss a couple of relevant concepts in the computation of the annotations relevant to functions, which are not issues in the original SPARK. The functions are `Target_Rate`, which is called from `Pitch_AP`, and `Target_ROC`, which is called from `Target_Rate`. The code for these functions and the annotations obtained with our tool are presented in Figs. 19 and 20, respectively.

```
function Target_Rate ( Present_Altitude  :  Instruments . Feet;
                       Target_Altitude   :  Instruments . Feet;
                       Climb_Rate  :  Instruments . Feetpermin )
                  return Degreespersec
--# derives @result
--#         from {},
--#              Climb_Rate ,
--#              Present_Altitude ,
--#              Target_Altitude
--#  ;
is
   Target_Climb_Rate  :  Floorfpm;
   Floor_Climb_Rate   :  Floorfpm;
   Result             :  Degreespersec ;
begin
   Target_Climb_Rate  := Target_ROC( Present_Altitude , Target_Altitude );
   if  Climb_Rate > Floorfpm ' Last then
      Floor_Climb_Rate := Floorfpm ' Last;
   elsif Climb_Rate < Floorfpm ' First then
      Floor_Climb_Rate := Floorfpm ' First;
   else
      Floor_Climb_Rate := Climb_Rate;
   end if;
   --# assert Floor_Climb_Rate  in Floorfpm and
   --#        Target_Climb_Rate  in Floorfpm:
   Result := Degreespersec( (Target_Climb_Rate — Floor_Climb_Rate) / 12);
   if (Result > 10) then
      Result := 10;
   elsif (Result < −10) then
      Result := −10;
   end if;
   return Result;
end Target_Rate ;
```

Fig. 19 Results of running tool on function AP.Altitude.Target_Rate from the autopilot code base

```
function Target_ROC( Present_Altitude  :  Instruments . Feet;
                     Target_Altitude   :  Instruments . Feet)
                  return Floorfpm
--# derives @result
--#         from {}
--#              when (( Target_Altitude — Present_Altitude ) / 10 < Floorfpm ' First
--#                 and not (( Target_Altitude — Present_Altitude ) / 10 > Floorfpm ' Last )),
--#              {}
--#              when (( Target_Altitude — Present_Altitude ) / 10 > Floorfpm ' Last ),
--#              Target_Altitude ,
--#              Present_Altitude
--#  ;
is
   Result : Instruments . Feetpermin ;
begin
   Result := Instruments . Feetpermin ( Integer ( Target_Altitude — Present_Altitude ) / 10);
   if (Result > Floorfpm ' Last ) then
      Result := Floorfpm ' Last;
   elsif Result < Floorfpm ' First then
      Result := Floorfpm ' First;
   end if;
   return Result;
end Target_ROC;
```

Fig. 20 Results of running tool on function AP.Altitude.Target_ROC from the autopilot code base

To conclude, we look at two SPARK functions that lie at the bottom of this call chain in Autopilot. This will reveal a couple of relevant concepts in the computation of the annotations relevant to functions, which are not issues in the original SPARK. The main thing to observe in these functions is the annotations. In SPARK, functions do not have `derives` annotations. Because functions are not allowed to have side-effects, the required information is implicit in the declaration of arguments and return value: the return value simply depends on all parameters and globals declared in the function. However, in our case, to provide a means to capture conditional flows, we need to explicitly introduce flow contracts as shown in Figs. 19 and 20.

In order to be able to specify conditional information flow for functions, we need to be able to talk about the value computed by the function. We define a special variable `@result`, which denotes the value returned by the function, and we compute dependences for this special variable. In the case of `Target_Rate`, we do not have conditional information flow specifications, but in the case of `Target_ROC`, we do have a couple of conditional flows. However, these two conditional flows are *constant flows* and as such they disappear in `Target_Rate`. The main point we want to present here is the need for information flow specifications for SPARK functions and the way we implement those in our adaptation of SPARK.

7 Related Work

The theoretical framework for the SPARK information flow framework is provided by Bergeretti and Carré [12] who present a compositional method for inferring and checking dependences [14] among variables. That approach is flow sensitive, whereas most security type systems [7, 33] are flow *insensitive* as they rely on assigning a security level ("high" or "low") to each variable. Chapman and Hilton [13] describe how SPARK information flow contracts could be extended with lattices of security levels and how the SPARK Examiner could be enhanced to check conformance of flows to particular security levels. Those ideas could be applied directly to provide security levels of flows in our framework. Rossebo et al.[26] show how the existing SPARK framework can be applied to verify various *unconditional* properties of a MILS Message Router. Apart from SPARK, there exist several tools for analyzing information flow properties, notably Jif (Java + information flow) which is based on [23] and Flow Caml [28].

The seminal work on agreement assertions is [1], whose logic is flow sensitive, and comes with an algorithm for computing (weakest) preconditions, but the approach does not integrate with programmer assertions. To address that, and to analyze heap-manipulating languages, the logic of [4] employs *three* kinds of primitive assertions: agreement, programmer, and region (for a simple alias analysis). But, since those can be combined only through conjunction, programmer assertions are not smoothly integrated, and it is not possible to capture *conditional* information flows. That was what motivated Amtoft and Banerjee [3] to introduce conditional agreement assertions for a heap-manipulating language. This paper integrates that approach into the SPARK setting (where the lack of heap objects enables us to omit the "object flow invariants" of [3]) for practical industrial development, adds interprocedural contract-based compositional checking, adds an algorithm for computing loop invariants (rather than assuming they are provided by the user), and provides an implementation as well as reports on experiments.

A recently popular approach to information flow analysis is *self-composition*, first proposed by Barthe et al. [10] and later extended by, e.g., Terauchi and Aiken [31] and (for heap-manipulating programs) Naumann [24]. Self-composition works as follows: for a given program S, a copy S' is created with all variables renamed (primed); with the observable variables say x, y, then noninterference holds provided the sequential composition $S; S'$ when given precondition $x = x' \wedge y = y'$ also ensures postcondition $x = x' \wedge y = y'$. This is a property that can be checked using existing verifiers like BLAST [20], Spec# [9], or ESC/Java2 [15]. Darvas et al. [16] use the key tool for interactive verification of noninterference; information flow is modeled by a dynamic logic formula, rather than by assertions as in self-composition.

When it comes to *conditional* information flow, the most noteworthy existing tool is the slicer by Snelting et al. [30] which generates *path conditions* in program dependence graphs for reasoning about end-to-end flows between specified program points/variables. In contrast, we provide a contract-based approach for *compositional* reasoning about conditions on flows with an underlying logic representation that can provide external evidence for conformance to conditional flow properties. We have recently received the implementation of the approach in [30], and we are currently investigating the deeper technical connections between the two approaches.

Finally, we have already noted how our work has been inspired by and aims to complement previous ground-breaking efforts in certification of MILS infrastructure [18, 19]. While the direct theorem-proving approach followed in these efforts enables proofs of very strong properties beyond what our framework can currently handle, our aim is to dramatically reduce the labor required, and the potential for error, by integrating automated techniques directly on code, models, and developer workflows to allow many information flow verification obligations to be discharged earlier in the life cycle.

8 Conclusion

We have presented what we believe to be an effective and developer-friendly framework for specification and automatic checking of conditional information flow properties, which are central to verification and certification of information applications built according to the MILS architecture. The directions that we are pursuing are inspired directly by challenge problems presented to us by industry teams using SPARK for MILS component development. The initial prototyping and evaluation of our framework has produced promising results, and we are pressing ahead with evaluating our techniques against actual product codebases developed at Rockwell Colins. A crucial concern in this effort will be to develop design and implementation methodologies for (a) exposing and checking conditional information flows and (b) specifying and checking security levels of data along conditional flows. We believe that our framework will nicely integrate with work on conditional declassification/degrading.

While our framework already supports many of the language features of SPARK and the extension to almost all other features (e.g., records) is straightforward, the primary remaining challenge is the effective treatment of arrays, which are often used in SPARK to implement complex data structures. Rockwell Collins developers are facing significant frustrations because SPARK treats arrays as atomic entities, i.e., it does not support even unconditional specification and checking of flows in/out of specific array components. This report contains some first steps toward building a theory for a more precise handling of conditional information flow for array components, but we still need to integrate it with the treatment of while loops and procedure calls, incorporate it into our implementation, and develop appropriate enhancements of SPARK contract notations.

Acknowledgments This work was supported in part by the US National Science Foundation (NSF) awards 0454348, 0429141, and CAREER award 0644288, the US Air Force Office of Scientific Research (AFOSR), and Rockwell Collins. The authors gratefully acknowledge the assistance of Rod Chapman and Trevor Jennings of Praxis High Integrity Systems in obtaining SPARK examples and running the SPARK tools. The material in this chapter originally appeared in the Proceedings of FM'08, LNCS 5014.

References

1. Amtoft T, Banerjee A (2004) Information flow analysis in logical form. In: 11th static analysis symposium (SAS), LNCS, vol 3148. Springer, Berlin, pp 100–115
2. Amtoft T, Banerjee A (2007a) A logic for information flow analysis with an application to forward slicing of simple imperative programs. Sci Comp Prog 64(1):3–28
3. Amtoft T, Banerjee A (2007b) Verification condition generation for conditional information flow. In: 5th ACM workshop on formal methods in security engineering (FMSE), a long version, with proofs, appears as technical report CIS TR 2007-2, Kansas State University, Manhattan, KS, pp 2–11
4. Amtoft T, Bandhakavi S, Banerjee A (2006) A logic for information flow in object-oriented programs. In: 33rd Principles of programming languages (POPL), pp 91–102
5. Amtoft T, Hatcliff J, Rodriguez E, Robby, Hoag J, Greve D (2007) Specification and checking of software contracts for conditional information flow (extended version). Technical report SAnToS-TR2007-5, CIS Department, Kansas State University. Available at http://www.sireum.org
6. Amtoft T, Hatcliff J, Rodríguez E (2009) Precise and automated contract-based reasoning for verification and certification of information flow properties of programs with arrays. Technical report, Kansas State University. URL http://www.cis.ksu.edu/~edwin/papers/TR-esop10.pdf, available from http://www.cis.ksu.edu/ edwin/papers/TR-esop10.pdf
7. Banerjee A, Naumann DA (2005) Stack-based access control and secure information flow. J Funct Program 2(15):131–177
8. Barnes J (2003) High integrity software – the SPARK approach to safety and security. Addison-Wesley, Reading, MA

9. Barnett M, Leino KRM, Schulte W (2004) The Spec# programming system: an overview. In: Construction and analysis of safe, secure, and interoperable smart devices (CASSIS), pp 49–69

10. Barthe G, D'Argenio P, Rezk T (2004) Secure information flow by self-composition. In: Foccardi R (ed) CSFW'04. IEEE, New York, NY, pp 100–114

11. Bell D, LaPadula L (1973) Secure computer systems: mathematical foundations. Technical report, MTR-2547, MITRE Corp

12. Bergeretti JF, Carré BA (1985) Information-flow and data-flow analysis of while-programs. ACM TOPLAS 7(1):37–61

13. Chapman R, Hilton A (2004) Enforcing security and safety models with an information flow analysis tool. In: SIGAda'04, Atlanta, Georgia. ACM, New York, NY, pp 39–46

14. Cohen ES (1978) Information transmission in sequential programs. In: Foundations of secure computation. Academic, New York, NY, pp 297–335

15. Cok DR, Kiniry J (2004) ESC/Java2: uniting ESC/Java and JML. In: Construction and analysis of safe, secure, and interoperable smart devices (CASSIS), pp 108–128

16. Darvas A, Hähnle R, Sands D (2005) A theorem proving approach to analysis of secure information flow. In: 2nd International conference on security in pervasive computing (SPC 2005), LNCS, vol 3450. Springer, Berlin, pp 193–209

17. Goguen JA, Meseguer J (1982) Security policies and security models. In: IEEE symposium on security and privacy, pp 11–20

18. Greve D, Wilding M, Vanfleet WM (2003) A separation kernel formal security policy. In: 4th International workshop on the ACL2 prover and its applications (ACL2-2003)

19. Heitmeyer CL, Archer M, Leonard EI, McLean J (2006) Formal specification and verification of data separation in a separation kernel for an embedded system. In: 13th ACM conference on computer and communications security (CCS'06), pp 346–355

20. Henzinger TA, Jhala R, Majumdar R, Sutre G (2003) Software verification with blast. In: 10th SPIN workshop, LNCS, vol 2648. Springer, Berlin, pp 235–239

21. Jackson D, Thomas M, Millett LI (eds) (2007) Software for dependable systems: sufficient evidence? National Academies Press, Committee on certifiably dependable software systems, National Research Council

22. Kaufmann M, Manolios P, Moore JS (2000) Computer-aided reasoning: an approach. Kluwer, Dordrecht

23. Myers AC (1999) JFlow: practical mostly-static information flow control. In: POPL'99, San Antonio, Texas. ACM, New York, NY, pp 228–241

24. Naumann DA (2006) From coupling relations to mated invariants for checking information flow. In: Gollmann D, Meier J, Sabelfeld A (eds) 11th European symposium on research in computer security (ESORICS'06), LNCS, vol 4189. Springer, Berlin, pp 279–296

25. Owre S, Rushby JM, Shankar N (1992) PVS: a prototype verification system. In: Proceedings of the 11th international conference on automated deduction (Lecture notes in computer science 607)

26. Rossebo B, Oman P, Alves-Foss J, Blue R, Jaszkowiak P (2006) Using SPARK-Ada to model and verify a MILS message router. In: Proceedings of the international symposium on secure software engineering

27. Rushby J (1981) The design and verification of secure systems. In: 8th ACM symposium on operating systems principles, vol 15, Issue 5, pp 12–21

28. Simonet V (2003) Flow Caml in a nutshell. In: Hutton G (ed) First APPSEM-II workshop, pp 152–165

29. Sireum website. http://www.sireum.org

30. Snelting G, Robschink T, Krinke J (2006) Efficient path conditions in dependence graphs for software safety analysis. ACM Trans Softw Eng Method 15(4):410–457

31. Terauchi T, Aiken A (2005) Secure information flow as a safety problem. In: 12th Static analysis symposium, LNCS, vol 3672. Springer, Berlin, pp 352–367

32. Vanfleet M, Luke J, Beckwith RW, Taylor C, Calloni B, Uchenick G (2005) MILS: architecture for high-assurance embedded computing. CrossTalk: J Defense Softw Eng 18:12–16

33. Volpano D, Smith G, Irvine C (1996) A sound type system for secure flow analysis. J Comput Security 4(3):167–188

Model Checking Information Flow

Michael W. Whalen, David A. Greve, and Lucas G. Wagner

1 Introduction

In order to describe the secure operation of a computer system, it is useful to study how information propagates through that system. For example, an unintended propagation of information between different components may constitute a *covert channel* that can be used by an attacker to gain access to protected information. We are therefore interested in determining how and when information may be communicated throughout a system. At Rockwell Collins, we have spent several years modeling *information flow* problems to support precise formal analyses of different kinds of software and hardware models.

In this chapter, we describe an analysis procedure that can be used to check a variety of information flow properties of hardware and software systems. One of the properties that can be checked is a form of *noninterference* [5, 19–21] that is defined over system traces. Informally, it states that a system input does not interfere with a particular output if it is possible to vary the trace of that input without affecting the output in question.

Although great strides have been made in the development of formal analysis tools over the last few years, there have been relatively few instances reported of their successful application to industrial problems outside of the realm of hardware engineering. In fact, software and system engineers are often completely unaware of the opportunities that these tools offer. One of the goals of our analysis was that it could be completely automated and directly applicable to the tools and languages used by engineers at Rockwell Collins, such as MATLAB Simulink® [11] and Esterel Technologies SCADE Suite™ [4]. These tools are achieving widespread use in the avionics and automotive industry and can also be used to describe hardware designs. The graphical models produced by these tools have straightforward formal semantics and are amenable to formal analysis. Furthermore, it is often the case that software and/or hardware implementations are generated directly from these models, so the analysis model is kept synchronized with the actual system artifact.

M.W. Whalen (✉)
Rockwell Collins, Inc., Bloomington, MN, USA
e-mail: mike.whalen@gmail.com

D.S. Hardin (ed.), *Design and Verification of Microprocessor Systems for High-Assurance Applications*, DOI 10.1007/978-1-4419-1539-9_13,
© Springer Science+Business Media, LLC 2010

Our analysis is based on annotations that can be added directly to a Simulink or SCADE model that describe specific sources and sinks of information. After this annotation phase, the translation and model checking tools can be used to automatically demonstrate a variety of information flow properties. In the case of noninterference, they will either prove that there is no information flow between the source and sinks or demonstrate a source of information flow in the form of a counterexample.

The result returned by the model checker must be justified by a general claim regarding the soundness of the analysis and the annotated model. To justify our analyses, we first define a kind of trace equivalence. This trace equivalence is just a form of the GWVr1 characterization defined earlier in Chap. 9 [6]. We then define syntax and semantics for a synchronous dataflow language and provide an information flow semantics for the language. Next, we demonstrate that this information flow semantics characterizes (i.e., enforces) the trace equivalence and define noninterference as a dual property of the information flow characterization. The information flow semantics is then directly reflected into a "flow model" that is emitted as part of the translation and conjoined with the original model. We finally show that model checking this conjoined model yields the same result as executing the flow model semantics.

The organization of the rest of the chapter is as follows: Sect. 2 introduces the concepts involved through the use of a motivating shared buffer example. Section 3 describes an abstract formalization of information flow through trace equivalence, presents the syntax and semantics for a simplified dataflow language, and proves an *interference theorem*, that is, the information flow semantics preserves the trace equivalence. Section 4 demonstrates how *noninterference* can be defined as a corollary of the interference theorem. Section 5 describes how this formalization is realized in the Gryphon tool suite. Section 6 describes how the tools can be used to analyze *intransitive interference*. Section 7 describes connections between the formalization in this chapter and the GWV formulation from Greve [6]. Section 8 describes applications of the analysis: the shared buffer model and also a large-scale model of the Rockwell Collins Turnstile high-assurance guard. Section 9 presents future directions for the analysis and concludes.

2 A Motivating Example

To motivate our presentation, we use an example of a shared buffer model, shown in Fig. 1. In this model, secret and unclassified information both pass through a shared buffer. In order to prevent leakage of secret information, this buffer is coordinated by a scheduler (bottom of the figure) that mediates access to the buffer. On the left, there are two input processes for secret and unclassified input. On the right, there are two output processes for secret and unclassified output.

When the scheduler is in the WAITING state, a write request from either input process will result in that process obtaining the buffer. The process will continue

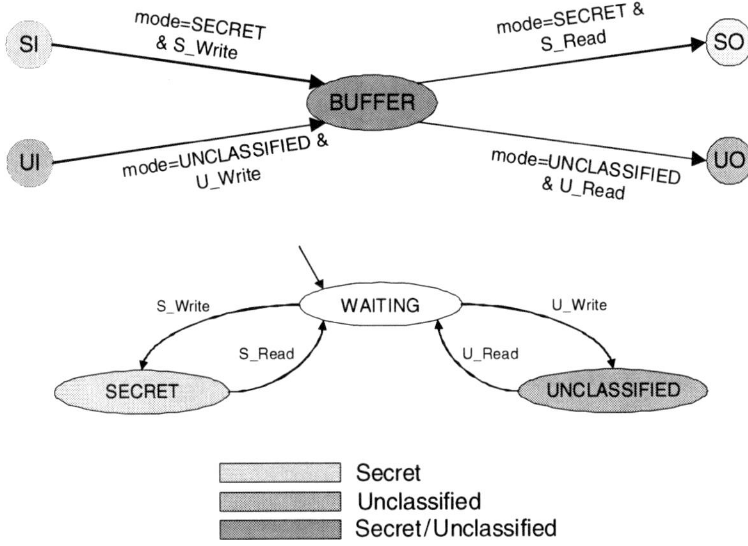

Fig. 1 Shared buffer architecture

to control the buffer until a corresponding read from the buffer is completed. The controller is designed to ensure that the secret data is only allowed to be consumed by the secret output and symmetrically that the unclassified data is only consumed by the unclassified output.

Given this system, we would like to determine whether or not there is information flow between the secret processes and the unclassified processes. In other words, is it possible for the unclassified processes to glean information of any kind from the secret processes and vice versa? This information sharing is usually called *interference*; *noninterference* is the dual idea expressing that no information sharing occurs. In this example, the potential for interference exists via the scheduler. Unclassified processes can perceive the state of the buffer (whether they are able to read and write from it) via the scheduler, which is affected by the secret processes.

If we decide that this interference is allowable, we would like to be able to determine whether there are any other sources of interference between the secret and unclassified processes. An analysis which does not account for the current system state will probably decide that there is the potential for interference, since both kinds of processes use a shared buffer. We would like a more accurate analysis that accounts for the scheduler state in order to show that there is no interference through the shared buffer.

This example demonstrates important features of the analysis that we will describe in the next sections:

- *Conditional information flow*. We would like the analysis to account for enough of the system state to allow an accurate analysis (e.g., that no information flows from a secret input to unclassified output through the shared buffer)

- *"Covert" information flow.* The scheduler does not directly convey information from secret processes to unclassified processes, yet its state allows information about the secret processes to be perceived. The analysis should detect this interference.
- *Intransitive information flow.* If we are willing to allow information flow through the scheduler, there should be a mechanism to allow us to tag this information path as "allowable" and determine if other sources of flow exist. In the noninterference literature, this is generally described as *intransitive noninterference* [5, 19, 20]. The meaning of *intransitive* has to do with the nature of information flows. Since the scheduler depends on the secret input and the unclassified output depends on the scheduler, a *transitive* analysis would assert that the unclassified output depends on the secret input. However, we would like to be able to tag certain mediation points (e.g., downgraders or encryptors) as "allowed" sources of information flow.

2.1 Shared Buffer Simulink Model

A Simulink model of the shared buffer example is shown in Fig. 2. The inputs to the model are shown on the left: we have the requests to use the buffer from the four processes (the secret input/output process and the unclassified input/output processes) as well as the input buffer data from the secret and unclassified input processes. The scheduler subsystem determines access to the buffer, while the buffer subsystem uses the scheduler state to determine which process writes to the shared buffer.

The information flow analysis is performed in terms of a set of *principal variables*. These variables are the variables that we are interested in tracking

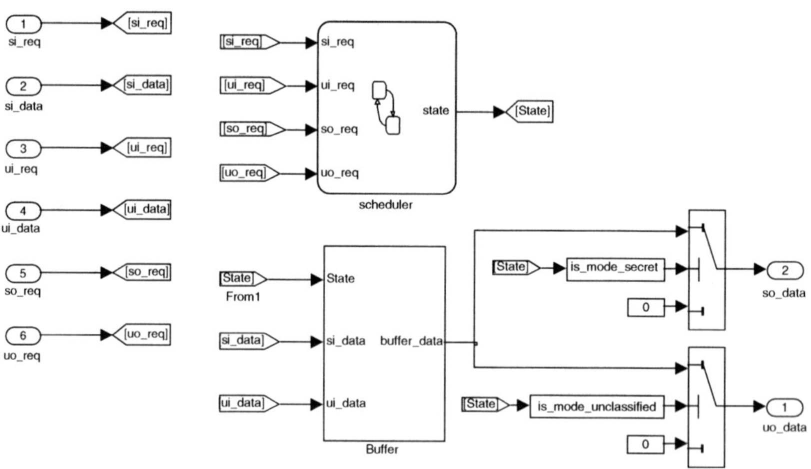

Fig. 2 Shared buffer example in Simulink

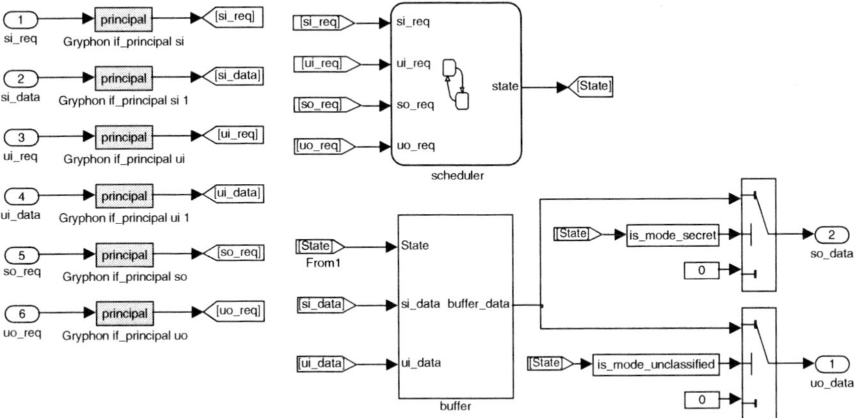

Fig. 3 Annotated Simulink model

through the model. We always track the input variables to the model, and we sometimes track computed variables internal to the model. To perform the analysis, the Simulink model is annotated to add the principal variables as shown in Fig. 3.

Once we have annotated the model, we use the *Gryphon* tool set [24] to automatically construct an information flow model that can be model checked on a variety of model checking tools including NuSMV [8], SAL [23], and Prover [16]. The analysis process extends the original model with a *flow model* that operates over sets of principal variables. Each computed variable in the original model has a *flow variable* in the flow model that tracks its dependencies in terms of the principal variables.

For model checking, sets of principal variables are encoded as bit sets, and checking whether information flow is possible is the same as determining whether it is possible that one of the principal bits is set. For the model above, the translation generates the following bit set for the principals:

```
Principal bit vector: {
    si maps to bit: 0,
    so maps to bit: 1,
    ui maps to bit: 2,
    uo maps to bit: 3 }
```

Now we can write properties over output variables. For example, suppose we want to show that the secret output data is unaffected by the unclassified input or output principal. In this case, we could write:

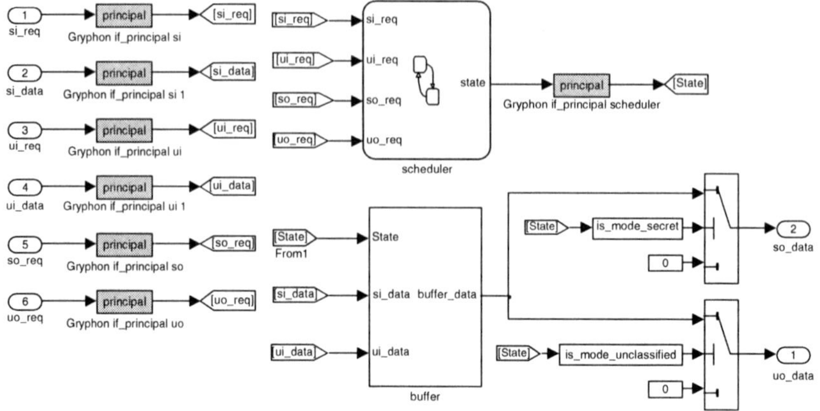

Fig. 4 Annotated Simulink model with intransitive flow

```
LTLSPEC G(!(gry_IF_so_data[ui_idx] |
gry_IF_so_data[uo_idx]));
```

gry_IF is the prefix used for the flow variables, so the analysis checks whether there is flow to the *so_data* output from the *ui* principal or the *uo* principal. These principals correspond to flow from the *ui_req*, *ui_data*, and *uo_req* input variables.

As described earlier, this property is violated, because there is information flow from the unclassified processes to the secret output through the scheduler. NuSMV generates a counterexample that we can examine to determine how the information leak occurred.

After analyzing the problem, we decide that the flow of information through the scheduler state is allowable. We would now like to search for additional sources of flow. By adding an additional principal for the scheduler state, as shown in Fig. 4, we can ignore the flows from the *ui* and *uo* principals that occur through the scheduler. After rerunning the analysis, the model checker finds no other sources of information flow.

3 Information Flow Modeling for Synchronous Dataflow Languages

Languages such as Simulink [11] and SCADE [4] are examples of *synchronous dataflow languages*. The languages are *synchronous* because computation proceeds in a sequence of discrete instants. In each instant, inputs are perceived and states and outputs are computed. From the perspective of the formal semantics, the computations are instantaneous. The languages are *dataflow* because they can be understood as a system of assignment equations, where an assignment can be computed as

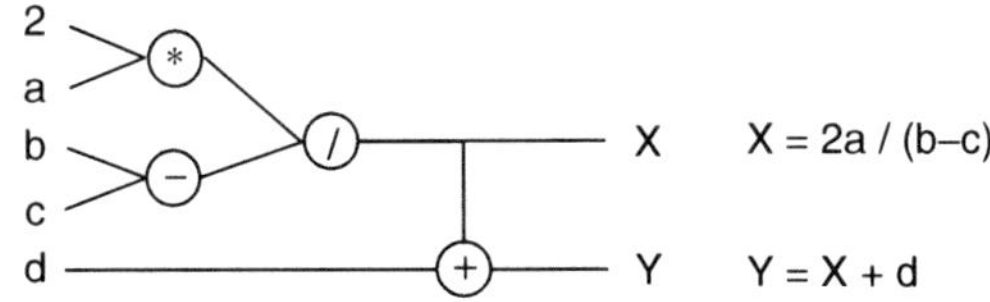

Fig. 5 Graphical and textual presentation of a set of equations

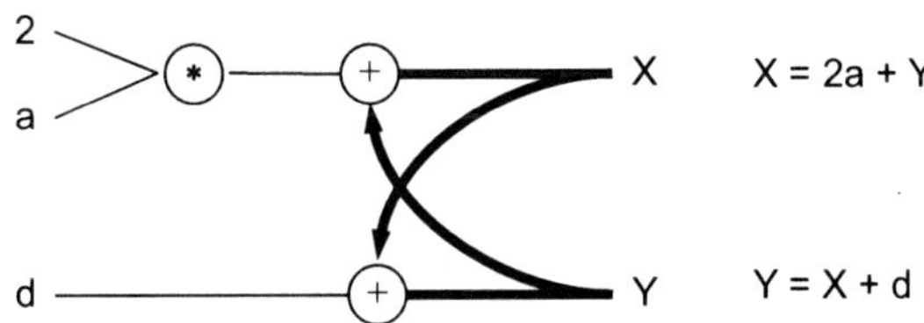

Fig. 6 Cyclic set of equations

soon as the equations on which it is dependent are computed. The equations can be represented either textually or graphically. As an example, consider a system that computes the values of two variables, X and Y, based on four inputs: a, b, c, and d, as shown in Fig. 5.

The variables (often referred to as *signals*) in a dataflow model are used to label a particular computation graph. Therefore, it is incorrect to view the equations as a set of constraints on the model: a set of equations shown in Fig. 6 is not a valid model because X and Y mutually refer to one another. This is shown in Fig. 6, where the bold lines indicate the cyclic dependencies. Such a system may have no solution or infinitely many solutions, so cannot be directly used as a deterministic program. If viewed as a graph, these sets of equations have *data dependency cycles* and are considered incorrect.

However, in order for the language to be useful, we must be able to have mutual reference between variables. To allow benign cyclic dependencies, we create a step-delay operator (i.e., a latch) using the comma operator. For example, $\{X = 2a/Y;\ Y = 1, (X + d))\}$ defines a system where X is equal to $2a$ divided by the current value of Y, while Y is initially equal to 1, and thereafter equal to the *previous* value of X plus d.

There are several examples of textual dataflow languages, including Lustre [7], Lucid Synchrone [3], and Signal [9], that differ in terms of structuring mechanisms, computational complexity (i.e., whether recursion is allowed), and *clocks* that define the rates of computation for variables. Our analysis is defined over the Lustre language. Lustre is the kernel language of the SCADE tool suite and also the internal language of the Rockwell Collins Gryphon tool suite. Lustre is also sufficient to model the portions of the Simulink/Stateflow languages that are suitable for hardware/software codesign.

3.1 Modeling Information Flow

When describing information flow, we are often attempting to define a *noninterference* relation of some kind. There have been several formulations of noninterference [5,19–21] involving transition systems and process algebras, which have focused on noninterference in terms of a trace of actions (inputs) fed into some machine that generates outputs. The idea of noninterference is simple: a security domain u does not interfere with domain v if no action performed by u can influence subsequent outputs of v.

In the formulation of [20], noninterference is demonstrated by removing actions from the trace T (call it T$'$) and showing that under certain conditions the final output of the machine is the same. However, for synchronous dataflow languages such as Lustre or Simulink, characterizing the "removable" inputs is difficult, as each input variable is assigned a value in each step; one must define predicates over the cross product of the input variables. Characterizing the "action" of a model with potentially tens or hundreds of outputs presents similar difficulties.

Instead, following Greve in an earlier chapter [6], we would like to define a notion of noninterference on individual variables within a model in terms of correspondences between two traces. In our formulation, a trace is a sequence of model states, each state containing the assignments to all variables within the model. We define a set of *principal variables* as a superset of the inputs and then define an *Interferes* function for any variable c that describes the set of principals that could possibly affect the value of c. We determine the correctness of the *Interferes* set in terms of trace correspondence. The *Interferes* set is correct if given any variable c and traces π_0 and π_1 and if the traces agree on all the variables of *Interferes(c)*, then they will agree on c. In other words, the variables in *Interferes(c)* are sufficient to determine the value of c at any step. Equivalently, any principal variable outside the *Interferes* set cannot affect the value of c.

Formalized in the PVS notation [22], the theorem that we are proving is as follows:

```
InterferenceTheorem: LEMMA

 FORALL (p: Program, gt1:gtrace, st1,st2:strace):
    FORALL (idx:index):
       WFp(p) & St(p,st1) & St(p,st2) &
       IFt(p,st1,gt1) &
       vtraceEquivSet(DepSet(idx,gt1),st1,st2) =>
                  liftv(idx,st1) = liftv(idx,st2)
```

This theorem states that if two traces are equivalent (*vtraceEquivSet*) on the dependencies computed for a variable *idx* by our *Interferes* set (*DepSet(idx,gt1)*), then two traces agree on the value of *idx*. The details of the theorem and steps in the proof will be explained in the following sections.

How this is used in practice is that the user suggests what is believed to be a noninterfering principal variable for some variable *c* and a model checker is used to determine whether or not this variable interferes with (i.e., affects) *c*.

3.2 Using PVS

PVS [15,22] is a mechanized theorem prover based on classical, typed higher order logic. Specifications are organized into (potentially parameterized) theories, which are collections of type and function definitions, assumptions, axioms, and theorems. The proof language of PVS is composed of a variety of primitive inference procedures that may be combined to construct more powerful proof strategies.

Normally in PVS the proof process is performed interactively, and the proof script encoding the entire proof is not visible to the user. In our development, we used the *ProofLite* [14] extension to PVS in order to embed the proofs as comments into the PVS theories. To make the theories shorter and easier to understand, we omit the ProofLite scripts in this chapter. However, the interested reader is encouraged to visit http://extras.springer.com, and enter the ISBN for this book, in order to view the complete scripts.

3.3 Traces and Processes

The semantics of synchronous dataflow languages are usually defined in terms of *traces* that describe the behavior of the system over time. These traces are formalized in the language of the PVS theorem prover in Fig. 7. We are interested in two kinds of traces. First, we are interested in the trace of values produced by the execution of the system. We define the set of values that can be assigned to variables using the opaque type *vtype*.[1] The execution traces are mappings from instants in time to states, where states map variables to values, and are defined by the *strace* and *state* types, respectively. The variables in our model correspond to indices in Greve's formulation, and we use the term *index* to identify a variable in a trace.

Second, we are interested in tracing the dependencies of a variable in terms of a set of other variables (in GWV terms, the information flow graph). These traces map instants in time to graph states, where each graph maps an index (i.e., variable) to sets of indices. At each instant, for a given variable *v*, the graph captures a set of

---

[1] Opaque types in PVS allow one to define a type as an unspecified set of values.

variables that are necessary for computing *v*. These traces are defined by the *gtrace* and *graphState* types, respectively.

Note that our states are defined over an infinite set of variables *nat*. In a real system, we would have a finite set, but this can be modeled by simply ignoring all variables above some maximum index. This change does not affect the formalization or the proofs.

Next, we define *processes* that constrain the traces in Fig. 8. The processes are built from expressions: an (unspecified) set of unary and binary operators, constant,

```
Traces: THEORY
BEGIN


  index: TYPE = nat
  time: TYPE = nat
  vtype: TYPE+


  state : TYPE = [ index -> vtype ]
  strace: TYPE = [ time -> state ]


  get(i: index, s: state): vtype = s(i)


  graphState: TYPE = [ index -> set[index] ]
  gtrace: TYPE = [ time -> graphState ]
END Traces
```

Fig. 7 Traces theory

```
ProcessExprTypes: THEORY
BEGIN
  IMPORTING Traces

  BopType: TYPE+
  UopType: TYPE+

  BopEx(Bop: BopType, v1,v2: vtype): vtype
  UopEx(Uop: UopType, v0: vtype): vtype
  isTrue(v0: vtype): bool
END ProcessExprTypes


ProcessExpr: DATATYPE
BEGIN
  IMPORTING ProcessExprTypes
```

Fig. 8 Processes and programs

```
    Constant(value : vtype): Constant?
    Variable(name  : index):  Variable?
    ITE(test: ProcessExpr, thn: ProcessExpr,
        els: ProcessExpr): ite?
    Bop(OpB: BopType, a1: ProcessExpr,
        a2: ProcessExpr): Bop?
    Uop(OpU: UopType, a0: ProcessExpr): Uop?
END ProcessExpr

ProcessAssignment: DATATYPE
BEGIN
  IMPORTING ProcessExpr
  Gate (gexpr: ProcessExpr): Gate?
  Latch(v0: vtype, lexpr: ProcessExpr): Latch?
  Input: Input?
END ProcessAssignment

Program: THEORY
BEGIN
  IMPORTING ProcessAssignment
  IMPORTING IndexSet[index]

  Program: TYPE = [ index -> ProcessAssignment ]

  StatesP(p: Program): set[index] =
    (LAMBDA (v: index): Latch?(p(v)))

  InputsP(p: Program): set[index] =
    (LAMBDA (v: index): Input?(p(v)))

  GatesP(p: Program): set[index] =
    (LAMBDA (v: index): Gate?(p(v)))

  De(e: ProcessExpr): RECURSIVE set[index] =
    CASES e OF
      Constant(value): Empty,
      Variable(name):  singleton(name),
      ITE(test,thn,els): De(test) + De(thn) +
        De(els),
      Bop(OpB,a1,a2):  De(a1) + De(a2),
      Uop(OpU,a0): De(a0)
    ENDCASES
  MEASURE e by <<

  belowSet(n: nat, s: set[nat]): bool =
    FORALL (i: nat): member(i,s) => (i < n)
```

Fig. 8 (continued)

```
   Ae(v: index, a: ProcessAssignment): ProcessExpr =
     CASES a OF
       Gate (gexpr)    : gexpr,
       Latch(v0,lexpr): lexpr,
       Input           : Variable(v)
     ENDCASES;

   WFp(p: Program) : bool =
     FORALL (v: index):
       belowSet(v, De(Ae(v,p(v))) & GatesP(p))

   WFPrograms : TYPE = { p : Program | WFp(p) }

 END Program
```

Fig. 8 (continued)

variable, and conditional (if/then/else) expressions. We next partition the indices into gates, latches, and inputs. Gates are computed from the current values of other variables, while latches are computed from the previous values of other variables. Latches also have an initial value which is their value in the first step of a trace. Inputs are not computed and assumed to be externally provided.

The processes described in Fig. 8 define a simple *synchronous dataflow language*, such as Simulink or SCADE. For the purposes of this discussion, the structuring mechanisms of these languages (nodes and subsystems) as well as the clocking mechanisms for variables can be thought of as syntactic sugar.

In general, a set of simultaneous equations may yield zero or multiple solutions. We want a program to be *functional*, given a particular input trace. In order to ensure that the assignments yield functional traces, we need a strict ordering on gate assignments. Since indices are defined as naturals, it suffices to define an ordering such that the assignment expression for a variable may only refer to gate indices that are strictly smaller than the index being assigned. Note that *only* gate indices are restricted – it is possible to write benign cyclic dependencies involving latches.

The *Ae* function returns the assignment expression associated with a particular index. For inputs, *Ae* just returns a variable expression referring to the input. The *De* predicate defines the dependencies of an expression and *WFp* defines the functional well-formedness constraint on programs. Note that this predicate also forms a basis for inducting over the gates within the program that we will use for several of the proofs.

3.4 Semantic Rule Conventions

We define different kinds of semantics for the values produced by a program and also for the information flow. The semantic functions that are introduced follow a naming convention to make them easier to follow and to relate to one another. The form of the semantics functions is as follows:

```
<TYPE><syntax><OPTIONAL RESTRICTION>
```

For example, the *Se* function defines the value-semantic function for expressions, and the *IFsG* function defines the information-flow function for states with respect to gates.

The <TYPE>s of semantics that will be used in the following discussion are as follows:

S: Value semantics for traces
D: Syntactic dependencies
DS: Dependencies based on syntax and current state
IF: Information flow dependencies

The <syntax>es that will be discussed are the following:

e: Expressions
i: Indices (assignments)
s: States
t: Traces

The <OPTIONAL RESTRICTION>s restrict the semantic functions at a particular syntactic level to:

I: Inputs
G: Gates
L: Latches

3.5 Value Trace Semantics

We next create semantic functions for the expressions and programs in Fig. 8. Following [1] and [12], the semantics are defined in terms of *trace conformance*, as shown in Fig. 9. We state that a trace conforms to a program if the values computed by the assignment expressions for the gates and latches correspond to the values in the trace. The *Se* function computes a value from a Process expression. The *SsG* predicate checks conformance between the gate assignments and a state, and the *SsL* predicates check conformance between the latch assignments and the trace. The *St* predicate defines trace conformance over both gates and latches.

3.6 Creating an Accurate Model of Information Flow

Now we can create a semantics that tracks information flow through the model, as shown in Fig. 10. This semantics maps indices to the set of indices used when computing the value of the index. For expressions, we create two different semantics; the first tracks the indices that are immediately used within the computation of the expression; the second traces the indices back to *principal variables*, which are

```
ProcessSemantics: THEORY
BEGIN
  IMPORTING Program

  Se(e: ProcessExpr, s: state): RECURSIVE vtype =
    CASES e OF
      Constant(value): value,
      Variable(name): s(name),
      ITE(test,thn,els):
        IF isTrue(Se(test,s)) THEN Se(thn,s)
        ELSE Se(els,s) ENDIF,
      Bop(OpB,a1,a2):  BopEx(OpB,Se(a1,s),Se(a2,s)),
      Uop(OpU,a0): UopEx(OpU,Se(a0,s))
    ENDCASES
  MEASURE e by <<

  Si(p: Program)(i: index, s0: state): vtype =
    CASES p(i) OF
      Gate (gexpr)     : s0(i),
      Latch(v0,lexpr)  : Se(lexpr,s0),
      Input            : s0(i)
    ENDCASES

  SsG(p: Program, s0: state): bool =
    FORALL (v: index): Gate?(p(v)) =>
      (s0(v) = Se(Ae(v,p(v)),s0))

  SsL0(p: Program, s0: state): bool =
    FORALL (v: index): Latch?(p(v)) =>
      (s0(v) = v0(p(v)))

  SsLn(p: Program, s0,s1: state): bool =
    FORALL (v: index): Latch?(p(v)) =>
      (get(v,s1) = Si(p)(v,s0))

  St(p: Program, st: strace): bool =
    FORALL (n: nat):
      IF (n = 0) THEN
        SsL0(p,st(0)) & SsG(p,st(0))
      ELSE
        SsLn(p,st(n-1),st(n)) & SsG(p,st(n))
      ENDIF

END ProcessSemantics
```

Fig. 9 Process trace semantics

the actual concern of the information flow analysis. For the moment, we consider the inputs as the principal variables. We expand this notion when we talk about *intransitive interference* in Sect. 6.

The only difference between the *DSe* and *IFe* semantics in Fig. 10 is in the behavior of the Variable branch. For the *IFe* semantics, a set of *principal variables* are provided. If a referenced variable is a principal variable, then we return it as a dependency; if it is not, then we return the dependencies of that variable. The effect of this rule is to backchain through the intermediate variables so that dependencies

```
ProcessIndexSets: THEORY
BEGIN
  IMPORTING ProcessSemantics
  IMPORTING MemberRules[index]

  DSe(e: ProcessExpr, s0: state):
      RECURSIVE set[index] =
    CASES e OF
      Constant(value): Empty,
      Variable(name):  singleton(name),
      ITE(test,thn,els):
        IF isTrue(Se(test,s0)) THEN
          SDe(test,s0) + SDe(thn,s0)
        ELSE
          SDe(test,s0) + SDe(els,s0)
        ENDIF,
      Bop(OpB,a1,a2):  SDe(a1,s0) + SDe(a2,s0),
      Uop(OpU,a0): SDe(a0,s0)
    ENDCASES
  MEASURE e by <<

  IFe(e: ProcessExpr, principal: set[index],
      s0: state, g0: graphState): RECURSIVE
      set[index] =
    CASES e OF
      Constant(value): Empty,
      Variable(name):
        IF principal(name) THEN
          singleton(name)
        ELSE
          g0(name)
        ENDIF,
      ITE(test,thn,els):
        IF isTrue(Se(test,s0)) THEN
          IFe(test,principal,s0,g0) +
          IFe(thn,principal,s0,g0)
        ELSE
          IFe(test,principal,s0,g0) +
          IFe(els,principal,s0,g0)
        ENDIF,
      Bop(OpB,a1,a2):  IFe(a1,principal,s0,g0) +
         IFe(a2,principal,s0,g0),
      Uop(OpU,a0): IFe(a0,principal,s0,g0)
    ENDCASES
  MEASURE e by <<

  IFsI(p: Program, s0: state, g0: graphState): bool=
    FORALL (v: index): Input?(p(v)) =>
      (g0(v) = IFe(Ae(v,p(v)),InputsP(p),s0,g0))
```

Fig. 10 Process index semantics

```
     IFtI(p: Program, st: strace, gt: gtrace): bool =
       FORALL (t: time) : IFsI(p, st(t), gt(t))

     IFsG(p: Program, s: state, g: graphState): bool =
       FORALL (v: index): Gate?(p(v)) =>
         (g(v) = IFe(Ae(v,p(v)),InputsP(p),s,g))

     IFtG(p: Program, st: strace, gt: gtrace): bool =
       FORALL (t: time) : IFsG(p, st(t), gt(t))

     IFsL0(p: Program, g0: graphState): bool =
       FORALL (v: index):
         Latch?(p(v)) => g0(v) = Empty

     IFsLn(p: Program, s0: state,
           g0,g1: graphState): bool =
       FORALL (v: index): Latch?(p(v)) =>
         (g1(v) = IFe(Ae(v,p(v)),InputsP(p),s0,g0))

     IFtL(p: Program, st: strace, gt: gtrace): bool =
       FORALL (n: nat):
         IF (n = 0) THEN
           IFsL0(p,gt(0))
         ELSE
           IFsLn(p,st(n-1),gt(n-1),gt(n))
         ENDIF

     IFt(p: Program, st: strace, gt: gtrace): bool =
       IFtG(p,st,gt) & IFtL(p,st,gt) & IFtI(p,st,gt)

     tracePair : TYPE = [# s: strace, g: gtrace #];

     tp_ok(p: Program, tp: tracePair) : bool =
       IFt(p, s(tp), g(tp)) AND St(p, s(tp)) ;
```

Fig. 10 (continued)

are always a subset of the principal variables. The *DSe* semantics, on the other hand, return the immediate dependencies (i.e., the indices of all variables referenced in the assignment expression).

Note that both the *DSe* and *IFe* semantics are state dependent: For if/then/else expressions, the set of dependencies depends on the if-test; only dependencies for the used branch are returned. This feature allows conditional dependencies to be tracked within the model.

After defining the expression semantics, we define the IF semantics on states and programs, matching the structure of the *S* definitions in Fig. 9. At the bottom of Fig. 10, we define trace pairs as a type and define trace pair conformance to a program based on both semantics.

3.7 PVS Proof of Trace Equivalence (InterferenceTheorem)

We can now state the interference theorem that should be proven over the trace pairs. Informally, we would like to state that for a particular index *idx*, if the inputs referenced in an information flow trace for *idx* (*DepSet*) have the same values in two state traces (*vtraceEquivSet*), then the two traces will have the same values for *idx*. Formally, this obligation is expressed in Fig. 11. Note that there is an asymmetry in the interference theorem: we define two execution traces (*st1* and *st2*) but only one graph trace (*gt1*). The graph trace (*gt1*) corresponding to an execution trace (*st1*) for a given index *idx* characterizes the signals that must match for any other execution trace (in this case *st2*) to match *st1* for signal *idx*. It is equivalent to use a graph trace based on *st2*.

To prove this theorem, we have to build a hierarchy of equivalences shown in Fig. 12. This graph does not show all of the connections between proofs (e.g., which theorems are instantiated in the proofs of other theorems), but it provides a good overview of the structure of the proof. Ultimately, we are interested in proving the final theorem, which defines a relationship between traces as described by the information flow semantics *IF* and the value semantics *S*. In order to prove this theorem, we define an intermediate flow semantics based on state dependencies (*DS*). Whereas the information flow semantics unwinds the dependencies from outputs to inputs implicitly through the use of the graph state and graph trace, the *DS* flow semantics unwind the graph explicitly and therefore provide an easier basis for inductive proof.

```
vtrace: TYPE = [ time -> vtype ]

liftv(i: index, st: strace): vtrace =
  (LAMBDA (t: time): st(t)(i))

vtraceEquivSet(set: set[index],st1,st2: strace):
  bool =
  FORALL (i: index): member(i,set) =>
    liftv(i,st1) = liftv(i,st2)

DepSet(x: index, gt: gtrace): set[index] =
  (lambda (i: index):
    (EXISTS (t: time): member(i,gt(t)(x))))

InterferenceTheorem: LEMMA
  FORALL (p: Program, gt1:gtrace, st1,st2:strace):
    FORALL (idx:index):
      WFp(p) & St(p,st1) & St(p,st2) &
      IFt(p,st1,gt1) &
      vtraceEquivSet(DepSet(idx,gt1),st1,st2) =>
        liftv(idx,st1) = liftv(idx,st2)
```

Fig. 11 Interference theorem

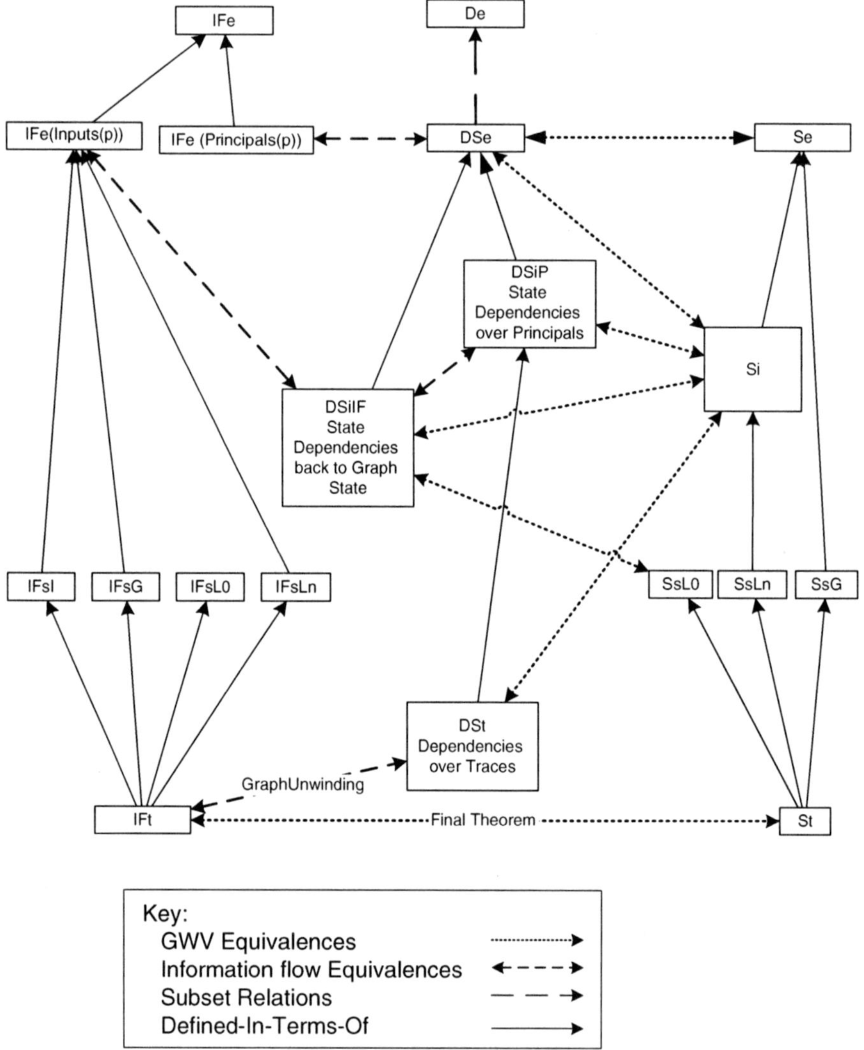

Fig. 12 Proof graph for final theorem

The "rows" of the proof graph correspond to a level in the evaluation hierarchy. Reading from top to bottom, we talk about equivalences in terms of expressions, then in terms of indices (assignments), then states, and, finally, traces. The "columns" correspond to the different semantics. On the left is the information flow (*IF*) semantics, in the middle is the *DS* semantics, and on the right is the value (*S*) semantics. One semantics bridges the *IF* and *DS* semantics (*DSiIF*).

There are two different kinds of theorems that are proved between the semantics. The first are equivalences between the different flow semantics (e.g., that two flow semantics yield the same set of dependencies). The second are GWV-style theorems, in the same style as [6]. These state that if the values of the dependent indices for a piece of syntax $\sum$ are equal within two states or traces $s1$ and $s2$, then the value produced by evaluating $\sum$ over $s1$ and $s2$ will be equal.

In our analysis, we prove GWVr1-style theorems. GWVr1 is less expressive than GWVr2 but it is simpler to formulate. The additional expressive power in GWVr2 is necessary to describe dynamic memory, but the synchronous models that we analyze in this chapter do not use dynamic memory, so GWVr1 is sufficiently expressive for our purposes. The connection between the formulation in this chapter and [6] is explored further in Sect. 7.

3.7.1 Expression Equivalence Theorems

In Fig. 13, we begin the process of proving the final theorem by describing some lemmas over expressions. These will form the basis of the later proofs over larger pieces of syntax.

The *DSe_subset_De* lemma states that the state-aware dependency function (*DSe*) returns a subset of the indices referenced by the syntactic dependency function (*De*). We appeal to this lemma (through another lemma: *WFg_to_WFgDSe*) to establish a basis for induction for some of the proofs involving equivalence of gate assignments.

The *Compose* function is used to look up each of the entries in a set in the graph state. It performs the same function as the Direct Interaction Allowed (DIA) function in Greve's formulation [6]. It is used to map from a set of immediate dependencies to their dependencies.

The *IFe_to_DSe_Property* lemma defines the first mapping between the state-based *DS* dependency semantics and the *gtrace*-based *IF* dependency semantics. Remember from Sect. 3.4 that the *IFe* semantics are defined in terms of a set of *principals*: if a variable is principal, then we look up its dependencies in the graph state. This property creates an equivalence between these semantics by looking up (via *Compose*) the nonprincipal variables from the *DSe* semantics.

3.7.2 Program Well-Formedness Theorems

In Fig. 14, we define a bridge between the program well-formedness constraint *WFp* and state dependencies (*DSe*). This bridge will allow us to use the *WFp* predicate in reasoning about GWV equivalences involving state dependencies. We define a *WFgDSe* predicate that defines well-formedness in terms of the *DSe* and show that *WFp* implies the (more accurate) *WFgDSe* predicate.

3.7.3 GWV Equivalence Theorems

Now, we can start proving GWV-style equivalence properties. These state that if the values of the dependent indices for a piece of syntax $\sum$ match within two states or

```
DSe_subset_De: LEMMA
   FORALL (e: ProcessExpr, s0: state):
     subset?(DSe(e,s0),De(e))


Compose(uset: set[index], g0: graphState):
     set[index] =
   (lambda (z: index):
    (EXISTS (m: index):
      member(m,uset) & member(z,g0(m)))))

member_Compose: LEMMA
   FORALL (i: index, uset: set[index],
          g0: graphState):
     member(i,Compose(uset,g0)) =
       (EXISTS (m: index):
         member(m,uset) & member(i,g0(m)))

IFe_to_DSe_Property(e: ProcessExpr): bool =
   FORALL (principal: set[index], s0: state,
          g0: graphState):
     IFe(e,principal,s0,g0) =
       LET uset: set[index] = DSe(e,s0) IN
         (uset & principal) +
          Compose(uset & (not(principal)),g0)

IFe_to_DSe_proof: LEMMA
   FORALL (e: ProcessExpr): IFe_to_DSe_Property(e)

IFe_to_DSe: LEMMA
   FORALL (e: ProcessExpr, principal: set[index],
          s0: state, g0: graphState):
     IFe(e,principal,s0,g0) =
       LET uset: set[index] = DSe(e,s0) IN
         (uset & principal) +
          Compose(uset & (not(principal)),g0)
```

Fig. 13 Expression equivalence proofs

traces $s1$ and $s2$, then the value produced by the evaluating $\sum$ over $s1$ and $s2$ will match. The idea is that we will start from the *immediate* dependencies of an expression and progressively unwind the dependencies toward the inputs. This unwinding occurs in two stages as follows:

- First we unwind to the *principals*, which (for the purposes of the proof) are the states and inputs. Another way of looking at this first unwinding is unwinding back to the "beginning" of the step. This is the definition of the *DSiP* dependencies.
- Next, we unwind the dependencies back to the inputs by examining the graph trace over time. This is the definition of the *DSt* dependencies.

```
   Principals(p: Program): set[index] =
     StatesP(p) + InputsP(p)

   WFg(p: Program): bool =
     FORALL (v: index):
       belowSet(v,De(Ae(v,p(v))) - Principals(p))

   Principals_Gates_partition : LEMMA
     FORALL (p: Program):
        (GatesP(p) = complement(Principals(p)))

   Principals_Gates_subset_equiv : LEMMA
     (FORALL (s: set[index], p: Program) :
       (s & GatesP(p)) = (s - Principals(p)))

   WFp_to_WFg : LEMMA
     FORALL (p: Program) :   (WFp(p) = WFg(p))

   WFgDSe(p: Program, s0: state): bool =
     FORALL (v: index):
       belowSet(v,DSe(Ae(v,p(v)),s0) - Principals(p))

   WFg_to_WFgDSe: LEMMA
     FORALL (p: Program, s0: state):
       WFg(p) => WFgDSe(p,s0)

END ProcessIndexSets
```

Fig. 14 Well-formedness predicates for programs

We also map these state-based equivalences that are computed via explicit unwindings of dependencies to the *IF* equivalences, which implicitly unwind the dependencies using the graph states. This is accomplished by using the *DSiIF* dependency relation. This will be the key lemma to show the equivalence of the *IF* and *DS* formulations.

Figure 15 shows the dependency proof for the *DSe* dependencies. There are two equivalences: the first over evaluation of expressions and the second over evaluation of indices.

Figure 16 shows the proofs for the next level of unwinding: showing that if the *principal* variables are the same for two states, then the results produced for an index will be the same. This step removes the gates from the dependency calculation.

Figure 17 shows the proofs of the next level of unwinding to the dependencies of the states. The definition of the *DSiIF* predicate is particularly important as it bridges between the graph-trace-based *IF* semantics and the state-based *DS* semantics. Like the *DSiP* semantics, it backtraces through the gates to reach dependencies based on states and inputs. The distinction is that it then looksup the

```
ProcessInterference: THEORY
BEGIN
  IMPORTING ProcessIndexSets
  IMPORTING GWV_EquivSetRules[index,state,vtype,get]

  StateEquivSet(s:set[index],s1,s2: state): bool =
    equivSet(s,s1,s2)

  GWVr1_Se_DSe: LEMMA
    FORALL (e: ProcessExpr):
      FORALL (in1, in2: state):
        StateEquivSet(DSe(e, in1), in1, in2) =>
          (Se(e, in1) = Se(e, in2))

  GWVr1_Si_DSe: LEMMA
    FORALL (p: Program):
      FORALL (i: index, in1, in2: state):
        SsG(p,in1) & SsG(p,in2) &
          StateEquivSet(DSe(Ae(i,p(i)), in1),
                        in1, in2) =>
            Si(p)(i,in1) = Si(p)(i,in2)
```

Fig. 15 GWVr1 for DSe dependencies

state dependencies in the graph state. This means that the dependencies computed by *DSiIF* will match the dependencies computed by the IF relation, as demonstrated by the *IFe_to_DSiIF* lemma. This is a key lemma in proving the unwinding theorem over state dependency traces *DSt* and information flow traces *IFt*.

Finally, in Fig. 18, we map dependencies to inputs across a multistep trace. First, we prove a lemma that is sufficient for the proof of latch assignment at step zero (*GWVr1_Si_SsL0*). This lemma will be used to provide the base case for latches in the *GWVr1_Si_DSt* proof.

Next, in Fig. 19, we have to define a graph unwinding theorem, which maps between our state-dependency-based formulation *DSt* and our graph-dependency-based formulation *IFt*. This is performed in two steps. First, we show that the *DSiIF* formulation matches the result returned by *IFe*. Next, we define the unwinding theorem which demonstrates that *DSt* and *IFt* yield the same dependencies.

3.7.4 Proof of InterferenceTheorem

Now we have finally assembled the pieces necessary to prove the trace theorem that was proposed in Fig. 8 in Sect. 3.7. The proof is shown in Fig. 20. We state that the information flow *characterizes* the execution of a model, if it satisfies the *InterferenceTheorem*.

```
DSiP(p: Program,s0: state)(x: index) :
RECURSIVE set[index] =
  LET uset: set[index] = DSe(Ae(x,p(x)),s0) IN
  LET pri : set[index] = Principals(p) IN
    (uset & pri) +
      (lambda (z: index):
        (EXISTS (m: index):
          m < x &
          member(m,uset & not(pri)) &
          member(z,DSiP(p,s0)(m))))
MEASURE x

DSiP_contains_only_Principals: LEMMA
  FORALL (x: index, p: Program, s0: state):
    subset?(DSiP(p,s0)(x),Principals(p))

DSiP_def: LEMMA
  FORALL (p: Program,s0: state,x: index):
    WFg(p) =>
      DSiP(p,s0)(x) =
        LET uset: set[index] = DSe(Ae(x,p(x)),s0)
        IN
        LET pri : set[index] = Principals(p)
        IN
          (uset & pri) +
          Compose(uset & not(pri),DSiP(p,s0))
GWVr1_Si_DSiP: LEMMA
  FORALL (p: Program):
    FORALL (i: index, s1,s2: state):
      WFg(p) & SsG(p,s1) & SsG(p,s2) &
        StateEquivSet(DSiP(p,s1)(i),s1,s2) =>
          Si(p)(i,s1) = Si(p)(i,s2)
```

Fig. 16 GWVr1 for principal dependencies

4 Interference to Noninterference

A nearly immediate corollary of the interference theorem is a noninterference theorem, shown in Fig. 21. If a variable *unclass* does not depend on a variable *secret* in any legal trace of the system (as defined by *tp_ok*), then we say that *secret* does not interfere with *unclass*. This is demonstrated by the *Non_Interference* lemma; in this lemma, we state that any two traces whose inputs differ only by *secret* will yield the same values for *unclass*.

```
GProgram: TYPE = { p : Program | WFg(p) }

DSiIF(p: GProgram, s0: state, g0: graphState)
   (x: index): RECURSIVE set[index] =
  LET uset  : set[index] = DSe(Ae(x,p(x)),s0) IN
  LET ins   : set[index] = InputsP(p) IN
  LET dff   : set[index] = StatesP(p) IN
  LET gates : set[index] = GatesP(p)   IN
    (uset & ins) +
      Compose(uset & dff, g0) +
        (lambda (z: index):
        (EXISTS (m: index):
          member(m,uset & gates) &
          member(z,DSiIF(p,s0,g0)(m)))))
MEASURE x

DSiIF_to_DSiP: LEMMA
  FORALL (p: Program, s0: state, g0: graphState):
    WFg(p) =>
      FORALL (x: index):
        DSiIF(p,s0,g0)(x) =
          (InputsP(p) & DSiP(p,s0)(x)) +
            Compose(StatesP(p) & DSiP(p,s0)(x),g0)
```

Fig. 17 GWVr1 for state-input dependencies

5 Model Checking Information Flow

Up to this point, we have defined formal notions of interference and noninterference over traces for a simple synchronous dataflow language and shown that an *information flow semantics* can be used to demonstrate noninterference. However, we have not yet proposed a mechanism for computing noninterference relations using the model checker using a temporal logic such as LTL [2].

In order to use a model checker to analyze the notion of noninterference proposed in Sect. 4, we must do two things. First, we must formalize noninterference in a temporal logic such as LTL that is understood by model checkers. Second, we must encode the model and information flow semantics into the notation of the model checker. The syntax (Fig. 8) and execution semantics (Fig. 9) of our language were chosen in part because they correspond to a subset of the syntax and semantics supported by several popular model checkers including NuSMV [8], SAL [23], and Prover [16]. The translation of the execution model and semantics is therefore immediate.

To support analysis of information flow, however, we have to encode the *IF* semantics in the syntax of the model checker. We call this encoding the *information flow model*. Then we can analyze a *hybrid model*, containing both the original program and the information flow model in order to reason about flow properties.

```
GWVr1_Si_SsL0: LEMMA
  FORALL (p: Program):
    FORALL (i: index, s1,s2: state):
      WFg(p) & SsL0(p,s1) & SsL0(p,s2) &
      SsG(p,s1) & SsG(p,s2) &
      StateEquivSet(InputsP(p) &
      DSiP(p,s1)(i),s1,s2) =>
          Si(p)(i,s1) = Si(p)(i,s2)

DSt(p: Program, st: strace, t: time)(i: index):
RECURSIVE set[index] =
  IF (t = 0) THEN
    InputsP(p) & DSiP(p,st(t))(i)
  ELSE
    LET uset: set[index] = DSiP(p,st(t))(i) IN
      (uset & InputsP(p)) +
      Compose(not(InputsP(p)) & uset,
              DSt(p,st,t - 1))
  ENDIF
MEASURE t

subset_Compose: LEMMA
  FORALL (a: index, x: set[index], g: graphState):
    member(a,x) => subset?(g(a),Compose(x,g))

vtrace: TYPE = [ time -> vtype ]

vtrace_extensionality: LEMMA
  FORALL (i: index, s1,s2: vtrace):
    (s1 = s2) =
      FORALL (t: time): s1(t) = s2(t)

AUTO_REWRITE+ vtrace_extensionality

liftv(i: index, st: strace): vtrace =
  (LAMBDA (t: time): st(t)(i))

vtraceEquivSet(set: set[index],st1,st2: strace):
bool =
  FORALL (i: index): member(i,set) =>
    liftv(i,st1) = liftv(i,st2)

GWVr1_Si_DSt: LEMMA
  FORALL (p: Program, st1,st2: strace):
    FORALL (t: time, i:index):
      WFg(p) & St(p,st1) & St(p,st2) &
        vtraceEquivSet(DSt(p,st1,t)(i),st1,st2) =>
          Si(p)(i,st1(t)) = Si(p)(i,st2(t))
```

Fig. 18 GWVr1 theorems for trace dependencies

```
IFe_to_DSiIF: LEMMA
  FORALL (p: GProgram, s0: state, g0: graphState):
    IFsG(p,s0,g0) & WFg(p) =>
      FORALL (x: index):
        IFe(Ae(x,p(x)),InputsP(p),s0,g0) =
          DSiIF(p,s0,g0)(x)

Graph_Unwinding: LEMMA
  FORALL (p: Program, st: strace, gt: gtrace):
    FORALL (t: time, v: index):
      WFg(p) & IFt(p,st,gt) =>
        IFe(Ae(v,p(v)),InputsP(p),st(t),gt(t)) =
          DSt(p,st,t)(v)
```

Fig. 19 The graph unwinding theorem demonstrating equivalence between *IFt* and *DSt* semantics

```
DepSet(x: index, gt: gtrace): set[index] =
  (lambda (i: index): (EXISTS (t: time):
  member(i,gt(t)(x)))))
InterferenceTheorem: LEMMA
 FORALL (p: Program, gt: gtrace, st1,st2: strace):
    FORALL (i:index):
      WFp(p) & St(p,st1) & St(p,st2) &
      IFt(p,st1,gt) &
        vtraceEquivSet(DepSet(i,gt),st1,st2) =>
          liftv(i,st1) = liftv(i,st2)
```

Fig. 20 Proof of the InterferenceTheorem

5.1 *Formalizing Noninterference in LTL*

We first assume Rushby's formalization of LTL [2] in PVS presented in [6]. We now prove in Fig. 22 that a noninterference assertion over a graph state machine follows from a particular LTL assertion, in the same way as in Greve [6].

5.2 *Creating the Information Flow Model*

Recall that the *IF* semantics correspond to graph traces (*gtrace*) that are composed of a sequence of graph states (*gstate*). Each *gstate* maps program variables to a finite set of *Principal* variables. The information flow semantics from the previous section are then encoded as set manipulations. The information flow model is then the set of assignments to the information flow variables.

```
ProcessNonInterference: THEORY
BEGIN
  IMPORTING ProcessInterference

  Never_Interferes(p: Program, secret: index,
                   unclass: index) : bool =
    FORALL (x: tracePair):
      tp_ok(p, x) =>
        (FORALL (t: time):
          not(member(.secret, g(x)(t)(unclass))))

  Inputs_Match_Except_Secret(p: Program,
         st1, st2: strace, secret: index) : bool =
    FORALL (t: time, idx: index):
      ((member(idx, InputsP(p)) AND
       (idx /= secret)) =>
         st1(t)(idx) = st2(t)(idx))

  Non_Interference : LEMMA
    FORALL (p: Program, secret: index,
          unclass: index, st1,st2: strace):
      (member(secret, InputsP(p)) &
       WFg(p) & St(p, st1) & St(p, st2) &
       Never_Interferes(p, secret, unclass)) &
       Inputs_Match_Except_Secret(p, st1, st2,
           secret)
      =>
         liftv(unclass, st1) = liftv(unclass, st2)

END ProcessNonInterference
```

Fig. 21 Process Noninterference

The mechanism for creating the information flow variable assignments is a set of
transformation rules that are applied to the syntax of *ProcessExpr* and *ProcessAs-
sign* datatypes defined in Fig. 8. The transformation rules generate a slightly richer
expression syntax (shown in Fig. 23) that contains two additional variables. The first
expression, *IF_Variable*, allows reference variables in the information flow graph
state. The second, *SingletonSet*, takes an index and generates a singleton set containing that index.

We can now reflect the information flow semantics into an extended program
ProgramExt that contains assignments for both the state and graph traces, as shown
in Fig. 24.

The hybrid model in Fig. 24 contains assignments both for the state variables (*st*)
and the graph variables (*gr*). The syntax of the state assignments does not change;
however, the strong typing of PVS requires that we define a transformation to

```
ProcessLTL: THEORY
BEGIN

  IMPORTING ProcessInterference

  GState : TYPE = [# g: graphState, s: state #]

  IMPORTING ltl[GState]

  P : Program
  P_inputs : TYPE = {x: index | Input?(P(x)) }

  split(x: sequence[GState]) : tracePair =
   (# s := LAMBDA (t: time): s(x(t)),
      g := LAMBDA (t: time): g(x(t)) #)

  merge(x: tracePair) : sequence[GState] =
   (LAMBDA (t: time):
      (# s := s(x)(t), g := g(x)(t) #) )

  GSTrace : TYPE =
    { x : sequence[GState] | tp_ok(P, split(x)) }

  Non_Interference(secret: P_inputs, unclass: index)
    (gs: GState) : bool =
      (not (member(secret, g(gs)(unclass))))

  % only consider well-formed models
  reduction: LEMMA
    WFg(P)   =>
      FORALL (secret: P_inputs, unclass: index):
        (FORALL (s: GSTrace):
          (s |=
            G(Holds(
              Non_Interference(secret,unclass)))))
      =>
        (FORALL (s: tracePair):
          tp_ok(P,s)  =>
            (FORALL (t: time):
              (not(member(secret,
                            g(s)(t)(unclass))))))

  END ProcessLTL
```

Fig. 22 Connection to LTL

map from the *ProcessExpr* and *ProcessAssignment* datatypes into the *ExprExt* and *AssignExt* datatypes, respectively. This is performed by the *IDe* and *IDa* functions, respectively.

The mapping of the information flow *IF* semantics into syntax that can be interpreted is performed by the *TR* functions. These functions create new syntax based

```
ExprExt: DATATYPE
BEGIN
  IMPORTING ProcessTypes
  Constant(value : vtype): Constant?
  Variable(sname  : index):  Variable?
  ITE(test: ExprExt, thn: ExprExt, els: ExprExt):
      ite?
  Bop(OpB: BopType, a1: ExprExt, a2: ExprExt): Bop?
  Uop(OpU: UopType, a0: ExprExt): Uop?
  IF_Variable(ifname  : index):  IF_Variable?
  SingletonSet(varSet: set[index], prname : index) :
      SingletonSet?
END ExprExt

AssignmentExt: DATATYPE
BEGIN
 IMPORTING ExprExt
  Gate (gexpr: ExprExt): Gate?
  Latch(v0: vtype, lexpr: ExprExt): Latch?
  Input: Input?
END AssignmentExt
```

Fig. 23 Extended process syntax

on an original program that manipulates index sets. It is instructive to compare the syntax created by the *TRe* function with the definition of the *IFe* semantics originally defined in Fig. 10 and shown again in Fig. 25. Note the similarities between the semantic definitions in *IFe* and the syntax generated by the *TRe* function.

The compositional equivalence between the syntactic rule and the semantic rule can be proven, but we do not demonstrate it in this chapter. To do so would require some further elucidation of sets-as-*vtype* elements as well as an algebraic formulation of the union binary operator over *vtype* elements to show its equivalence to the standard set-union operator. We plan to do this in future work.

The model encoding tool in the Rockwell Collins *Gryphon* tool suite implements the transformation defined by the *TR* rules. It operates over the Lustre language [7]. Lustre includes a superset of the expressions described in the *TR* rules, such as expressions for creating and manipulating composite datatypes including arrays, records, and tuples. It also accounts for Lustre's notion of modularity, called the *node*, which corresponds to Simulink *subsystems*. The complete rules for rewriting Lustre programs are described in a Rockwell Collins technical report that is available by visiting http://extras.springer.com and entering the ISBN for this book.

For encoding the set of principals for model checking tools, we use bitvectors. The models that we attempt to analyze will always consist of a finite number of variables, and therefore, the principal variables form a finite set. We encode this set as a bitvector containing one bit per principal signal. The *Union* and *SingletonSet* operations are encoded as *bit_or* operators and bitvector constants, respectively.

```
TransformIF : THEORY
BEGIN
  IMPORTING Program, AssignmentExt

  Union : BopType
  EMPTYSET : vtype
  principal_index : [set[index], index -> vtype]

  IDe(e: ProcessExpr): RECURSIVE ExprExt =
  CASES e OF
    Constant(value): Constant(value),
    Variable(name):  Variable(name),
    ITE(test,thn,els):
      ITE(IDe(test), IDe(thn), IDe(els)),
    Bop(OpB,a1,a2):  Bop(OpB, IDe(a1), IDe(a2)),
    Uop(OpU,a0): Uop(OpU, IDe(a0))
  ENDCASES
  MEASURE e by <<

  IDa(a: ProcessAssignment) : AssignmentExt =
     CASES a OF
       Gate(gexpr) : Gate(IDe(gexpr)),
       Latch(v0, lexpr) : Latch(v0,IDe(lexpr)),
       Input : Input
     ENDCASES

  TRe(e: ProcessExpr, Pr: set[index]):
  RECURSIVE ExprExt =
    CASES e OF
      Constant(value): Constant(EMPTYSET),
      Variable(name):
        IF Pr(name) THEN
          SingletonSet(Pr, name)
        ELSE
          IF_Variable(name)
        ENDIF,
      ITE(test,thn,els):
        Bop(Union,
          ITE(IDe(test),TRe(thn, Pr),TRe(els, Pr)),
          TRe(test, Pr)),
      Bop(OpB,a1,a2):
        Bop(Union, TRe(a1, Pr), TRe(a2, Pr)),
      Uop(OpU,a0): TRe(a0, Pr)
```

Fig. 24 Hybrid model definitions

5.3 *From Principals to Domains*

Our implementation allows multiple variables to be mapped to the same principal
identifier (id). This identifier can be thought of as a *security domain* [5, 20]. For
the purposes of analysis, this can reduce the number of bits necessary for a model

```
      ENDCASES
    MEASURE e by <<

    TRa(a: ProcessAssignment, Pr: set[index]) :
    AssignmentExt =
      CASES a OF
        Gate(gexpr) : Gate(TRe(gexpr, Pr)),
        Latch(v0, lexpr) :
          Latch(EMPTYSET, TRe(lexpr, Pr)),
        Input : Input
      ENDCASES

    AssignSet: TYPE = [index -> AssignmentExt ]
    ProgramExt: TYPE =
        [# st: AssignSet, gr: AssignSet #]

    TRp(p: Program) : ProgramExt =
      (# st := (LAMBDA (idx: index) : IDa(p(idx))),
         gr := (LAMBDA (idx: index) :
                  TRa(p(idx), InputsP(p))) #)

  END TransformIF
```

Fig. 24 (continued)

```
    IFe(e: ProcessExpr, principal: set[index],
        s0: state, g0: graphState): RECURSIVE
        set[index] =
      CASES e OF
        Constant(value): Empty,
        Variable(name):
            IF principal(name) THEN
                singleton(name)
            ELSE
                g0(name)
            ENDIF,
        ITE(test,thn,els):
          IF isTrue(Se(test,s0)) THEN
            IFe(test,principal,s0,g0) +
            IFe(thn,principal,s0,g0)
          ELSE
            IFe(test,principal,s0,g0) +
            IFe(els,principal,s0,g0)
          ENDIF,
        Bop(OpB,a1,a2):  IFe(a1,principal,s0,g0) +
           IFe(a2,principal,s0,g0),
        Uop(OpU,a0): IFe(a0,principal,s0,g0)
      ENDCASES
    MEASURE e by <<
```

Fig. 25 Another presentation of the IFe function

checking analysis, which improves performance. It also coarsens the analysis, as it is no longer clear from a counterexample which of the variables mapped to the principal id is responsible for information flow.

5.4 Adding Control Variables

The implementation allows variables to be designated as control variables. The intuition is that an operand of an AND or OR gate sometimes acts as a mask for the other operand (toward FALSE and TRUE, respectively). In this instance, we would like to consider the information flow from the other variable into the gate only if the control variable has the appropriate value. This feature allows for slightly more accurate analysis in some models. It is a conservative extension because the semantics of AND and OR gates are semantically the same as the following if/then/else structure:

```
Y = C and E  ⇔  Y = if C then E else false;
Y = C or E  ⇔  Y = if C then true else E ;
```

Y is semantically equivalent in both cases, and the soundness of the flow analysis follows from the existing proof of if/then/else expressions in Fig. 12. Note that the condition variable for if/then/else (C) is always used for the information flow analysis, so if *both* variables in a Boolean expression are control variables, the following is generated:

```
Y = C₀ and C₁   ⇔
Y = if C₀ then (if C₁ then C₀ else false) else
(if C₁ then C₀ else false)
```

After applying the syntactic *TRe* transformation to the right-hand side of the equivalence and simplifying, this yields the "standard" information flow expression for the original binary expression: *Bop(Union, TRe(a1, Pr), TRe(a2, Pr))*.

6 Intransitive Interference and Noninterference

We have defined a considerable amount of infrastructure for determining which variables can *interfere* with a particular computed variable within a model. In the approach that we have pursued in the previous sections of this chapter, all interference relations are *transitive*. That is, if variable A interferes with variable B and B interferes with C, then A interferes with C. However, there are several systems in which we are willing to allow certain kinds of interference across security domains, as long as it is mediated in some way. The reasoning for allowing this interference is well explained by Roscoe and Goldsmith [19]:

> It seems intuitively obvious that the relation must be transitive: how can it make sense for A to have lower security level than B, and B to have lower level than C, without A having lower level than C? But this argument misses a crucial possibility, that some high-level users are trusted to *downgrade* material or otherwise influence low-level users. Indeed, it has been argued that no large-scale system for handling classified data would make sense without some mechanism for downgrading information after some review process, interval (e.g., the U.K. 30-year rule) or defined event (the execution of some classified mission plan, for example). Largely to handle this important problem, a variety of extended theories proposing definitions of "intransitive noninterference" have appeared, though we observe that this term is not really accurate, as it is in fact the *interference* rather than the *noninterference* relation which is not transitive. Perhaps the best way to read the term is as an abbreviation for "noninterference under an intransitive security policy."

There have been several formulations of intransitive interference based on state machines [20], process algebras [19], and event traces [10].

6.1 Formulating Intransitive Interference

Our model is entirely defined in terms of variables. Operations such as encryption or downgrading are implemented as subsystems (sets of variables) within a larger model whose output is another variable within the model. Therefore, it is natural to think of extending the set of principal variables P from only the inputs to include internal variables that define the mediation points of interest. Since the definition of *Noninterference* requires only that the principal variables agree, these intermediaries are easily incorporated into our definition.

For example, in the shared buffer model, we are willing to allow information to flow through the scheduler. By adding the scheduler state to P, we restrict ourselves to reasoning over traces in which the scheduler states match. From the perspective of reasoning, it is straightforward to parameterize the proofs over a superset of the inputs and reprove the *InterferenceTheorem* and *NoninterferenceTheorem* defined in Sects. 3 and 4.

6.1.1 The Problem of Implicit Functional Dependencies

Unfortunately, this formulation of "correctness" allows unintended covert information flows around the mediation point as long as they can be *functionally derived* from an input variable.

Figure 26 presents a Simulink model that illustrates the problem. Output O is a record type that contains two fields B and C. Field B is the output of a subsystem that encrypts the input variable (A); field C is a simple pass-through of A. Suppose that the output of the encryptor B is functionally derived from input A. That is, two traces on B agree *only when* the traces on A also agree. In this case, according to the interference theorem, we can adjudge output O to be dependent only

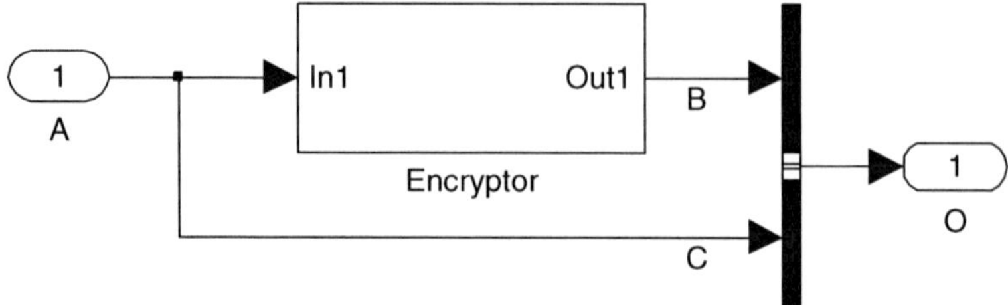

Fig. 26 Simulink model containing a bypass

on B, even though there is clearly a flow that bypasses B. The problem is that the encryptor variable is *functionally derived* from a single input A, so the equivalence on B forces a corresponding equivalence on the input A. In other words, requiring a trace equivalence on a computed principal variable may cause an implicit equivalence on another principal variable. These implicit equivalences allow an attacker to bypass the desired mediation variable.

6.1.2 An Overly Conservative Formulation

An approach that could be considered for intransitive interference reframes the problem: given a program P involving a computed principal variable c, we construct a program P' in which c is an input and assert that all traces must agree on P'. P' has at least as many traces as P, as the value of c is unconstrained with respect to the other variables in P'. The additional traces distinguish variables that bypass the computed principal as there is no longer a functional connection between the computed variable and the inputs.

Unfortunately, treating states as inputs leads to overly conservative analyses involving traces that are impossible in the original program. Consider the shared buffer model from Sect. 2. If a new model is created in which the scheduler output is instead a system input, then the scheduler can no longer correctly mediate access to the shared buffer and so information flow occurs through the buffer. The flow analysis will (correctly) state that there is information flow through the buffer, but the flagged traces are not possible in the original model.

6.2 *Modeling Intransitive Interference Using Graph Cuts*

The analysis approach that is used in Gryphon is to model intransitive information flow through *cuts* in the information flow graph. That is, we define a new principal variable in the information flow graph by cutting the edges that define the dependencies of that computed variable. To implement the change in the information flow (IF) semantics defined in Sect. 3, we add the internal variable indices to the set of inputs that are used in the *IFe, IFi, IFs*, and *IFt* relations. The definition of the program P is left unchanged.

This modified graph model is sufficient to correctly characterize both a program P and a modified program P' in which a principal variable c is treated as an input. In other words, this formulation is sensitive to the *structure* of the computation of the system execution traces as well as the *functional result*. The original program P is analyzed, so there are no problems introduced by the additional traces of P', but we (correctly) characterize models such as the one described by Fig. 26 as containing direct information flows from input variable A to output O.

Illustrations of a transitive flow model and an intransitive model using graph cuts are shown in Fig. 27. Recall that the *hybrid model* that is generated for model checking is composed of both a *functional model* (the original system) and an *information flow model*, which is an encoding of the *IF* semantics as described in Sect. 5. In Fig. 27, the functional model is shown at the top of the figure. In the middle is a *transitive* information flow model. At the bottom is an *intransitive* information flow model.[2] Each model is presented both graphically on the left and in terms of equations on the right. In this figure, the principal bitvector for a variable X is notated as X.

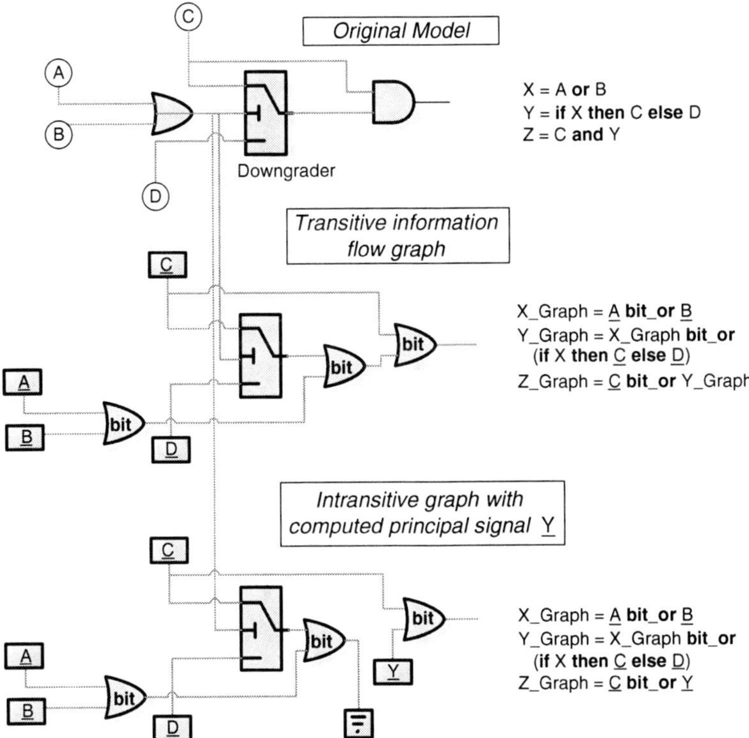

Fig. 27 Transitive vs. intransitive flow graphs

Suppose variable Y (the switch gate) acts as a downgrader for variable D. We would like to state that the output (Z) depends on input D only when mediated through the downgrader. Given the transitive formulation of information flow in the middle of Fig. 27, it is not possible to make this claim. However, the intransitive graph at the bottom of Fig. 27 breaks the information flow graph for each use of variable Y, replacing the input flows through the computed definition of Y with a new principal signal Y. Given this new graph, it is possible to prove that no information flows from D to Z that is not mediated by Y. On the other hand, note that with this intransitive graph, a noninterference proof would still not be possible for variable C as it has a flow to Z that bypasses Y.

We currently do not have a strong theorem (such as the *InterferenceTheorem*) that we can prove about intransitive dependencies. Further, we conjecture that it is not possible to functionally characterize such dependencies using trace semantics. Instead, the structure of the computation function must be examined – the property is *intrinsic* to the structure.

7 Connections to GWV

In the current chapter and the earlier chapter by Greve [6], we have presented two quite similar formulations of information flow modeling. The formulation in Greve's chapter is more abstract and describes information flow over arbitrary functions using flow graphs. It then describes how these functions can be composed and how multistep state transition systems can be encoded. Two different formulations (GWVr1 and GWVr2) are presented. The GWVr2 formulation is capable of modeling dynamic information flows, in which storage locations are created and released during the computation of the function, but this additional capability comes at a cost of some additional complexity.

In this chapter, we have modeled information flow specifically for synchronous dataflow languages. The basis for this approach was modeling GWV-style equivalences using a model checker. However, the approach was originally justified by manual proofs over trace equivalences due to the first author's familiarity with this style of formalization for synchronous dataflow languages. The mechanized proofs in this chapter reflect the manual proofs.

As a basis for formalization, the trace equivalence allows a very natural style of presentation. It provides a nice abstraction of the computation and information flow analysis in that a total computation order for the assignments of the semantic and flow analyses is not required. Instead, we can talk about *conformance* to some existing trace. Also, since the entire trace is provided, we can describe latch conformance by examining the previous state in the trace.

```
GWVr1_Connection[
     (importing ProcessInterference)
     P: WFPrograms]: THEORY
BEGIN
  IMPORTING ProcessInterference

  valid_tp : TYPE = {tp: tracePair | tp_ok(P, tp)}

  st_liftv(i: index, tp: tracePair) : vtrace =
    liftv (i, s(tp))

  IMPORTING GWVr1[index, valid_tp, vtrace, st_liftv,
                  index, valid_tp, vtrace, st_liftv]

  step_id(tp: valid_tp) : valid_tp = tp;

  gtrace_graph(tp: valid_tp)(idx: index) :
  GraphEdge[index] =
     Compute(DepSet(idx, g(tp)))

  precondition(tp: valid_tp) : bool = true ;

  inputEquivSet_to_vtraceEquivSet : LEMMA
    (FORALL (is: set[index], tp1, tp2: tracePair) :
        Input.equivSet(is, tp1, tp2) =>
        vtraceEquivSet(is, s(tp1), s(tp2)))

   GraphIsGWVr1 : LEMMA
      GWVr1(step_id)(precondition, gtrace_graph);

  END GWVr1_Connection
```

Fig. 28 Connection to GWVr1 theorem

7.1 *From InterferenceTheorem to GWVr1*

From the *InterferenceTheorem*, it is straightforward to map directly into the GWVr1 theorem presented in Greve's chapter [6], as shown in Fig. 28.

GWVr1 is defined as a proof obligation over a transition function from an input state to an output state. The fragment of the GWVr1 theory required for the proof is shown in Fig. 29.

The *index*, *state*, *value*, and *get* parameters to the theory define the indices of discourse, the state, the values that can be stored at indices, and the "getter" function to look up a value for the inputs and outputs of the transition function. In our case, the types of inputs and outputs are the same: we are looking at traces. To format our

```
GWVr1 [INindex, INState, INvalue: TYPE,
       getIN: [[INindex, INState] -> INvalue],
       OUTindex, OUTState, OUTvalue: TYPE,
       getOUT: [[OUTindex, OUTState] -> OUTvalue]

GWVr1 [INindex, INState, INvalue: TYPE,
       getIN: [[INindex, INState] -> INvalue],
       OUTindex, OUTState, OUTvalue: TYPE,
       getOUT: [[OUTindex, OUTState] -> OUTvalue]
      ]: THEORY

BEGIN

  IMPORTING GWV_Graph[INindex,OUTindex]
  IMPORTING GWV_Equiv[INindex,INState,INvalue,getIN]
    AS Input
  IMPORTING GWV_Equiv[OUTindex,OUTState,OUTvalue,
                      getOUT] AS Output

  StepFunction:  TYPE = [ INState -> OUTState ]
  GraphFunction: TYPE = [ INState -> graph ]
  PreCondition:  TYPE = [ INState -> bool ]

  GWVr1(Next: StepFunction)
    (Hyps: PreCondition, Graph: GraphFunction): bool=
      FORALL (x: OUTindex, in1,in2: INState):
        Input.equivSet(DIA(x,Graph(in1)),in1,in2) &
          Hyps(in1) & Hyps(in2) =>
            Output.equiv(x,Next(in1),Next(in2))
```

Fig. 29 Fragment of GWVr1 theory

trace equivalences as a GWVr1 theorem, we create a theory parameterized by an arbitrary well-formed program. The GWV index values are simply our index type, the state is the trace pair containing both the execution state and the information flow state, values map to our *vtype*, and the *get* function returns a variable trace from the state trace.

The proof to GWVr1 merely involves reshaping the *InterferenceTheorem* into the expected arguments for GWVr1. Our *StepFunction* is simply the identity; we already have the entire trace. The *GraphFunction* returns the trace dependency set for a variable of interest; this is the same set used by the *InterferenceTheorem*. No hypotheses are necessary, so we create a trivial precondition function. We introduce a lemma *inputEquivSet_to_vtraceEquivSet* to map between the set equivalence functions used by *InterferenceTheorem* and *GWVr1*, then can establish the *GraphIs-GWVr1* lemma with very little difficulty using the *InterferenceTheorem* as a lemma.

Although the trace formulation provides a nice level of abstraction for describing synchronous dataflow languages, in this chapter we have duplicated some of the infrastructure that had already been established in [6] with respect to function composition, mapping from interference to noninterference, and justifying LTL theorems in terms of trace equivalence. It would be possible to reformalize the synchronous language semantics defined in Sect. 3 in order to better utilize the GWV infrastructure, but we leave this for future work.

8 Using Gryphon for Information Flow Analysis

We now demonstrate the information flow analysis in the Rockwell Collins *Gryphon* tool suite. *Gryphon* is an analysis framework designed to support model-based development tools such as Simulink/Stateflow and SCADE. Model-based development (MBD) refers to the use of domain-specific, graphical modeling languages that can be executed and analyzed before the actual system is built. The use of such modeling languages allows the developers to create a model of the system, execute it on their desktop, analyze it with automated tools, and use it to automatically generate code and test cases.

As MBD established itself as a reliable technique for software development, an effort was made to develop a set of tools to enable the practitioners of MBD to formally reason about the models they created. Figure 30 illustrates MBD development process flow.

8.1 Model-Based Development ToolChain

The following sections briefly describe each component of the MBD toolchain.

8.1.1 Simulink, Stateflow, MATLAB

Simulink, Stateflow, and MATLAB are products of The MathWorks, Inc. [11] Simulink is an interactive graphical environment for use in the design, simulation, implementation, and testing of dynamic systems. The environment provides a customizable set of block libraries from which the user assembles a system model by selecting and connecting blocks. Blocks may be hierarchically composed from predefined blocks.

8.1.2 Reactis

Reactis® [17], a product of Reactive Systems, Inc., is an automated test generation tool that uses a Simulink/Stateflow model as input and autogenerates test code for the verification of the model. The generated test suites target specific

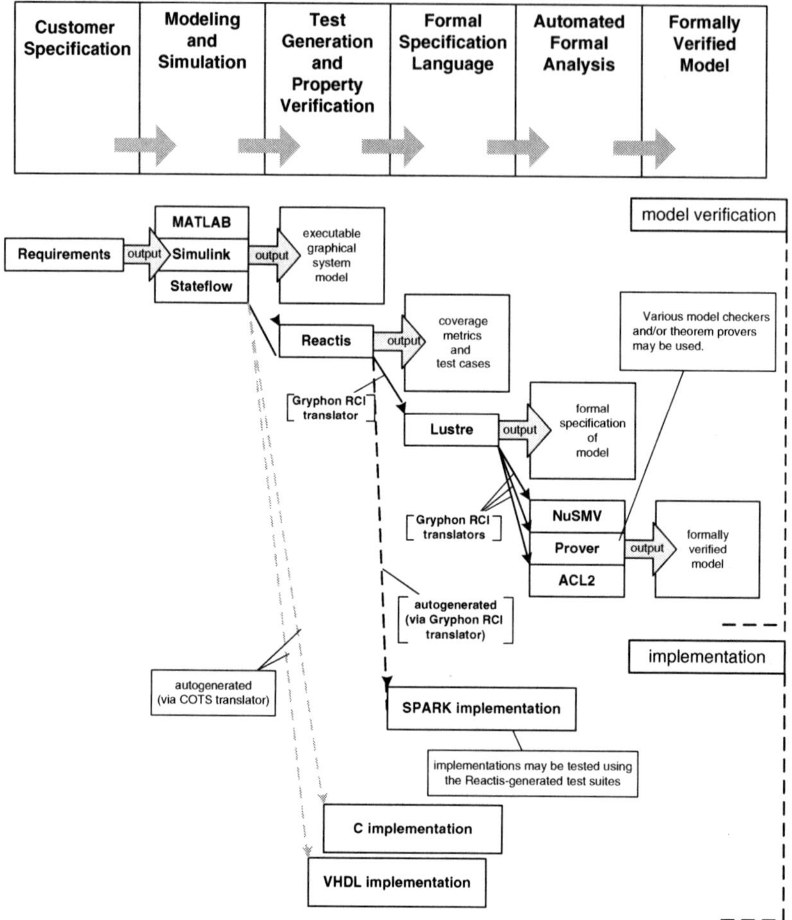

Fig. 30 Model-based development process flow

levels of coverage, including state, condition, branch, boundary, and modified condition/decision coverage (MC/DC). Each test case in the generated test suite consists of a sequence of inputs to the model and the generated outputs from the model. Hence, the test suites may be used in testing of the implementation for behavioral conformance to the model, as well as for model testing and debugging.

8.1.3   Gryphon

Gryphon [24] refers to the Rockwell Collins tool suite that automatically translates from two popular commercial modeling languages, Simulink/Stateflow and SCADE [4], into several back-end analysis tools, including model checkers and theorem provers. Gryphon also supports code generation into Spark/Ada and C. An overview

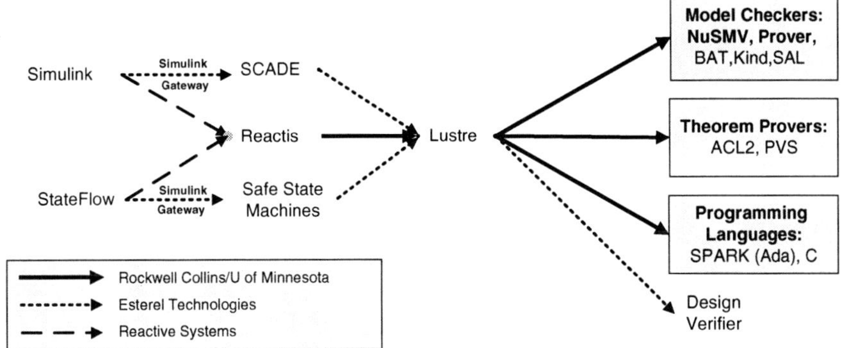

Fig. 31 Gryphon translator framework

of the Gryphon framework is shown in Fig. 31. Gryphon uses the Lustre [7] formal specification language (the kernel language of SCADE) as its internal representation. This allows for the reuse of many of the RCI proprietary optimizations.

8.1.4 Prover

Prover [16] is a best-of-breed commercial model checking tool for analysis of the behavior of software and hardware models. Prover can analyze both finite-state models and infinite-state models, that is, models with unbounded integers and real numbers, through the use of integrated decision procedures for real and integer arithmetic. Prover supports several proof strategies that offer high performance for a number of different analysis tasks including functional verification, test-case generation, and bounded model checking (exhaustive verification to a certain maximum number of execution steps).

8.1.5 Custom Code Generation

By leveraging its existing Gryphon translator framework, Rockwell Collins designed and implemented a toolchain capable of automatically generating SPARK-compliant Ada95 source code from Simulink/Stateflow models.

8.2 Modeling and Analyzing the Turnstile High-Assurance Guard Architecture

A large scale use of the Gryphon analysis was performed on the Rockwell Collins Turnstile high-assurance cross-domain guard [18]. A high-level view of

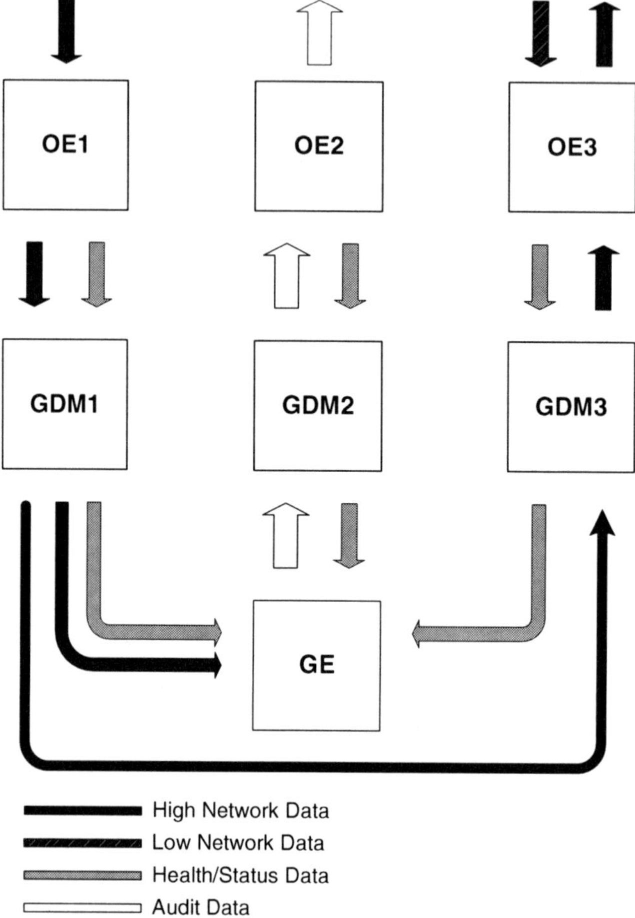

Fig. 32 Turnstile system architecture

the architecture is shown in Fig. 32. The offload engines (OEs) provide the external interface to Turnstile. The Guard Engine (GE) is responsible for enforcing the desired security policy for message transport. The guard data movers (GDMs) provide a high-speed mechanism to transfer messages under the direction of the GE. The GE is implemented on the EAL-7 AAMP7 microprocessor [25] and uses the partitioning guarantees provided by the AAMP to ensure secure operation.

In its initial implementation, Turnstile provides a "one-way" guard. It has a *high-side* OE (OE1 in Fig. 32) that submits messages (generates input) for the guard, a *low-side* OE (OE3 in Fig. 32) that emits messages if they are allowed to pass through the guard, and an *audit* OE (OE2 in Fig. 32) that provides audit functionality for the system.

The architectural analysis focused on the interaction between the GDMs, GE, and OEs. The OEs, GDMs, and GE do not share a common clock and both execute and communicate asynchronously. In the model, we *clock* each of the subsystems using a system input. This input is allowed to vary nondeterministically, allowing us to model all possible interleavings of system execution.

8.2.1 Representing the Turnstile Architecture in Simulink

The Simulink model of the Turnstile system architecture is shown in Fig. 33. The components were modeled at various levels of fidelity, depending on their relevance to the information flow problem:

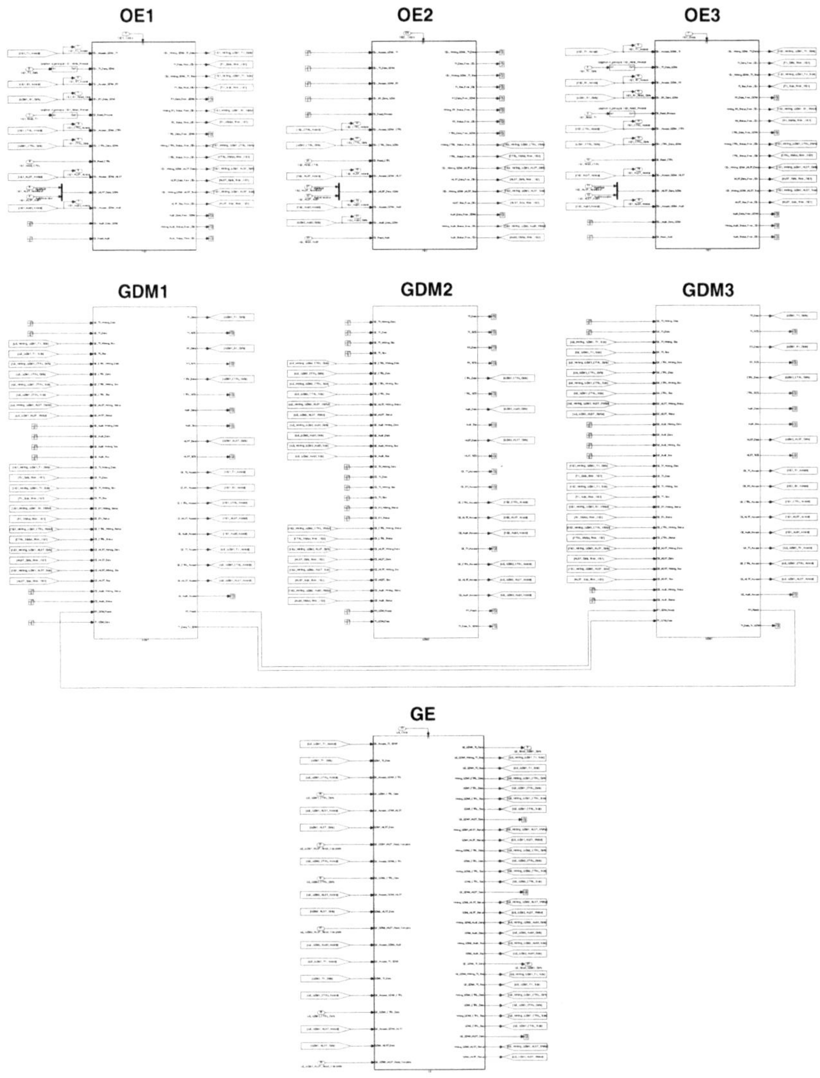

Fig. 33 Simulink turnstile model

- The GDMs are responsible for most of the data routing and are modeled to a high level of fidelity. All of the GDM channels (transmit, receive, audit, control, and health monitor) are modeled as well as the GDM-to-GDM and GDM-to-GE transfer protocols.
- The data routing portions of the GE were accurately modeled. The policy enforcement portions (the guard evaluator) were modeled nondeterministically: the GE component randomly chooses whether messages are dropped or propagated.
- The OEs were modeled at a fairly low level of fidelity. As the OEs are not trusted by the Turnstile architecture, we allow them to nondeterministically submit requests on all of the interfaces between OE and GDM. This approach allows us to model situations in which the OE violates the Turnstile communications protocols (which should cause the system to enter a fail-safe mode).

The principals of interest are those processes on the Offload Engines that interact with the outside world (the low and high networks): the reading and writing processes on OE1 and the reading and writing processes on OE3. To represent the arbitrary interleavings of the Turnstile processes, we used enabled (clocked) subsystems in Simulink. The GDMs run in synchrony at the basic rate of the model while the OEs and GE run at arbitrary intervals of the basic rate.

The model in Fig. 33 was translated via Gryphon into the model checkers NuSMV [8] and Prover [16]. With these tools we analyzed several of the information flows through the model. Since the OE has multiple inputs in our model (and in real life), we analyzed every input into the OEs for the possible presence of information from an unwanted source. In a one-way guard configuration, we are interested in determining whether there is backflow of information to the high-side network, that is, whether any GDM input into OE1 is influenced by the low-side (OE3) reading or writing principals. These properties can be encoded as shown in Fig. 34.

One of the back flow properties (shown in bold font) was violated in the architectural model. However, this was already a known source of back flow because of the implementation of the GDM transfer protocol that resulted from a quality of service

```
       l2h_tx1 = not gry_IF_OE1_TX_Access[p_oe3_writer] ;
   ►   l2h_tx2 = not gry_IF_OE1_TX_Access[p_oe3_reader] ;
       l2h_tx3 = not gry_IF_OE1_RX_Access[p_oe3_writer] ;
       l2h_tx4 = not gry_IF_OE1_RX_Access[p_oe3_reader] ;
       l2h_tx5 = not gry_IF_OE1_RX_Read_Data[p_oe3_writer] ;
       l2h_tx6 = not gry_IF_OE1_RX_Read_Data[p_oe3_reader] ;
       l2h_ctrl1 = not gry_IF_OE1_CTRL_Data[p_oe3_writer] ;
       l2h_ctrl2 = not gry_IF_OE1_CTRL_Data[p_oe3_reader] ;
       l2h_ctrl3 = not gry_IF_OE1_CTRL_Access[p_oe3_writer] ;
       l2h_ctrl4 = not gry_IF_OE1_CTRL_Access[p_oe3_reader] ;
       l2h_hlst1 = not gry_IF_OE1_HLST_Access[p_oe3_writer] ;
       l2h_hlst2 = not gry_IF_OE1_HLST_Access[p_oe3_reader] ;
```

Fig. 34 Backflow properties from "low-side" OE3 to "high-side" OE1

```
oe2_audit1 = not gry_IF_OE2_Audit_Access[p_oe3_writer] ;
oe2_audit2 = not gry_IF_OE2_Audit_Access[p_oe3_reader] ;
oe2_audit3 = not gry_IF_OE2_Audit_Data[p_oe3_writer] ;
oe2_audit4 = not gry_IF_OE2_Audit_Data[p_oe3_reader] ;
oe2_audit5 = not gry_IF_OE2_CTRL_Access[p_oe3_writer] ;
oe2_audit6 = not gry_IF_OE2_CTRL_Access[p_oe3_reader] ;
oe2_audit7 = not gry_IF_OE2_CTRL_Data[p_oe3_writer] ;
oe2_audit8 = not gry_IF_OE2_CTRL_Data[p_oe3_reader] ;
oe2_audit9 = not gry_IF_OE2_HLST_Access[p_oe3_writer] ;
oe2_audit10 = not gry_IF_OE2_HLST_Access[p_oe3_reader] ;
```

Fig. 35 Backflow properties from "low-side" OE3 to audit OE2

requirement levied on the Turnstile implementation. This requirement stated that a new message cannot be accepted until the previous message had been delivered. In the Turnstile architecture, the high-side writer is unable to transmit to the GDM until the low-side reader has finished consuming the last message. The low-side reader could potentially use this mechanism to transmit information (interfere) with the high-side network. The verification of the other properties demonstrates that the high-side OE is not, for example, influenced by the low-side writer.

Also, because the Audit OE may also be connected to the high network, we wanted to verify that no information from OE3 leaks out to the Audit network from any of the GDM inputs to OE2. These properties, which are all proven correct by the Prover model checker, are shown in Fig. 35.

Though much more complex, the Turnstile architectural model is conceptually similar to the shared buffer example. The GE acts as the scheduler between the GDMs, which are physically connected together and can be thought of as defining a "shared" resource. It is crucial to note that accurate *conditional* information flow is necessary to successfully analyze the Turnstile system architecture and many other industrial systems of interest. Since the GDMs are directly connected, an unconditional analysis of the architecture would not be able to demonstrate noninterference properties between the high- and low-side OEs. Only by considering the state of the system (especially the GE) can one demonstrate the security of the architecture.

9 Conclusion and Future Work

In this chapter, we have described an analysis procedure that can be used to check a variety of information flow properties of hardware and software systems, including *noninterference* over system traces. This procedure is an instantiation of the GWV-style flow analysis specialized for synchronous dataflow languages such as SCADE [4] and Simulink [11]. Our analysis is based on annotations that can be added directly to a Simulink or SCADE model that describe specific sources and sinks of information. After this annotation phase, the translation and model checking tools can be used to automatically demonstrate a variety of information flow properties. In the case of noninterference, they will either prove that there is no information flow between the source and a variable of interest or demonstrate a source of information flow in the form of a counterexample.

In order to justify the model checking analysis, we have presented a formalization of our approach in PVS and demonstrated a *NoninterferenceTheorem*. This theorem states that if our model checking analysis determines a system input that does not interfere with a particular output, then it is possible to vary the trace of that input without affecting the output in question. The analysis is both scalable and accurate and can be used to describe the following:

- *Conditional information flow*. The analysis is sensitive to the state of the model and can be used in situations in which multiple domains "share" a resource, such as the shared buffer model.
- *"Covert" information flow*. The analysis can detect flows due to (for example) contention for resources. These flows are ultimately manifest in the test expressions for conditionals, which are propagated to the output of the conditional.
- *Intransitive information flow*. The analysis can be used to define intransitive information flows, in which we are willing to allow information flows between domains as long as they occur through well-defined mediation points.

Our analysis is implemented in the *Gryphon* tool suite that supports several kinds of formal analysis of Simulink and Stateflow models. *Gryphon* has been used in several large-scale formal verification efforts [24], including a flow analysis of the Turnstile high-assurance cross-domain guard.

9.1 Future Work

There are several directions for future work given the framework that has been created. First, there are a variety of interesting properties beyond noninterference that can be formalized using temporal logic. For example, it is possible to begin talking about *rates* of information flow through a system by creating more interesting temporal logic formulations of flow properties. For example, one can state that flow occurs at most every ten cycles of evaluation (say), with the following Real-Time CTL (RTCTL) [2] property:

SPEC AG(gry_IF_output[P1] -> ABF[5, 23] (!gry_IF_output[P1]));

where "ABF" is the bounded future operator of RTCTL. This formula states that if flow occurs from principal $P1$ to variable *output* in the current steps, then no flow occurs from $P1$ to *output* over the next ten steps. In order to be informative, this obligation would have to be paired with some notion of *how much* information was being transmitted by a particular flow in an instant when flow occurs. It should be possible to annotate (manually or automatically) an information flow model with the flow rates along particular edges within the graph. Such an annotation could be used to overapproximate "acceptable" levels of information loss when strict noninterference is not possible (such as with the scheduler in the shared buffer example from Sect. 2.2).

Similarly, we may want to describe modal information flow properties. For example: *as long as the system is not in the self-test mode, then no information flows from A to B*. These properties are straightforward to specify in temporal logic, but precisely defining the meaning of these kinds of properties in a more general *InterferenceTheorem* would be a useful exercise.

It should be possible to partition the model checking analyses using compositional reasoning techniques such as those described in [12,13] for very large models. Determining the obligations over both the functional state and also the information flow graph should be an interesting exercise and may yield further insights into the relationship between a functional model and information flow graph.

There are several directions to extend the full formalization of the approach in PVS. First, we should formalize the proof of equivalence between the *IFe* semantics and the *information flow model* that is generated by the translation rules in Sect. 5. A more ambitious step would be to formalize the entire Lustre language in PVS including the clock operators and modularity constructs and demonstrate the correctness of the complete translation provided in the Gryphon toolsuite.

Finally, we would like to be able to compose the model checking results with results from theorem proving GWV-style theorems using a theorem prover such as PVS or ACL2. This would allow partitioning of very large problems into portions that can be analyzed with "the right tool for the job," using theorem proving where required (e.g., when complex dynamic data structures are involved) but using automated analysis using model checking where possible.

Acknowledgments We would like to thank the reviewers of early drafts of this paper, especially Matt Staats, Andrew Gacek, and Kimberly Whalen, for their many helpful comments and suggestions.

References

1. Bensalem S, Caspi P, Parent-Vigouroux C, Dumas C (1999) A methodology for proving control systems with Lustre and PVS. In: Proceedings of the seventh working conference on dependable computing for critical applications (DCCA 7), San Jose, January 1999
2. Clarke EM, Grumberg O, Peled DA (1999) Model checking. MIT, Cambridge, MA
3. Colaço JL, Girault A, Hamon G, Pouzet M (2004) Towards a higher-order synchronous dataflow language. In ACM fourth international conference on embedded software (EMSOFT'04), Pisa, Italy, September 2004
4. Esterel Technologies, Inc. SCADE Suite product description. http://www.esterel-technologies.com/products/scade-suite
5. Goguen JA, Meseguer J (1982) Security policies and security models. In: Proceedings of the 1982 IEEE symposium on security and privacy. IEEE Computer Society Press, Washington, DC, pp 11–20
6. Greve D (2010) Information security modeling and analysis. In: Hardin D (ed) Design and verification of microprocessor systems for high-assurance applications. Springer, Berlin, pp 249–299

7. Halbwachs N, Caspi P, Raymond P, Pilaud D (1991) The Synchronous dataflow programming language lustre. Proc IEEE 79(9):1305–1320
8. IRST http://nusmv.irst.itc.it/ The NuSMV model checker, IRST, Trento, Italy
9. Le Guernic P, Gautier T, Borgne ML, Maire CL (1991) Programming real-time applications with signal. Proc IEEE 79:1321–1336
10. MacLean J (1992) Proving noninterference and functional correctness using traces. J Comput Secur 1:37–57
11. The Mathworks, Inc., Simulink and stateflow product description. http://www.mathworks. com/Simulink, http://www.mathworks.com/products/stateflow/
12. McMillan K (1998) Verification of an implementation of Tomasulo's algorithm by compositional model checking. In: Proceedings of the tenth international conference on computer aided verification (CAV'98) Vancouver, Canada, June 1998
13. McMillan K (1999) Circular compositional reasoning about liveness. In: Advances in hardware design and verification: IFIP WG10.5 international conference on correct hardware design and verification methods (CHARME'99), pp 342–345
14. Munoz C (2009) ProofLite product description. http://research.nianet.org/~munoz/ProofLite
15. Owre S, Rushby JM, Shankar N (1992) PVS: a prototype verification system. In: 11th International conference on automated deduction (CADE), vol 607, Saratoga, NY, June 1992, pp 748–752
16. Prover Technologies, Inc. Prover SL/DE plug-in product description. http://www.prover. com/products/prover_plugin
17. Reactive Systems, Inc. Reactis product description. http://www.reactive-systems.com
18. Rockwell Collins Turnstile Product Page. http://www.rockwellcollins.com/products/gov/ airborne/cross-platform/information-assurance/cross-domain-solutions/index.html
19. Roscoe AW, Goldsmith MH (1999) What is intransitive noninterference? In: Proceedings of the 12th IEEE computer security foundations workshop, pp 228–238
20. Rushby J (1992) Noninterference, transitivity, and channel-control security policies. Technical report csl-92-2, SRI
21. Ryan PYA (1991) A CSP formulation of non-interference. In: Cipher. IEEE Computer Society Press, Washington, DC, pp 19–27
22. SRI, Incorporated. PVS specification and verification system. http://pvs.csl.sri.com
23. SRI, SAL home page. http://www.csl.sri.com/projects/sal/
24. Whalen M, Cofer D, Miller S, Krogh B, Storm W (2007) Integration of formal analysis into a model-based software development process. In: 12th International workshop on industrial critical systems (FMICS 2007), Berlin, Germany, July 2007
25. Wilding M, Greve D, Richards R, Hardin D (2010) Formal verification of partition management for the AAMP7G microprocessor. In: Hardin D (ed) Design and verification of microprocessor systems for high-assurance applications. Springer, Berlin, pp 175–191

Index

CPSIA information can be obtained at www.ICGtesting.com
Printed in the USA
238335LV00005B/65/P